Application of Artificial Neural Network in Agriculture

Guest Editors
Ray E. Sheriff
Department of
Computer Science
Edge Hill University
Ormskirk
United Kingdom

Chiew Foong Kwong
Department of Electrical and
Electronic Engineering
University of Nottingham
Ningbo China
Ningbo
China

Editorial Office
MDPI AG
Grosspeteranlage 5
4052 Basel, Switzerland

This is a reprint of the Special Issue, published open access by the journal *AgriEngineering* (ISSN 2624-7402), freely accessible at: www.mdpi.com/journal/agriengineering/special_issues/QBJA5BAQI7.

For citation purposes, cite each article independently as indicated on the article page online and using the guide below:

Lastname, A.A.; Lastname, B.B. Article Title. *Journal Name* **Year**, *Volume Number*, Page Range.

ISBN 978-3-7258-3122-7 (Hbk)
ISBN 978-3-7258-3121-0 (PDF)
https://doi.org/10.3390/books978-3-7258-3121-0

Application of Artificial Neural Network in Agriculture

Guest Editors

Ray E. Sheriff
Chiew Foong Kwong

MDPI

Basel • Beijing • Wuhan • Barcelona • Belgrade • Novi Sad • Cluj • Manchester

Contents

About the Editors

Ray E. Sheriff

Professor Ray E. Sheriff received a B.Eng. degree (Hons.) in Electrical and Electronic Engineering from the University of Leeds, Leeds, U.K., in 1986. He then worked within the satellite communications industry before taking up a Lectureship in 1991 with the University of Bradford, where he received his Doctorate in 1995 and Master of Business Administration in 2010. He is currently a Professor and Associate Head of the Department of Engineering within the Department of Computer Science, Edge Hill University.

He is a Fellow of the Institution of Engineering and Technology (FIET), a Senior Member of the Institute of Electrical and Electronics Engineers (SMIEEE), a Chartered Engineer with the Engineering Council UK (CEng), and a Fellow of the Higher Education Academy (FHEA).

His research interests include artificial intelligence, information technology, telecommunications engineering, and wireless systems.

Chiew Foong Kwong

Dr. C.F. Kwong is an Associate Professor and Head of the Department of Electrical and Electronic Engineering at the University of Nottingham Ningbo China (UNNC). He holds a PhD in Mobile Communications from Multimedia University Malaysia, an MSc in Personal Mobile and Satellite Communications from the University of Bradford, UK, and a BEng in Electrical and Electronics Engineering from the University of Hertfordshire, UK.

Dr. Kwong's academic journey is complemented by a wealth of practical experience in the field, with notable roles in the development of mobile communication technologies, satellite navigation, artificial intelligence applications, and wireless and mobile computing. His research interests include advanced mobile communications, smart mobility, the Internet of Things (IoT), and educational innovations in engineering. He is also the Director of the Next-Generation Internet of Everything Laboratory.

Dr. Kwong has served in leadership roles, including as the Head of the Department of Electrical and Electronic Engineering, Associate Dean of Admission and International Relations at UNNC, has extensive involvement in academic accreditation, and is currently a Member of the UK-based Institution of Engineering and Technology (IET) Academic Accreditation Committee. He has supervised numerous PhD students and significantly contributed to the academic community through research publications, conference involvements, and international collaborations.

He is a Senior Member of IEEE, a Chartered Engineer with the Engineering Council UK, and a Senior Fellow of the Higher Education Academy (SFHEA). Dr. Kwong's research achievements include securing major funding for projects, and his academic influence extends through his roles in various professional bodies and committees.

Preface

We are delighted to present the Reprint of our Special Issue titled "Application of Artificial Neural Network in Agriculture", which showcases fourteen high-quality research articles highlighting the use of artificial neural networks (ANNs) and related technologies when applied to agricultural practice and production.

With the need for farming to become more efficient and environmentally sustainable to meet the demands of a growing global population, the application of ANNs to inform and enhance agricultural practice and production is gathering pace.

The introduction of enabling technologies, such as edge computing, the Internet of Things, and mobile applications, is creating new opportunities to gather extensive digital datasets in real time, which can then be used by models with the ability to learn from and interpret this information, including ANNs, artificial intelligence, and machine learning.

In this Reprint, we highlight how ANNs may be employed for pest detection, plant disease detection, nutritional monitoring and prediction, the classification of the phenology of beans, and the optimisation of egg production. This Reprint also presents how ANNs may be used for predicting the power requirements of agricultural machinery and the dynamic behaviour forecasting of an indirect solar dryer.

The multidisciplinary nature of the articles presented will likely strongly appeal to researchers and scientists working in the areas of agricultural technology, communications engineering, computer science, data analytics, electronic engineering, information technology, image processing, and mathematical modelling.

We would like to express our gratitude to the many peer reviewers who have contributed their time, comments, and opinions to ensure that the presented articles are of the highest quality. We would also like to thank the editorial staff of *AgriEngineering* and our contact editor Mr Dušan Zečević for their continuous support throughout the publication process.

We hope you find our Reprint to be of interest and a source for future research in this increasingly important application of ANNs.

Ray E. Sheriff and Chiew Foong Kwong
Guest Editors

AgriEngineering

Article

TFEMRNet: A Two-Stage Multi-Feature Fusion Model for Efficient Small Pest Detection on Edge Platforms

Junlin Mu [1,†], Linlin Sun [1,2,†], Bo Ma [1], Ruofei Liu [1], Shuangxi Liu [1], Xianliang Hu [3], Hongjian Zhang [1,2,*] and Jinxing Wang [1,4,*]

1 College of Mechanical and Electronic Engineering, Shandong Agricultural University, Taian 271018, China; mjl19970127@163.com (J.M.); sunlinlin@sdau.edu.cn (L.S.); 2023010151@sdau.edu.cn (B.M.); 19860912870@163.com (R.L.); shuangxiliu168@163.com (S.L.)
2 Shandong Province Key Laboratory of Horticultural Machinery and Equipment, Taian 271018, China
3 Shandong Xiangchen Technology Group Co., Ltd., Jinan 250000, China; 2022010103@sdau.edu.cn
4 Shandong Provincial Engineering Laboratory of Agricultural Equipment Intelligence, Taian 271018, China
* Correspondence: zhanghongji_an@163.com (H.Z.); jinxingw@163.com (J.W.)
† These authors contributed equally to this work.

Abstract: Currently, intelligent pest monitoring systems transmit entire monitoring images to cloud servers for analysis. This approach not only consumes significant bandwidth and increases monitoring costs, but also struggles with accurately recognizing small-target and overlapping pests. To overcome these challenges, this paper introduces a two-stage multi-feature fusion small-target pest detection algorithm based on edge computing devices, termed TFEMRNet. The algorithm initially conducts semantic segmentation on an edge processor, followed by uploading the segmented images to a cloud server for target identification. Specifically, the semantic segmentation model TFENet incorporates a Multi-Attention Channel Aggregation (MACA) module, which integrates semantic features from EfficientNet-Pest and Swin Transformer, thereby enhancing the model's ability to extract features of small-target pests. Experimental results demonstrate that TFEMRNet surpasses models such as YOLOv11, Fast R-CNN, and Mask R-CNN on small-target pest datasets, achieving precision of 96.75%, recall of 96.45%, and an F1 score of 95.60%. Notably, the TFENet model within TFEMRNet attains an IoU of 91.63% in semantic segmentation accuracy, outperforming other segmentation models such as U-Net and PSPNet. These findings affirm TFEMRNet's superior efficacy in small-target pest detection, offering an effective solution for agricultural pest monitoring.

Keywords: pest identification; tiny object recognition; multi-stage detection approach; edge-based processing; feature integration

Citation: Mu, J.; Sun, L.; Ma, B.; Liu, R.; Liu, S.; Hu, X.; Zhang, H.; Wang, J. TFEMRNet: A Two-Stage Multi-Feature Fusion Model for Efficient Small Pest Detection on Edge Platforms. *AgriEngineering* **2024**, *6*, 4688–4703. https://doi.org/10.3390/agriengineering6040268

Academic Editors: Ray E. Sheriff and Chiew Foong Kwong

Received: 5 November 2024
Revised: 29 November 2024
Accepted: 2 December 2024
Published: 5 December 2024

1. Introduction

Pest infestation has been regarded as a severe disaster in apple production [1]. These pests reproduce rapidly, and if not promptly controlled, they can severely damage fruits and significantly reduce apple yields. Therefore, the rapid and accurate identification of pest species has been crucial for the management of apple pests and diseases [2,3]. Precise pest identification not only enhances control efficacy, but also reduces the use of pesticides, thereby minimizing environmental impact [4]. However, existing agricultural pest detection methods face numerous challenges in practical applications, primarily manifested in the following aspects.

Current apple pest detection methods face several significant challenges that hinder their effectiveness and scalability. Firstly, the accuracy of existing techniques is inadequate due to the high diversity of apple pest populations, which exhibit considerable variations in morphology, size, and coloration across different developmental stages. This diversity complicates the reliable distinction among various species and their growth phases, leading to difficulties in precise classification and preventing farmers from implementing effective control measures, thereby negatively impacting both the yield and quality of apples [5]. Secondly, the processing speed of traditional manual detection methods is slow and labor-intensive, making large-scale pest monitoring in extensive orchards time-consuming and inefficient. For instance, inspecting several hundred acres manually may take days or weeks, while pest reproductive rates can outpace detection efforts, resulting in increased losses [6]. Thirdly, advanced technologies and algorithms such as drones, remote sensing, and deep learning, although enhancing detection efficiency, impose high demands on computational power and energy consumption. In resource-constrained rural areas, the deployment and maintenance of such high-tech equipment are challenging and costly, leading to significant increases in both operational costs and energy usage [7]. These factors collectively impede the widespread adoption and effectiveness of current pest monitoring systems in apple orchards.

To address the aforementioned challenges, Intelligent Integrated Pest Management (IPM) has emerged by incorporating Artificial Intelligence (AI) and edge computing technologies, providing new solutions for pest monitoring. Edge computing enables real-time data processing at the device level, reducing reliance on cloud servers, lowering latency, and conserving bandwidth [8]. Intelligent IPM has proven highly effective in controlling economically important crop pests [9]. At the same time, the identification of small-target pests has become a significant research focus. These pests often cause direct damage to fruits, making them essential subjects for real-time identification [10]. Additionally, a top-down feature fusion algorithm has been proposed to improve feature utilization for small-target pests by integrating top-down feature fusion with prior bounding box clustering, thereby achieving precise localization of these pests [11]. A detection model named Skip DETR has been introduced, which enhances the detection performance of small objects by incorporating skip connections and spatial pyramid pooling layers to facilitate feature fusion across different network layers [12]. Furthermore, the novel multi-scale dense MD-YOLO method has improved feature map utilization and reduced information loss by integrating DenseNet modules and adaptive attention modules [13]. In summary, existing solutions for small-target pest detection have primarily focused on enhancing the feature utilization of images depicting small-target pests [14]. However, most of these approaches still face the following limitations: in the presence of interference from other pests, the semantic features of small-target pest images remain limited [15]. Although these models offer advantages in improving image feature utilization, the challenge of feature loss due to downsampling has not been adequately addressed, thereby hindering the accurate identification of small-target pests.

As previously discussed, this research seeks to bridge the gap between pest recognition accuracy and the limitations imposed by resource constraints in deploying models on edge devices by introducing a lightweight pest detection algorithm named TFEMRNet. This approach involves conducting semantic segmentation directly on the edge device and subsequently transmitting the segmented images to a cloud server for identification. This strategy effectively lowers the computational power demands and conserves bandwidth. To address the existing challenge of insufficient information utilization in the detection of small-target pests, a dual-encoder feature-enhanced semantic segmentation network has been developed. This network increases the visibility of small-target pests within the entire image, thereby enriching the semantic features associated with these pests. Furthermore, a parallel branch architecture has been incorporated, which simultaneously integrates features from the Swin Transformer and EfficientNet-Pest. This configuration facilitates the efficient capture of global dependencies and low-level spatial information in a streamlined manner, significantly improving the model's capacity to represent pest characteristics. The primary contributions of this study are summarized as follows:

1. We construct a dataset that includes pests such as *Grapholitha molesta* and *Helicoverpa armigera*, collected from intelligent pest monitoring devices. This dataset provides a critical foundation for advancing pest identification research and validating detection models.

2. We design a two-stage small-target pest recognition algorithm that reduces computing power requirements on edge devices and saves bandwidth.
3. We design a dual-encoder feature-enhanced semantic segmentation network (TFENet) based on feature fusion, addressing the issue of inadequate semantic feature utilization for small-target pests in high-resolution images. TFENet employs a parallel branch structure that integrates features from the Swin Transformer and EfficientNet-Pest, enabling efficient acquisition of global dependencies and low-level spatial information in a shallow manner.
4. We design a Multi-Attention Channel Aggregation module (MACA) which optimizes global information and local details using SE-Block and CBAM modules. Additionally, it enhances the model's feature extraction capability for small-target pests by interacting feature maps from the two branches through the Hadamard product.

The structure of this study is organized as follows: Section 2 outlines the materials used and the two-stage recognition methodology employed. Section 3 details the experimental hardware, the obtained results, and their discussion. Section 4 presents the conclusions, and finally, future work is described.

2. Materials and Methods

2.1. Image Acquisition

Grapholitha molesta is one of the primary small-target pests in orchards, causing damage to both fruits and leaves. Due to its small size, strong camouflage, high mobility, and lack of phototaxis, it is difficult to monitor. This study focuses on the recognition of small-target pests, using the oriental fruit moth as an example.

Pest monitoring equipment plays a vital role in tracking agricultural pests. The intelligent pest monitoring device utilized in this study for apple orchards is illustrated in Figure 1a. This apparatus consists of a solar power module, a central processing unit, an image acquisition module, and edge computing devices. The central processing unit includes an edge computing gateway that conducts initial image segmentation. To attract specific pest species, targeted pheromone baits are installed within the device, facilitating the trapping of designated pests. Multiple pheromone baits for various pest species can be deployed simultaneously, allowing for a single device to monitor different pests at the same time, thereby lowering monitoring expenses. Once captured, the pests fall onto a collection plate where a 12-megapixel high-definition camera periodically takes high-resolution images, which are then sent to the central processing unit. Following this, the data transmission module forwards the collected images to a cloud server through 4G/5G networks. An example of the captured images is presented in Figure 1b.

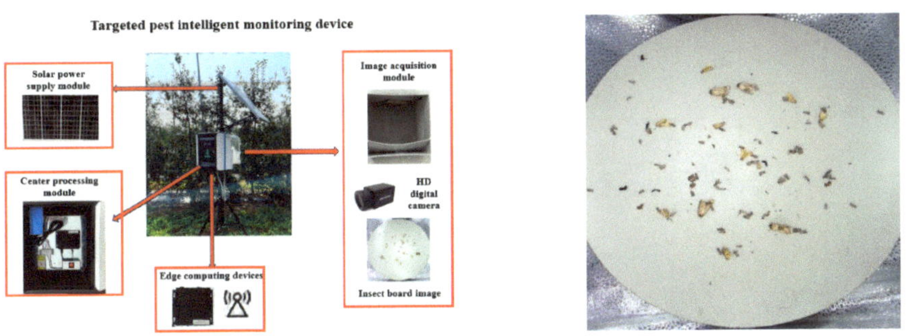

(a) (b)

Figure 1. Pest monitoring device and pest images. (**a**) For the intelligent monitoring device for targeted pests in apple orchards, (**b**) the pest images obtained using the intelligent monitoring device for targeted pests in apple orchards.

Edge computing is utilized in the proposed pest detection system to reduce latency and bandwidth usage by processing data locally on edge devices rather than relying on centralized cloud computing. This approach ensures faster response times, which is critical for real-time pest monitoring in agricultural environments. Additionally, edge computing alleviates the need for continuous data transmission to cloud servers, making it more cost-effective and energy-efficient for large-scale deployment in remote areas.

From 2021 to 2023, we extensively collected images using both the targeted pest trapping device and the intelligent pest monitoring device in apple orchards. Additionally, we obtained pest samples from the Shandong Jinan Agricultural Science Academy to simulate device environments and augment the sample data. A total of 452 images were collected for this study, comprising 326 images obtained from the intelligent monitoring device and 126 images sourced from the targeted trapping device. These images were captured using a high-definition camera, each possessing a resolution of 5760 × 4420 pixels.

The LabelMe software (LabelMe v5.2.0.) was used for manual annotation [16], generating labeled files with category and location information. The annotated images were utilized to construct the model's training, validation, and test datasets, partitioned in a 6:2:2 ratio, resulting in 1350 training images, 450 validation images, and 450 test images. To enhance the dataset, five data augmentation techniques were employed: brightness adjustment, noise addition, random point addition, mirroring, and translation [17]. Following data augmentation, the training set increased from 1350 to 5400 pest images, thereby providing a more robust foundation for network training.

2.2. *Two-Stage Recognition Method TFEMRNet*

The two-stage recognition network used in this study is named TFEMRNet, and its structure is shown in Figure 2. TFENet employs a parallel branch structure, with the blue frame representing TFENet and the green frame representing Mask R-CNN [18]. The outputs of TFENet and Mask R-CNN are displayed outside the frame. TFENet integrates features from Swin Transformer [19] and EfficientNet-Pest [20] in a parallel manner, allowing for the efficient acquisition of global dependencies and low-level spatial information. Additionally, we introduced a new fusion module called the multi-attention channel aggregation module (MACA), which effectively merges multi-level features from the two branches.

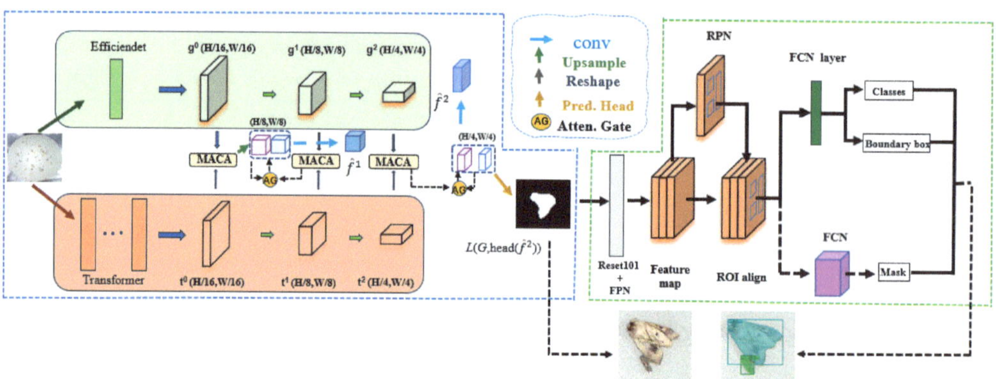

Figure 2. Overall structure diagram of TFEMRNet. The blue box on the left represents the TFENet structural framework, the green box on the right represents the MaskRCNN structural framework, and outside the box are the TFENet segmentation results and MaskRCNN detection results, respectively. In the figure, H represents the feature map height, W represents the feature map width, and Head represents the prediction head.

Specifically, TFEMRNet combines semantic segmentation and object detection. In the first stage, semantic segmentation is performed. The segmentation results are used to crop target areas from the original images, obtaining image fragments containing pests. We selected three outstanding semantic segmentation networks, TFENet, PSPNet, and UNet, and tried efficient feature extraction backbone networks for each, using EfficientNet and BiFPN for efficient multi-scale feature fusion. For instance, with UNet, we replaced the original VGG16 with EfficientNet as the backbone network. The second stage of TFEMRNet involves object detection, where the image fragments from semantic segmentation are input into the Mask R-CNN model to obtain pest identification and classification results. Mask R-CNN detects pests in the image fragments and provides pest species information.

The first stage of TFEMRNet, the TFENet semantic segmentation model, is detailed in Figure 3. Swin Transformer and EfficientNet serve as encoders for the TFENet segmentation network, with their feature maps fused via the multi-attention channel aggregation module. The semantic information and spatial context of pest images are crucial for improving the segmentation accuracy of small-target pests. Traditional CNNs struggle to model the global semantic information of pest images due to the fixed size of convolution kernels. In contrast, the Swin Transformer captures global semantic information through a self-attention mechanism, efficiently capturing both low-level spatial features and high-level semantic information without relying on excessively deep network architectures. This mitigates issues like gradient vanishing and insufficient feature reuse, addressing the loss of high-level semantic features in small-target pest images. However, the Swin Transformer has limitations in modeling fine-grained features, potentially causing spatial inductive bias when processing local information. To better integrate the strengths of both encoders and resolve the precise segmentation of small-target pests, we designed a segmentation model that combines features from the Swin Transformer and EfficientNet-pest. This integration improves the model's capacity for feature representation.

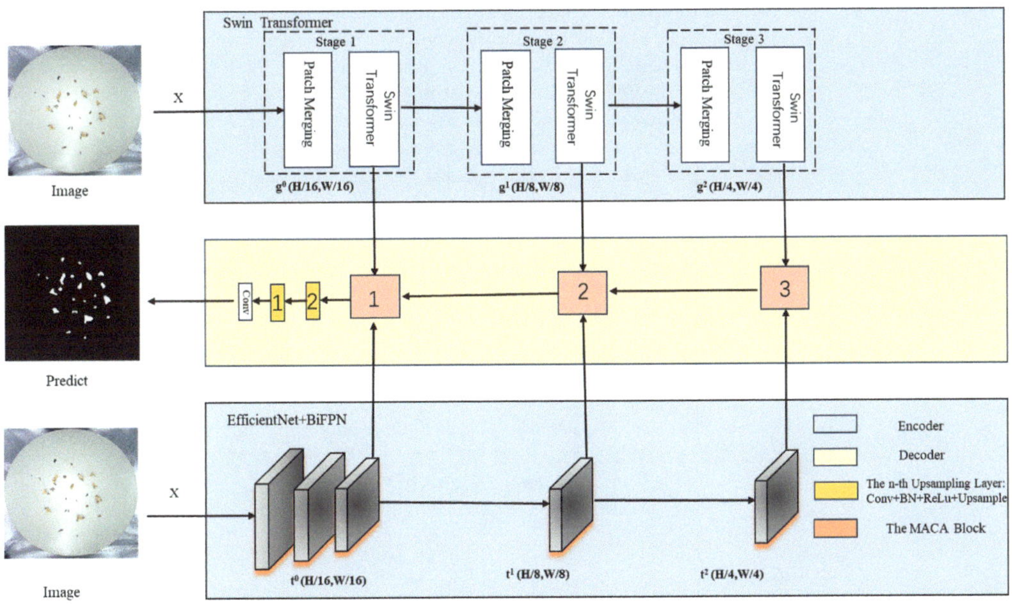

Figure 3. The overall structure of TFENet, with Swin Transformer and EfficientNet-Pest feature extraction networks in the upper and lower gray boxes and Multi Attention Channel Aggregation (MACA) modules in the orange boxes. The left side shows the input image and prediction image.

In the TFENet semantic segmentation network, the features extracted by the Swin Transformer and EfficientNet-pest are input into the MACA (Multi-Attention Channel Aggregation) module for feature fusion. This fusion process uses Attention Gates (AG) skip connections to combine features, ultimately generating the final segmentation map through a simple head. The output segmentation map is subsequently resized to match the original image resolution. This feature fusion network architecture effectively addresses the segmentation challenges posed by small-target pests in high-resolution images, thereby enhancing the detection performance of the recognition network for these pests.

Ultimately, the two-stage recognition strategy offers an effective approach for identifying small-target pests in high-resolution images. Initially, it employs a dual-phase recognition algorithm framework to address the challenges associated with detecting small-target pests within detailed imagery. In TFEMRNet, to prevent feature loss caused by image compression, the prediction map generated using the TFENet semantic segmentation network is restored to the original image resolution. Subsequently, corresponding segments are extracted from the original image to serve as object detection samples. Additionally, the accuracy of semantic segmentation impacts the integrity of the object detection samples. To ensure that the pest regions do not exceed the image boundaries, we appropriately expand the prediction boxes obtained from semantic segmentation to be the minimum enclosing rectangles of the pest regions predicted by the semantic segmentation model. This adjustment method keeps the centroid position unchanged and expands the side length of the bounding box by 1.2 times.

2.2.1. Swin Transformer Module

Swin Transformer is a Transformer architecture suitable for visual tasks that effectively extracts multi-scale image features through a hierarchical structure and sliding window mechanism. The hierarchical structure is similar to a convolutional neural network, allowing the model to gradually capture the details and global information of the image, while the sliding window performs a self-attention calculation locally through a fixed-size window, significantly reducing the computational cost.

In order to achieve cross-window information fusion, the Swin Transformer introduces the Shifted Window mechanism, that is, window sliding, to realize information interaction between windows. At the same time, the multi-head self-attention mechanism helps the model focus on details in different areas, thereby improving the detection capabilities of small targets and complex backgrounds. The overall architecture of this integration is depicted in Figure 4.

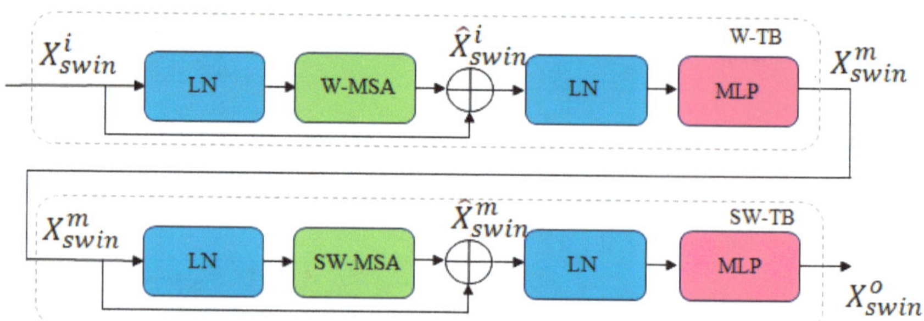

Figure 4. Swin Transformer module structure diagram.

2.2.2. EfficientNet-Pest Module

EfficientNet has provided high accuracy for small-target objects), aligning with the requirements of this study. As illustrated in Figure 5, EfficientNet consists of one stem and seven blocks. The function of the stem structure is to adjust the number of channels through convolution. Each block comprises multiple inverted bottleneck convolution (MBConv) modules. Effective feature layers P3–P5 have been extracted through these blocks, followed by two downsampling operations to obtain P6 and P7. Subsequently, P3–P7 have been input into the BiFPN for multi-scale feature fusion. During this process, each BiFPN layer has performed bidirectional feature fusion, both top-down and bottom-up, to enhance feature extraction for small targets in high-resolution images. The outputs of BiFPN, P3_out, P4_out, and P5_out have served as the input feature maps for the multi-attention channel aggregation module, corresponding to $g^0 = R^{H/16 \times W/16 \times 256}$, $g^1 = R^{H/8 \times W/8 \times 128}$, and $g^2 = R^{H/4 \times W/4 \times 64}$, respectively.

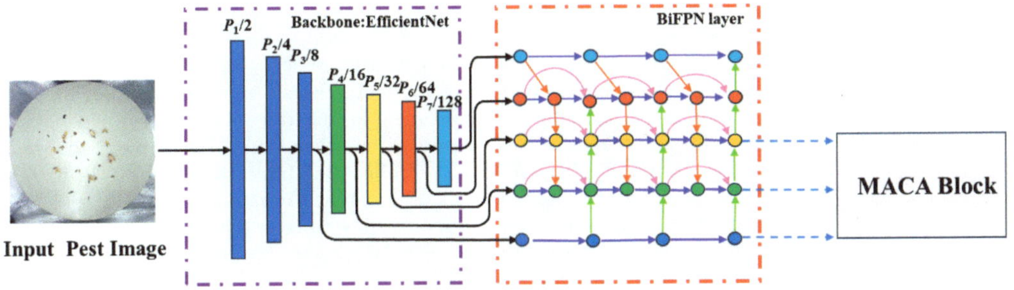

Figure 5. EfficientNet-Pest module structure diagram.

2.2.3. MACA

Upon analyzing high-resolution images, it was observed that small-target pests possess fewer pixel features. This limitation may lead to the extraction of relatively sparse feature information during the detection process, thereby significantly diminishing detection accuracy. To mitigate this issue, this study introduces the Multi-Attention Channel Aggregation (MACA) module. The MACA enhances the extraction of feature information for small-target pests by integrating encoded features from EfficientNet-Pest and the Swin Transformer. The architecture of this module is depicted in Figure 6.

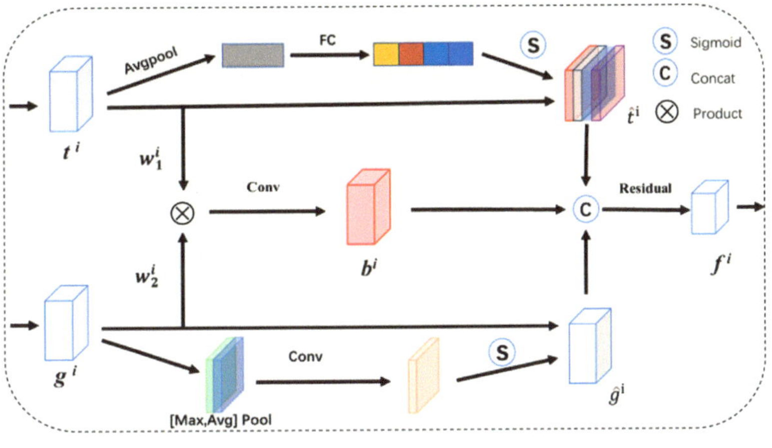

Figure 6. MACA structure diagram.

The module integrates the self-attention mechanism and multimodal fusion mechanism to obtain the fused features f^i, where i is 0, 1, or 2, through the following steps. The calculation formulas are presented in Equations (1)–(4):

$$\hat{t}^i = channelattn(t^i) \quad (1)$$

$$\hat{g}^i = channelattn(\hat{g}^i) \quad (2)$$

$$\hat{b}_i = Conv(t^i W_1^i \otimes g^i W_1^i) \quad (3)$$

$$f^i = Residual([\hat{b}_i, \hat{t}^i, \hat{g}^i]) \quad (4)$$

In the MACA, the symbols $W_1^i \in R^{Di \times Li}$ and $W_2^i \in R^{Ci \times Li}$ denote the element-wise product. This module employs a channel attention mechanism using an SE block to capture global contextual information from the Swin Transformer. In contrast, a CBAM block is utilized for spatial attention, enhancing localized features in pest-relevant areas and suppressing non-essential background content. The feature maps from both branches are then fused through element-wise multiplication. Interactive features $\hat{b}_i$, together with refined attention features $\hat{t}^i$ and $\hat{g}^i$, are combined within a residual block. The output features fif^ifi effectively retain both comprehensive and localized details at the present spatial scale. Finally, AG skip connections incorporate f^i to generate the final segmentation, as described in Equations (5) and (6).

$$\hat{f}^0 = f^0 \quad (5)$$

$$\hat{f}^{i+1} = Conv([Up(\hat{f}^i), AG(\hat{f}^{i+1}, Up(\hat{f}^i))] \quad (6)$$

The MACA amalgamates both local and global features from pest images, thereby enhancing the network's capability to extract features from small-target pests and markedly increasing the segmentation accuracy for these pests.

2.2.4. Loss Function

The TFENet semantic segmentation network uses an end-to-end training strategy that combines a weighted IoU loss alongside a binary cross-entropy term, expressed as $L = L_{IoU}^w + L_I^w$.

In this configuration, boundary pixels are assigned higher weights to enhance their significance. Additionally, to facilitate improved gradient propagation, deep supervision is incorporated by providing extra guidance to the Swin Transformer branch. Consequently, the overall training loss is formulated as presented in Equation (8):

$$L = \alpha L(G, head(\hat{f}^2)) + \gamma L(G, head(t^2)) + \beta L(G, head(f^0)) \quad (7)$$

In this context, α, β, and γ are adjustable hyperparameters set to 0.4, 0.4, and 0.2, respectively, based on prior experience derived from similar studies in pest detection optimization [21]. These values were selected to balance the contributions of precision, recall, and detection robustness in accordance with the findings reported in previous research.

3. Results and Discussion

3.1. Experimental Setup

The experiments were conducted on a 64-bit Windows 10 operating system using a workstation equipped with an Intel® Xeon® Silver 4210R processor (Santa Clara, CA, USA) and an NVIDIA GeForce RTX 2080 Ti GPU (Santa Clara, CA, USA) to expedite the experimental process. Additionally, the system was outfitted with 128 GB of RAM and a 2 TB SSD to ensure efficient data processing and model training.

Training of TFEMRNet was performed using Python version 3.8 and developed based on the PyTorch 1.10 framework. To fully leverage GPU acceleration, the experimental environment was configured with CUDA 10 and cuDNN version 7.4.1.5. Furthermore, essential deep learning libraries and toolkits such as NumPy 1.19.5, Pandas 1.1.5, and Scikit-learn 0.24.2 were utilized to ensure the smooth progression of the experiments.

During the experimental process, fixed random seeds were set to guarantee the reproducibility of results. Specifically, the Python random seed was set to 42, the NumPy random seed to 42, and the PyTorch random seed to 42. Additionally, torch.backends.cudnn.deterministic was set to True and torch.backends.cudnn.benchmark to False to ensure consistent outcomes across experiments. Moreover, fixed shuffle parameters and data splitting methods were employed during data loading to further enhance the reproducibility and reliability of the experiments.

Through the above detailed environmental configurations and random seed settings, the stability and reproducibility of the experimental results were ensured, providing a solid foundation for the training and evaluation of the TFEMRNet model.

3.2. Evaluation Metrics

In this study, Intersection over Union (IoU) was utilized as the primary evaluation metric for the semantic segmentation phase. Based on extensive prior research experience in the fields of image processing and object detection, IoU was selected as the main evaluation metric because it effectively measures the degree of overlap between the segmentation results and the ground truth labels, and it has been proven to possess high reliability and consistency in numerous related studies. During the application of the object detection algorithm to the image fragments obtained in the initial segmentation phase, the original test set labels could not be directly applied to the new samples due to alterations in image dimensions. Drawing on previous methods for handling changes in image size, this study innovatively employs the location information derived from semantic segmentation to precisely isolate the target regions from the original images, thereby generating new labeled samples suitable for the subsequent object detection phase. This approach not only resolved the challenges posed by label migration due to size changes, but also enhanced the accuracy and efficiency of object detection.

For evaluating the object detection performance, precision, recall, and F1-score were adopted as the key metrics. Precision measures the proportion of correctly identified pests out of all detected pests, with higher precision indicating a lower rate of false positives. Recall assesses the proportion of correctly detected pests out of all actual pests, where higher recall signifies fewer missed detections. The F1-score, calculated as the harmonic mean of precision and recall, acts as an essential measure of the model's overall performance accuracy. The selection of these evaluation metrics is based on the need for a comprehensive assessment of model performance in prior research, ensuring that the model's performance across different aspects is thoroughly reflected. Specifically, for the identification of *Grapholitha molesta* in this research, all evaluation metrics were applied to this particular pest species. The mathematical formulations for IoU, precision, recall, and F1-score are presented in Equations (8) through (11):

$$IOU = \frac{Area\ of\ Overlap}{Area\ of\ Union} \tag{8}$$

$$recall = \frac{TP}{TP + FN} \tag{9}$$

$$precision = \frac{TP}{TP + FP} \tag{10}$$

$$F_1 = 2 \times \frac{recall \times precision}{recall + precision} \tag{11}$$

3.3. Model Training

The initial phase of semantic segmentation involves classifying each pixel within an image, necessitating substantial computational resources. To accommodate hardware constraints, input images were resized to a default resolution of 512 × 384 pixels. During model training, the initial learning rate was set to 0.0001, and the batch size was configured to 16. To ensure the effectiveness of the proposed enhancements in this study, all other hyperparameters were maintained consistently with the original TFENet. Each method was trained iteratively for 100 epochs, with the model being saved after each epoch. The model demonstrating the best performance on the validation set was selected as the final model for each method.

3.4. *Performance Advantages of Two-Stage Segmentation Recognition Strategy*

This research employs a two-stage recognition approach to identify small-target pests. To assess the effectiveness of the proposed method, comparative experiments were conducted against conventional recognition techniques, including Mask R-CNN, YOLOv11, and Fast R-CNN. The experimental outcomes, presented in Table 1, demonstrate that the proposed method attains the highest recognition accuracy of 96.75%, accompanied by a recall rate of 96.45% and an F1-score of 95.60%.

Table 1. Performance comparison with state-of-the-art object detection models.

Method	Backbone	Precision (%)	Recall (%)	F1 (%)
TFEMRNet	EfficientNet	96.75	96.45	95.60
MaskRCNN	VGG16	71.92	72.82	72.37
YOLOv11	Darknet53	75.43	74.45	74.68
FastRCNN	VGG16	71.60	72.56	72.08

The P–R curves for the four network models on the custom dataset are presented in Figure 7. In the P–R curves, the horizontal axis represents precision, and the vertical axis represents recall. The P–R curve is obtained by sequentially connecting points with different recall and precision rates. A P–R curve that bulges towards the top right corner indicates the stronger detection capability of the model. As shown in Figure 7, TFEMRNet exhibits better detection performance on this dataset compared to the other three models.

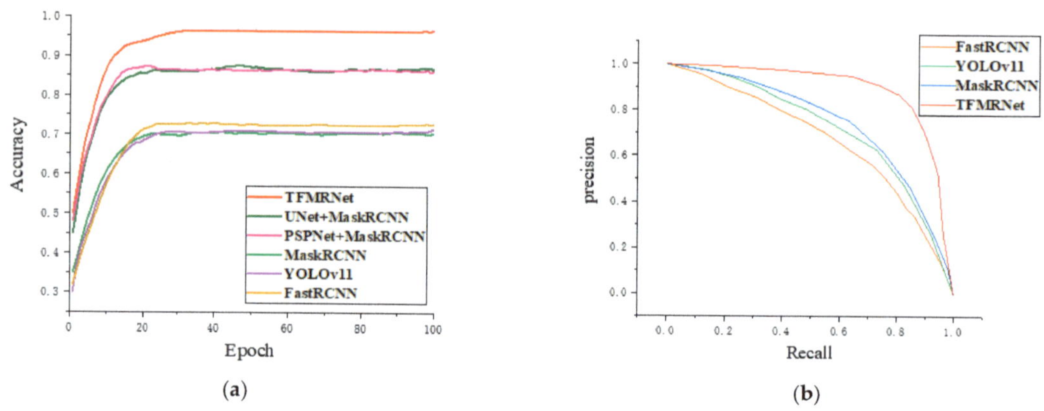

Figure 7. Compare the accuracy of the module with the PR curve. (**a**) For comparison of accuracy of variable algorithms, (**b**) for P–R curves of each algorithm.

According to the results in Figure 8, most pest images have been accurately identified using YOLOv11, but some *Grapholitha molesta*, as indicated by the red circles in Figure 8, have not been detected using the advanced object detection models. The green circles indicate incorrectly detected objects. Traditional object detection models suffer from significantly reduced recognition accuracy in the presence of pest adhesion or overlap. The substantial pixel-scale difference between the *Grapholitha molesta* and other pests makes it difficult for detectors to identify them. Moreover, the small proportion of *Grapholitha molesta* in the images, coupled with their few pixels, further complicates their recognition.

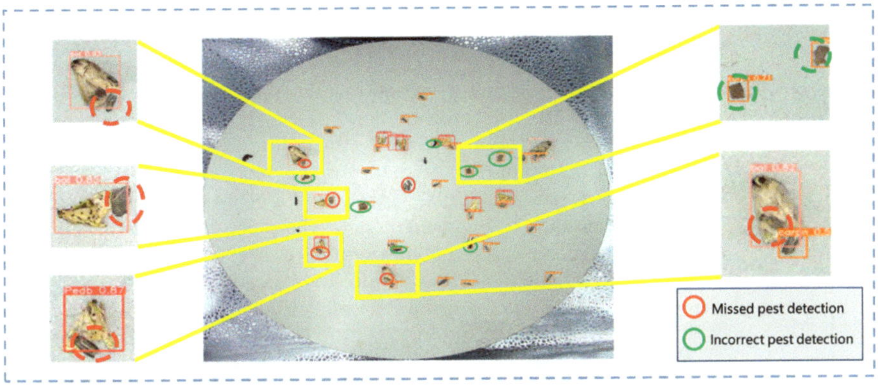

Image detection results of targeted pest intelligent monitoring device

Figure 8. The detection results of YOLOv11 model for two types of device images. The red circle represents missed detection objects, while the green circle represents missed detection objects.

To address these issues, the proposed strategy first uses a semantic segmentation algorithm to locate the target areas containing pests and then recognizes the target areas. This recognition strategy significantly mitigates the problem of low recognition accuracy in the presence of interference. Using images captured with an intelligent monitoring device targeting pests in apple orchards as an example, Figure 9 displays the recognition results of advanced object detection models and the two-stage object detection model under interference conditions. Figure 9a shows the original images to be detected, with red boxes numbered 1–5 indicating image segments with interference. Figure 9b presents the detection results of the image segments numbered 1–5 using TFEMRNet, YOLOv11, FastRCNN, and MaskRCNN models.

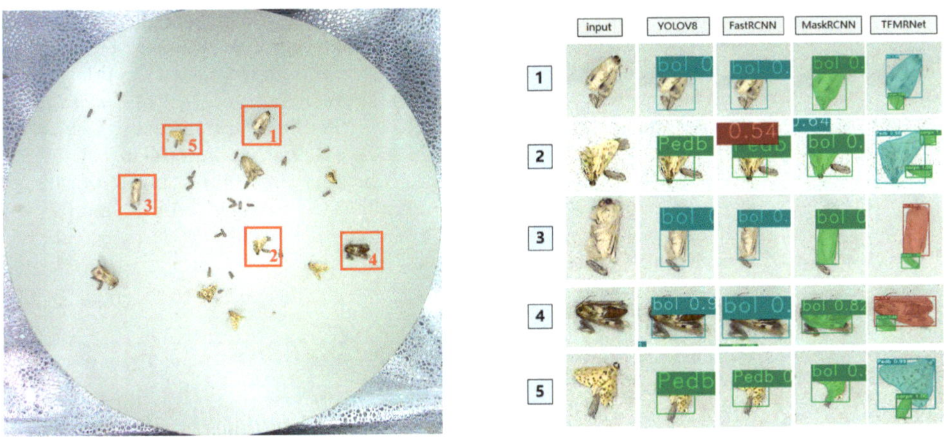

(a) (b)

Figure 9. Performance comparison of the model under interference, where (**a**) represents the original image with Numbers 1–5. The red boxes represent image segments with interference present. (**b**) Recognition results of four object detection models.

The results demonstrate that YOLOv11, FastRCNN, and MaskRCNN exhibit poor recognition performance under interference conditions, while TFEMRNet accurately identifies the pests, indicating that the proposed method can accurately recognize *Grapholitha molesta* in high-resolution images under various conditions.

3.5. Performance Advantages of TFENet Semantic Segmentation Model

The two-stage object detection results are not limited to TFEMRNet; all methods based on two-stage object recognition can enhance detection accuracy for adhered and overlapping objects. However, for *Grapholitha molesta,* not all semantic segmentation models can accurately segment pests from the original images. In this study, the backbone feature extraction networks of TFENet, Unet, and PSPNet were replaced with EfficientNet-pest and compared with their original models. The experimental results are shown in Figure 10. Figure 10b displays the prediction results of TFENet, Figure 10c shows the prediction results of the Unet network, and Figure 10d presents the prediction results of the PSPNet network. The red areas in Figure 10b indicate the objects missed by Unet and PSPNet.

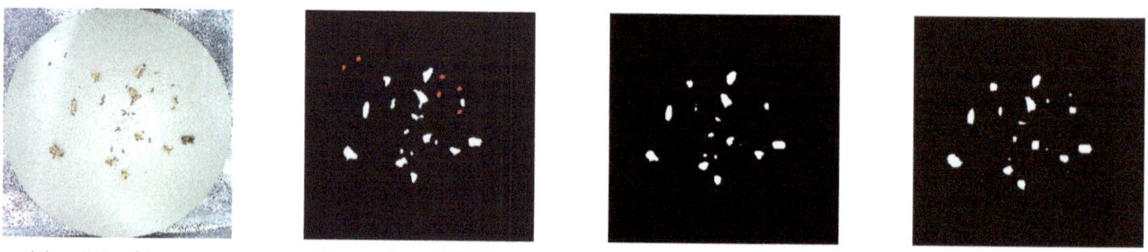

(**a**) original image (**b**) results of TFENet (**c**) results of Unet (**d**) results of PSPNet

Figure 10. Performance comparison of semantic segmentation models, where (**a**) represents the original image of the device and (**b–d**) represent the predicted results of three semantic segmentation models. The red area in (**b**) represents the missed detection objects in other semantic segmentation models.

Compared to TFENet, Unet and PSPNet exhibited missed detections, which is unacceptable in practical applications. These missed detections may result from information loss due to image compression, potentially leading to missing objects in the target detection phase and consequently reducing the overall recognition accuracy of the model. In contrast, the improved TFENet demonstrated superior performance in small-target segmentation, showing a significant advantage in the segmentation accuracy of *Grapholitha molesta.*

An ablation study was conducted on the *Grapholitha molesta* dataset, with the results presented in Table 2. Using TFENet as the baseline, the introduction of the MACA model led to a 10% improvement in Intersection over Union (IoU), no significant enhancement in precision, and a 9% increase in recall. These findings indicate that the MACA model enhances the segmentation performance for *Grapholitha molesta,* particularly excelling in reducing missed detections.

Table 2. Performance of six semantic segmentation models.

Method	MACA	IOU	Precision	Recall	Inference Times (s)	Model Size
TFENet	-	0.82	0.93	0.76	0.0191	135 MB
TFENet	√	0.92	0.96	0.85	0.0213	139 MB
UNet	-	0.85	0.95	0.76	0.0253	147 MB
PSPNet	-	0.86	0.85	0.79	0.0301	156 MB

3.6. Performance Advantages of TFEMRNet Under Multiple Interference Conditions

Through the analysis of recognition results, it has been found that in high-resolution images obtained using the intelligent monitoring device targeting pests in apple orchards, traditional object detection models perform reasonably well for recognizing isolated *Grapholitha molesta*. However, their performance significantly declines for pest images under mutual adhesion, pest images under occlusion, and pest images under overlap. These three interference conditions are illustrated in Figure 11.

(**a**) Pest images under mutual adhesion

(**b**) Pest images under occlusion

(**c**) Pest images under overlap

Figure 11. Example of pest images under three types of interference.

To further validate the performance of the TFEMRNet algorithm under interference conditions, a comparative experiment with YOLOv11, the model with the best detection performance, has been conducted. We have analyzed YOLOv11's recognition results for *Grapholitha molesta* in four scenarios using a test set of 50 images: independent pest images, pest images under mutual adhesion, pest images under occlusion, and pest images under overlap. The precision, recall, and F1-score parameters for these scenarios were obtained.

The statistical analysis reveals that TFEMRNet achieves the most significant improvement in detection performance under overlapping conditions, with precision increasing by 46.73% and recall increasing by 82.84%. This enhancement addresses the issue of small-target pest features being lost due to overlap in high-resolution images, which previously led to missed detections. In occlusion scenarios, where features of small-target pests are partially lost, TFEMRNet outperforms YOLOv11 with precision improving by 20.06% and recall improving by 71.15%. For mutual adhesion conditions, TFEMRNet improves precision by 21.63% and recall by 28.51% compared to YOLOv11. In the most frequent independent pest scenario, TFEMRNet achieves a precision improvement of 36.97% and a recall improvement of 11.68% compared to YOLOv11.

An analysis of Table 3 shows that in the independent pest scenario, the number of false positives (FP) is relatively high due to the similarity between impurity features and *Grapholitha molesta* features. In high-resolution images, YOLOv11 extracts fewer semantic features of *Grapholitha molesta*, making it difficult to distinguish them from impurities, resulting in low recognition accuracy. TFEMRNet, however, extracts more semantic features of *Grapholitha molesta*, effectively distinguishing them from impurities and thus improving recognition accuracy.

Table 3. Performance of two models under multiple interference conditions.

Method	Situations	TP	TP	FN	Precision (%)	Recall (%)	F1 (%)
YOLOv11	independent	564	241	78	70.06	87.85	77.95
	Occlusion	16	5	58	76.19	21.62	33.68
	adhesion	303	108	146	73.72	67.48	70.46
	overlap	3	4	45	42.85	6.74	10.91
TFEMRNet	independent	639	5	3	99.22	99.53	99.37
	Occlusion	68	3	6	95.77	91.89	93.79
	adhesion	431	21	18	95.35	95.99	95.67
	overlap	43	5	5	89.58	89.58	89.58

In occlusion and overlap scenarios, the number of false negatives (FN) is higher due to the loss of *Grapholitha molesta* features, which are not detected by YOLOv11. This loss of features in the extracted feature maps leads to YOLOv11's inability to detect *Grapholitha molesta* under these conditions. TFEMRNet outperforms YOLOv11 across all four scenarios because YOLOv11 suffers from the loss of high-level semantic features of *Grapholitha molesta*. TFEMRNet's two-stage recognition method overcomes the challenges faced by traditional object detection models in extracting small-target features from high-resolution images, thus addressing the issue of low recognition accuracy for *Grapholitha molesta* in various conditions.

This demonstrates that our model performs well in handling small-target pests in high-resolution images.

4. Conclusions and Future Work

Pest monitoring systems are integral to apple cultivation and broader agricultural practices. This study focuses on the identification of *Grapholitha molesta* by introducing a two-stage detection model named TFEMRNet. The model effectively identifies *Grapholitha molesta* in high-resolution images, achieving a detection accuracy of 96.75% and a recall rate of 96.45%. The proposed two-stage recognition approach enhances the semantic features of *Grapholitha molesta* during the object detection phase, thereby addressing the challenge of low detection accuracy. To elevate the performance of the semantic segmentation component, the TFENet module incorporates Swin Transformer and EfficientNet architectures to extract both global and local pest features. Consequently, the enhanced network attains an Intersection over Union (IoU) of 91.63% in image segmentation, successfully segmenting *Grapholitha molesta* and providing high-quality samples for the subsequent object detection phase.

TFEMRNet demonstrates excellent scalability through its modular design and efficient feature extraction capabilities, enabling it to handle larger datasets and adapt to the monitoring of various harmful biological species. Its two-stage architecture facilitates distributed computing and parallel processing, ensuring high performance even when confronted with massive image data. Additionally, the TFENet module incorporates a Multi-Attention Channel Aggregation (MACA) mechanism that integrates diverse semantic features, thereby enhancing the model's ability to recognize different pest characteristics and allowing it to flexibly adapt to the detection requirements of various species. Leveraging transfer learning and data augmentation techniques, TFEMRNet can rapidly adapt to new pest types and improve its generalization capabilities across diversified datasets. Furthermore, TFEMRNet supports multi-task learning, enabling the simultaneous detection of multiple pest species, which further enhances the model's efficiency and adaptability. In practical applications, TFEMRNet continuously optimizes and updates the model by integrating multi-source data and real-time monitoring feedback, ensuring its effective deployment across different agricultural environments. Future research will focus on developing a lightweight version of TFEMRNet suitable for mobile devices and expanding the dataset to include a broader range of pest species, thereby further advancing the modernization and precision of intelligent agricultural practices.

Furthermore, TFEMRNet has been deployed on a pest warning website for practical application. The experimental results underscore the superiority of TFEMRNet's pest detection architecture, demonstrating its potential for identifying other small-target pests. However, the current optimization objective function of TFEMRNet does not yet address model lightweighting. Future research will focus on developing a lightweight version of TFEMRNet suitable for mobile devices and expanding the dataset to include samples of peach small borers and citrus small borers for training and deployment. Additionally, we plan to integrate climate and soil moisture data to identify patterns of pest occurrence, thereby advancing the modernization of smart agriculture through comprehensive analysis and data processing.

Author Contributions: J.M.: Conceptualization, Methodology, Software, Investigation, Writing—original draft. L.S.: Validation, Visualization, Software, Formal analysis. B.M.: Investigation, Visualization, Data curation. R.L.: Conceptualization. S.L.: Writing—review and editing. X.H.: Conceptualization. H.Z.: Project administration, Funding acquisition, Resources. J.W.: Project administration, Funding acquisition, Resources. All authors contributed to the article and approved the submitted version. All authors have read and agreed to the published version of the manuscript.

Funding: This research was funded by the National Natural Science Foundation of China (32071908) (32472014), China Agriculture Research System (CARS-27), Shandong-Chongqing Science and Technology Collaboration Project, Young Talent of Lifting engineering for Science and Technology in Shandong, China (SDAST2024QTA050), Shandong Province "University Youth Innovation Team" Program (2023KJ160).

Data Availability Statement: We acknowledge the importance of data sharing for research reproducibility and collaboration. Due to privacy concerns, the dataset used in this study cannot be fully open-sourced. However, anonymized data may be made available upon reasonable request, subject to compliance with applicable privacy regulations and ethical considerations. Researchers interested in accessing the dataset are encouraged to contact the corresponding author.

Acknowledgments: We acknowledge the financial support provided by the National Natural Science Foundation of China (Grant Nos. 32071908 and 32472014), the China Agriculture Research System (CARS-27), the Shandong-Chongqing Science and Technology Collaboration Project, the Young Talent Lifting Engineering Program for Science and Technology in Shandong, China (Grant No. SDAST2024QTA050), and the Shandong Province "University Youth Innovation Team" Program (Grant No. 2023KJ160).

Conflicts of Interest: The authors declare that they have no known competing financial interests or personal relationships that could have appeared to influence the work reported in this paper.

References

1. Kobiljonovna, Y.S. Importance of biological control against apple pests. In Proceedings of the International Conference on Scientific Research in Natural and Social Sciences, Virtual, 22–24 April 2022; Volume 1, pp. 201–207. Available online: https://econferenceseries.com/index.php/srnss/article/view/78 (accessed on 7 October 2022).
2. Čirjak, D.; Miklečić, I.; Lemić, D.; Kos, T.; Živković, I.P. Automatic pest monitoring systems in apple production under changing climatic conditions. *Horticulturae* **2022**, *8*, 520. [CrossRef]
3. Abbaspour-Gilandeh, Y.; Aghabara, A.; Davari, M.; Maja, J.M. Feasibility of using computer vision and artificial intelligence techniques in detection of some apple pests and diseases. *Appl. Sci.* **2022**, *12*, 906. [CrossRef]
4. Ryalls, J.M.W.; Garratt, M.P.D.; Spadaro, D.; Mauchline, A.L. The benefits of integrated pest management for apple depend on pest type and production metrics. *Front. Sustain. Food Syst.* **2024**, *8*, 1321067. [CrossRef]
5. Cardim Ferreira Lima, M.; Damascena de Almeida Leandro, M.E.; Valero, C.; Pereira Coronel, L.C.; Gonçalves Bazzo, C.O. Automatic detection and monitoring of insect pests—A review. *Agriculture* **2020**, *10*, 161. [CrossRef]
6. Passias, A.; Tsakalos, K.-A.; Rigogiannis, N.; Voglitsis, D.; Papanikolaou, N.; Michalopoulou, M.; Broufas, G.; Sirakoulis, G.C. Insect Pest Trap Development and DL-Based Pest Detection: A Comprehensive Review. *IEEE Trans. AgriFood Electron.* **2024**, *2*, 323–334. [CrossRef]
7. Barbedo, J.G.A. Detecting and classifying pests in crops using proximal images and machine learning: A review. *AI* **2020**, *1*, 312–328. [CrossRef]
8. Bilal, K.; Khalid, O.; Erbad, A.; Khan, S.U. Potentials, trends, and prospects in edge technologies: Fog, cloudlet, mobile edge, and micro data centers. *Comput. Netw.* **2018**, *130*, 94–120. [CrossRef]

9. Demirel, M.; Kumral, N.A. Artificial intelligence in integrated pest management. In *Artificial Intelligence and IoT-Based Technologies for Sustainable Farming and Smart Agriculture*; IGI Global: Hershey, PA, USA, 2021; pp. 289–313. [CrossRef]
10. Wang, Q.-J.; Zhang, S.-Y.; Dong, S.-F.; Zhang, G.-C.; Yang, J.; Li, R.; Wang, H.-Q. Pest24: A large-scale very small object data set of agricultural pests for multi-target detection. *Comput. Electron. Agric.* **2020**, *175*, 105585. [CrossRef]
11. Lyu, Z.; Jin, H.; Zhen, T.; Sun, F.; Xu, H. Small object recognition algorithm of grain pests based on SSD feature fusion. *IEEE Access* **2021**, *9*, 43202–43213. [CrossRef]
12. Liu, B.; Jia, Y.; Liu, L.; Dang, Y.; Song, S. Skip DETR: End-to-end Skip connection model for small object detection in forestry pest dataset. *Front. Plant Sci.* **2023**, *14*, 1219474. [CrossRef] [PubMed]
13. Tian, Y.; Wang, S.; Li, E.; Yang, G.; Liang, Z.; Tan, M. MD-YOLO: Multi-scale Dense YOLO for small target pest detection. *Comput. Electron. Agric.* **2023**, *213*, 108233. [CrossRef]
14. Liu, J.; Wang, X. Plant diseases and pests detection based on deep learning: A review. *Plant Methods* **2021**, *17*, 22. [CrossRef] [PubMed]
15. Zhang, Z.; Rong, J.; Qi, Z.; Yang, Y.; Zheng, X.; Gao, J.; Li, W.; Yuan, T. A multi-species pest recognition and counting method based on a density map in the greenhouse. *Comput. Electron. Agric.* **2024**, *217*, 108554. [CrossRef]
16. Russell, B.C.; Torralba, A.; Murphy, K.P.; Freeman, W.T. LabelMe: A database and web-based tool for image annotation. *Int. J. Comput. Vis.* **2008**, *77*, 157–173. [CrossRef]
17. Kim, E.K.; Lee, H.; Kim, J.Y.; Kim, S. Data augmentation method by applying color perturbation of inverse PSNR and geometric transformations for object recognition based on deep learning. *Appl. Sci.* **2020**, *10*, 3755. [CrossRef]
18. He, K.; Gkioxari, G.; Dollár, P.; Girshick, R. Mask r-cnn. In Proceedings of the IEEE International Conference on Computer Vision, Venice, Italy, 22–29 October 2017; pp. 2961–2969. [CrossRef]
19. Liu, Z.; Ning, J.; Cao, Y.; Wei, Y.; Zhang, Z.; Lin, S.; Hu, H. Video swin transformer. In Proceedings of the IEEE/CVF Conference on Computer Vision and Pattern Recognition, New Orleans, LA, USA, 18–24 June 2022. [CrossRef]
20. Koonce, B.; Koonce, B.E. *Convolutional Neural Networks with Swift for Tensorflow: Image Recognition and Dataset Categorization*; Apress: New York, NY, USA, 2021. [CrossRef]
21. Zhang, Y.; Liu, H.; Hu, Q. Transfuse: Fusing transformers and cnns for medical image segmentation. In *Medical Image Computing and Computer Assisted Intervention—MICCAI 2021: 24th International Conference, Strasbourg, France, September 27–October 1, 2021, Proceedings, Part I 24*; Springer International Publishing: Berlin/Heidelberg, Germany, 2021; pp. 14–24. [CrossRef]

AgriEngineering

Article

A Convolutional Neural Network Algorithm for Pest Detection Using GoogleNet

Intan Nurma Yulita [1,*], Muhamad Farid Ridho Rambe [1], Asep Sholahuddin [1] and Anton Satria Prabuwono [2]

1 Department of Computer Science, Faculty of Mathematics and Natural Sciences, Universitas Padjadjaran, Sumedang 45363, Indonesia
2 Faculty of Computing and Information Technology in Rabigh, King Abdulaziz University, Rabigh 21911, Saudi Arabia; aprabuwono@kau.edu.sa
* Correspondence: intan.nurma@unpad.ac.id

Abstract: The primary strategy for mitigating lost productivity entails promptly, accurately, and efficiently detecting plant pests. Although detection by humans can be useful in detecting certain pests, it is often slower compared to automated methods, such as machine learning. Hence, this study employs a Convolutional Neural Network (CNN) model, specifically GoogleNet, to detect pests within mobile applications. The technique of detection involves the input of images depicting plant pests, which are subsequently subjected to further processing. This study employed many experimental methods to determine the most effective model. The model exhibiting a 93.78% accuracy stands out as the most superior model within the scope of this investigation. The aforementioned model has been included in a smartphone application with the purpose of facilitating Indonesian farmers in the identification of pests affecting their crops. The implementation of an Indonesian language application is a contribution to this research. Using this local language makes it easier for Indonesian farmers to use it. The potential impact of this application on Indonesian farmers is anticipated to be significant. By enhancing pest identification capabilities, farmers may employ more suitable pest management strategies, leading to improved crop yields in the long run.

Keywords: pest detection; GoogLeNet; Convolutional Neural Network; mobile application

Citation: Yulita, I.N.; Rambe, M.F.R.; Sholahuddin, A.; Prabuwono, A.S. A Convolutional Neural Network Algorithm for Pest Detection Using GoogleNet. *AgriEngineering* **2023**, *5*, 2366–2380. https://doi.org/10.3390/agriengineering5040145

Academic Editors: Ray E. Sheriff and Chiew Foong Kwong

Received: 27 September 2023
Revised: 2 December 2023
Accepted: 6 December 2023
Published: 8 December 2023

1. Introduction

Food is one of the most fundamental requirements for maintaining human health and vitality. Almost anywhere in Indonesia, a wide variety of crops may be grown due to the country's abundance of arable land [1]. When it comes to plant exports, Indonesia is a major player [2]. Due to its tropical location, Indonesia frequently experiences floods, droughts, and other forms of severe weather. The changing patterns of precipitation and temperature brought on by global warming also threaten plant growth and crop output. Traditional, inefficient farming practices are still used by some Indonesian farmers. Another factor that might prevent farmers from boosting output is a lack of access to cutting-edge agricultural technologies like organic fertilizer, improved seeds, enough irrigation, and safe pesticides. Social disparities in Indonesian society, which may be experienced by farmers in particular locations, may also have an impact on agricultural output. Small farmers' inability to run their farms efficiently is largely attributable to their lack of knowledge. Poor harvest outcomes are to be expected under these conditions.

The situation of farmers in Indonesia is strongly connected to the challenge of identifying pests in agricultural settings, which in turn can have an effect on crop yields that are not at their full potential [3]. The education level of many farmers in Indonesia is rather poor, and they have limited access to agricultural training and knowledge [4–6]. As a consequence of this, they might not have an appropriate understanding of how to properly detect and control agricultural pests. Smaller farms may not have the financial means to invest in or learn to utilize sophisticated software tools for pest identification. As a result, it may be challenging to see pests in their early stages. Without prompt and proper treatment, a pest infestation on one plant will spread to others in the area [7–11]. In a typical crop, a single rat may harm anywhere from 11 to 176 plants in a single night [12]. During the breeding season, a single rat can destroy up to 26 stalks in a single night, leaving behind only a few rows of plants. It takes a mouse only 21 days to have her own litter of 6–8 young. Obviously, the severity of this problem might increase dramatically if it is not addressed promptly and effectively. Depending on the surrounding environment and the care offered to the plant, there may be a wide variety of pests that target a single plant type [13].

Several ancillary technologies [14] are already in use in the agriculture industry, including drones. Farmers may utilize drones to keep an eye on crops from above and protect them from pests by spraying pesticides over large areas [15–18]. However, without special detecting equipment, drones cannot identify which pests are present on a farm, and not every farmer can afford to buy them. Pest detection apps for smartphones that are already ubiquitous in Indonesia are a necessity.

It is important to be able to identify the numerous plant pests that occur. Machine learning using supervised model approaches is one approach [19–25]. This technique needs data that already have a previous label. In machine learning, there are also various methods of image classification. One effective method is the use of CNNs. The Pest Region-CNN end-to-end detection model, otherwise known as Pest R-CNN, was created by Du et al. for the maize pest Spodoptera frugiperda [26]. It tracks their consumption of maize leaves using the Faster R-CNN model. The model classifies invasion intensity into juvenile, minor, moderate, and severe using UAV-acquired high-spatial-resolution RGB ortho-images. The model demonstrated the efficacy of deep learning object detection in pest monitoring with a mean average accuracy of 43.6% on the test dataset. Their model's detection accuracy improved by 12% and 19% over Faster R-CNN and YOLOv5, respectively. Jiao et al. developed a CNN-based multi-class pest identification approach for complicated scenarios [27]. Their research introduces adaptive feature fusion within feature pyramid networks to extract deeper pest characteristics. To reduce information loss in the highest-level feature map, an adaptive augmentation module was designed. Finally, a two-stage region-based CNN (R-CNN) refined the predicted bounding boxes to determine the pest types and locations in each image. Compared to SSD, RetinaNet, FPN, Dynamic R-CNN, and Cascade R-CNN, their method achieved 77.0% accuracy, which was the highest.

The key to lowering the amount of produce that is lost is finding plant pests as early, correctly, and effectively as possible. In this scenario, apps that run on mobile devices provide a significant and pertinent solution. With the help of smartphone applications, farmers are able to conveniently perform routine checks on their crops. They can capture images of plants and detect pests using their mobile devices, making it simpler for them to record and monitor everyday agricultural conditions. Mobile applications have been developed with user interfaces that are straightforward and simple to operate. Because of this, they may be used without trouble by farmers who have varied levels of education and prior experience with technology. A number of applications have been developed to help farmers detect pests. Prabha et al. developed an Android mobile application that utilizes a Deep CNN (DCNN)-based artificial intelligence system for the purpose of insect identification in maize production [28]. The utilization of CNN models on mobile devices yields advantages for all stakeholders, with particular benefits being offered to farmers and agricultural extension experts, since they enhance their accessibility. A mobile application for Android was developed for the purpose of detecting fall armyworm infection in maize crops. This application incorporated the advice provided by Tamil Nadu Agricultural University's Integrated Pest Management (TNAU IPM) capsules. This mobile software offers guidance on effectively managing the issue of fall armyworm infestation. Chen et al. presented an application for the identification of scale pests in Taiwan using image-based methods [29]. The Coccidae and Diaspididae species, known as mealybugs, are the predominant pests of scale insects in Taiwan. These species have the potential to cause significant harm to plants and pose a substantial threat to agricultural productivity. Therefore, the identification of scale pests holds significant importance within the agricultural sector of Taiwan. You Only Look Once v4 (YOLO v4), single-shot multi-box detectors (SSDs), and faster region-based CNNs (Faster R-CNNs) are some of the deep learning algorithms used to detect and precisely localize scale pests inside an image. A smartphone application has been created that utilizes a trained scale pest detection model. This application aims to assist in the identification of pests in agricultural settings, hence enabling farmers to effectively apply targeted pesticides and minimize crop losses.

Despite the existence of several related applications in other countries, there is a distinct necessity for tailored mobile-based applications designed specifically for Indonesian farmers. The compatibility between applications produced in other countries and the local language may not always be guaranteed. The development of a mobile application tailored to Indonesian farmers would facilitate the incorporation of local language into the program's content and user interface. This approach would enhance comprehension and acceptance among farmers, hence facilitating their utilization of the application. There is a poor general level of English proficiency in the whole population of Indonesia [30]. Farmers who have completed elementary and secondary schooling will also find this to be a difficult situation [31]. As a result, developing a mobile application tailored to the needs of Indonesian farmers has the potential to address the distinct obstacles encountered within the agricultural sector in Indonesia.

This study also employs a CNN architecture, namely GoogleNet, for the purpose of identifying pests within mobile applications. GoogleNet makes use of a technology called Inception blocks, which integrate a variety of convolutional algorithms and filter sizes into a single layer. This results in a reduction in the number of computing processes and parameters that are necessary, making the model simpler. When compared to other CNN architectures that were available at the time, such as AlexNet or VGG, this model featured a significantly lower total number of parameters [32,33]. This enabled a more effective utilization of the RAM of the mobile device. In addition, this model includes methods such as L2 regularization and dropout to avoid overfitting [34]. This can assist in retaining model performance on small datasets without the need for a high number of parameters by reducing the likelihood of the model being overly accurate. The fact that GoogleNet was developed with hardware capabilities in mind, such as the Single Instruction Multiple Data (SIMD) instructions that are present on many recent mobile devices [35], is another key aspect to take into consideration. This helps to maximize the usage of the resources available in the hardware. Traditional convolution is rendered obsolete as a result of the development of depth-wise separable convolution. Because of this, the number of operations and parameters that are necessary is decreased, which results in increased computing efficiency. The use of a large number of convolutional layers of relatively modest size helps to extract information in an effective manner and also enables dimensionality reduction. All of the aforementioned criteria contribute to GoogleNet's ability to achieve an optimal performance on mobile devices with limited resource capacities. Because of this, it is an excellent option for mobile apps, which place a premium on computing efficiency as well as model size. This aligns with the varied agricultural landscape of Indonesia.

2. Materials and Methods

2.1. Data Collecting

Insects and other organisms that cause damage to plants are the focus of this research. The information was obtained from the Kaggle website, which may be viewed at the following address on the internet: https://www.kaggle.com/simranvolunesia/pest-dataset (accessed on 21 February 2022). The data that were used cover nine different kinds of plant-destroying insects, including aphids, armyworm, beetles, bollworms, grasshoppers, mites, mosquitoes, and sawflies. Stem borers were also included in the list. There are two types of data in the dataset provided. The data included 2700 images, with each class contributing 300 images. These data were used to train the model, with 1800 data as training data, and the remaining 900 data as validation data. The data were distributed evenly in terms of number throughout each class so that the process of learning could go more smoothly with the weighting model. The second data group comprised the taking of 450 images, with each class contributing 50 images. This second data group was used to test the model.

2.2. Data Preprocessing

In the first step of the data preparation stage, the data were separated into two categories: training data and validation data. Following the distribution of the data, the images were subjected to preprocessing. Image preprocessing was performed so that the model was able to detect characteristics in images with greater ease as a result of the work that was performed. Image data relating to plants and pests were examined in this study. The data were eventually utilized as input data in the classification process once they had been processed via the preparation step. Data preprocessing consisted of two main stages, namely image resizing and image augmentation. Image resizing was performed to change the input image to a size of 224 × 224, with a size of 3 bits per pixel. Some deep learning operations are more efficient when the size of the image is a multiple of certain numbers, such as 32, 64, or 128. The GPU can process data more efficiently with an image size of 224 × 224 pixels [36]. Numerous deep learning models have been trained (pre-trained) with datasets containing images of size 224 × 224 [37]. Utilizing pre-trained models is facilitated by employing the same size during detection. Also, certain deep learning frameworks, such as TensorFlow, have pre-trained models that support 224 × 224 image sizes directly [38]. This facilitates the implementation of these models within applications.

Image augmentation was only carried out on the training data, because a lot of data were needed for the model training process [39–42]. The process was carried out to expand the variety of images that were used by the model using the module tf.image in TensorFlow. This stage was carried out to increase the amount of data by changing the shape of the image. The augmentation process carried out involved rotating the image, shifting the image based on the width and height of the image, applying shearing to the image, enlarging the image, and flipping the image horizontally.

- Width Shift

  This was carried out to shift the image width to the left and right to provide variations for the model to learn. The width shift range value given was 0.2.

- Height Shift

  This was carried out to shift the image height up and down to provide variations for the model to learn. The height shift range value given was 0.2.

- Shears

  This was carried out to tilt the image regarding the *x*-axis and *y*-axis to provide variations for the model to learn. The value of the shear range given was 0.2.

- Zoom

  This was carried out to enlarge or reduce the image to provide variations for the model to learn. The value of the given zoom range is 0.2.

- Horizontal Flip

  This involved flipping an image vertically or horizontally to provide variations for the model to learn. It was assisted using external tools in the form of an image data generator library that had been previously provided by Tensorflow to make the process easier.

2.3. Classification

One deep learning approach is CNN, which is based on artificial neural networks (ANN) [43–45]. The process data at both the input layer and the hidden layer then output the results to the output layer. The difference between CNNs and ANNs is that the former focuses more on recognition than the latter. A CNN has a structure consisting of various layers that form a stacked network [46,47]. Convolutional layers, pooling layers, and fully connected layers are the three main layers that produce a CNN after being combined. The CNN layer is very important. This layer continuously uses filters on the data in the input layer, and then produces the output. In the pooling layer process, the output convolutional layer has its dimensions and parameters reduced [48–50]. Maximum pooling takes the largest value from the pooling filter used. Average pooling takes the average value from the pooling filter used. The pooling process helps reduce the computational complexity of the model and prevents overfitting. Multilayer perceptron applications typically use fully connected layers. The goal of convolutional or pooling layers is to increase the dimensionality of data that can be classified linearly. To achieve this goal, data that have been previously processed in this layer will be flattened first before being inserted into this layer.

In 2014, Google collaborated with various universities to conduct research on GoogleNet, also known as Inception V1. The winner of ILSVRC 2014 is this architectural design. Compared to winning designs from previous years, this design is more accurate. This architecture uses global average pooling and 1 × 1 convolution in the middle. The main aim of this architecture is to determine how optimally the local sparse structure in a convolutional vision network can be estimated and encompassed in easily accessible thick components [51–55]. This architecture's 1 × 1 convolution concept helps reduce the number of parameters (loads and biases) present in the architecture significantly. At the end of the network, this architecture uses an elative global average pooling method. Overall, this structure consists of twenty-two layers. To maintain computational efficiency, this architecture can be used on devices. It has low computing capabilities, as depicted in Figure 1, which explains the details of the twenty-two architectural layers. This study used the CNN method. The architecture used was GoogLeNet, which was a development of LeNet and AlexNet and had twenty-two layers. The GoogLeNet model was implemented by training the model using previously obtained data. By using a pre-existing dataset, this model predicted images of plants.

Figure 1 demonstrates the GoogLeNet architecture that was utilized in the research. The subsequent section provides an outline of the subsequent steps:

1. The input used is an image measuring 224 × 224.
2. The image data undergo five phases of convolution prior to reaching the fully connected layer.
3. In the initial phase, the data are subjected to 7 × 7 convolution and 3 × 3 max pooling.
4. The data are then transit through two convolution layers with dimensions of 1 × 1 and 3 × 3, followed by a max pooling layer with dimensions of 3 × 3.
5. In the third stage, data enter the Inception model layer twice and are processed by max pooling with 3 × 3 dimensions.
6. In the fourth stage, the data are first processed by an Inception model layer and then by an auxiliary classifier. The data then pass through three layers of the inception model and then the second auxiliary classifier. Then, the data pass through an Inception model layer once more prior to max pooling with 3 × 3 dimensions.
7. In the fifth stage, the data travel through two layers of the Inception model and then undergo 7 × 7-dimensional average pooling. The data then proceed to the output after passing through the fully connected layer, which is also the sixth stage of the architecture.
8. At stages 2 through 5, a 1 × 1 convolution layer is used to reduce the dimensions of the output.

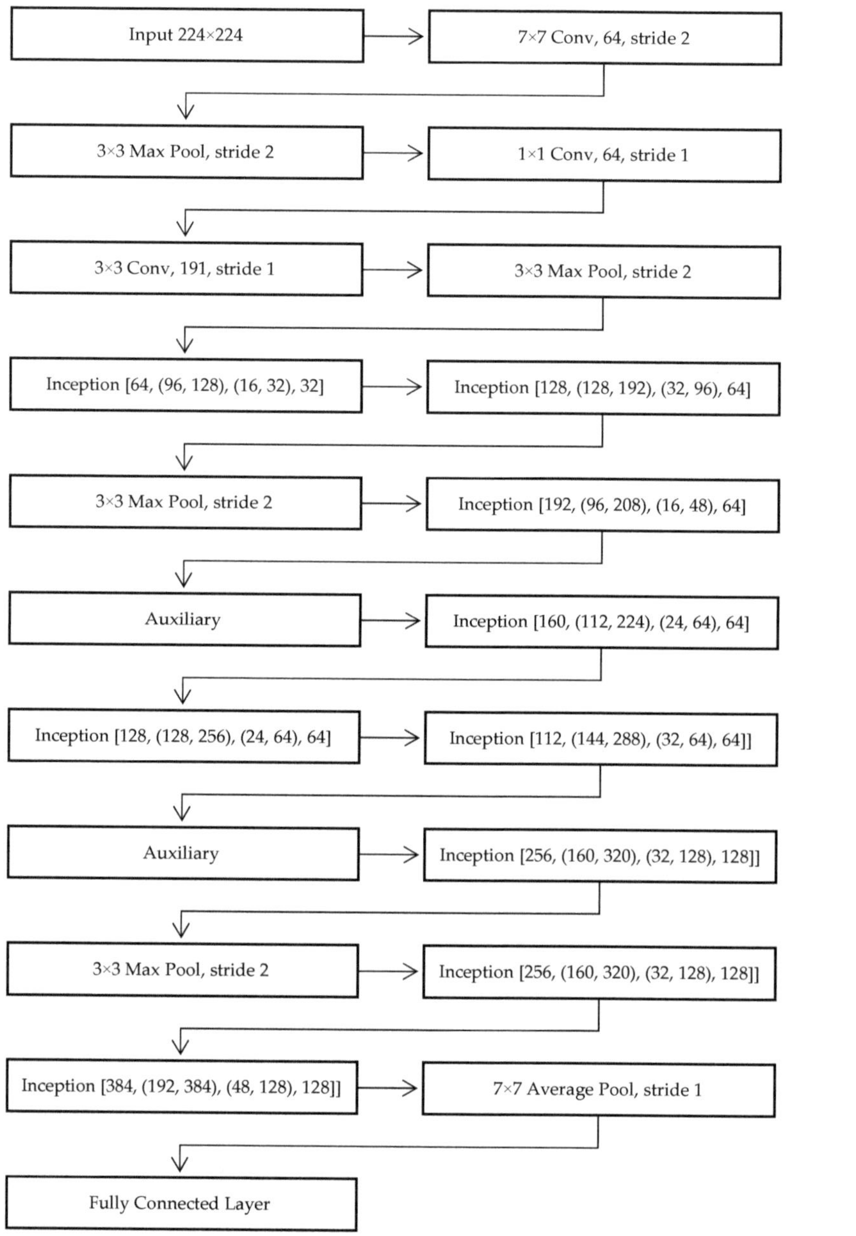

Figure 1. Details of each layer on GoogleNet.

An inception model and auxiliary classifier were developed for the GoogLeNet architecture in this study before creating the main model. Figure 2 illustrates the inception model's architecture. In addition, Figure 3 describes the auxiliary classifier, which was utilized by the GoogleNet model.

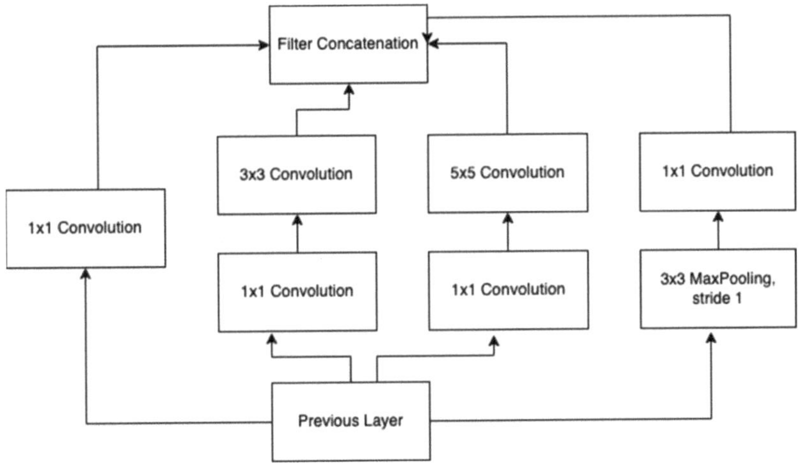

Figure 2. Architecture of the Inception Model.

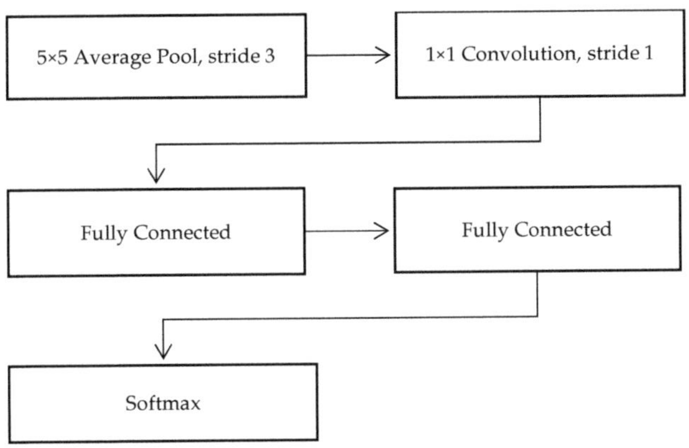

Figure 3. Architecture of Auxiliary Classifiers.

Before the model was implemented, hyperparameters were determined. The research was conducted by experimenting with multiple hyperparameters to determine the optimal pair of hyperparameters that yield the most accurate results from the model. This research employed the number of dense layers in the fully connected layer and various dropout values as hyperparameters. The research utilized these two categories of hyperparameters to find the optimal solution. The dropout values evaluated were 0.3, 0.4, and 0.5, and the number of dense layers was 1, 3, or 5. The hyperparameter determination procedure involved varying the number of dense layers and the dropout value on the fully connected layer to select the model with the highest accuracy. After determining the optimal hyperparameters, the GoogleNet model was applied to the applications created with the hyperparameter pair that yielded the highest accuracy.

In addition to the hyperparameters that were investigated, this study also set additional factors, including epochs, the batch size, optimizers, and the image size. The number of epochs was set to 100. The batch size, which determined the number of samples processed in each iteration, was specified as 8. The optimizer in this study was Stochastic Gradient Descent (SGD). The learning rate, denoting the step size at each iteration, was assigned a value of 0.00001. Lastly, the image size was 224 × 224. Hyperparameter changes involved a dense layer between the flatten layer and the output layer, as well as modifications to the values inside the dropout layer.

2.4. Evaluation

Model testing was carried out using data that were different from the training and validation data. This was to ensure that each model created made predictions on data that the model had not previously studied. Model validation was carried out via K-fold cross-validation (K = 3). This method helped the process of evaluating and determining hyperparameters, especially on smaller datasets. This method divided k datasets randomly. To determine the validation and training data, the rotation of each piece of data was performed [56–58]. The aim of this training was to produce different evaluation results for each fold performed. The hyperparameters were determined by taking the average value of these assessments. In addition, a confusion matrix was displayed in this study to simplify the test. The accuracy values were obtained using this matrix. This was the amount of data that had been classified correctly. During this procedure for testing models, the only model that was chosen was the one that had the highest quality validation values for each hyperparameter. This method was used to check the model with new data. The test results were used to assess the Android application system model created.

2.5. Implementation of Mobile Applications

The training process was carried out using Google Colab (Google LLC, Mountain View, CA, USA). After the model evaluation process was complete, the conversion stage was carried out. Of all the hyperparameter experiments carried out, only the best model was implemented in the application. Because the application to be developed was smartphone-based, this conversion process was necessary. Tensorflow Lite is a version of Tensorflow specifically made for mobile-based applications used on the Android operating system, and was used in this study. The disparities in the accuracy and duration of detection between training in Google Colab and testing on a smartphone might arise from several sources. Some examples include hardware disparities, deployment optimization, and dependencies on frameworks and libraries. In order to address this issue, we employ frameworks such as TensorFlow Lite, which is specifically developed to enhance the efficiency of models for deployment on mobile devices. It provides tools and techniques particularly designed to enhance the speed and efficiency of inference on mobile devices. It can strive to minimize the disparities in precision and response time between the training environment in Google Colab and the testing environment on a smartphone. This process was carried out to change the format of the model obtained, which was previously in h5 format, into TFlite format. After the conversion process was complete, the model was moved to the Android application directory. It carried out the detection process using the application diagram package that is presented in Figure 4. The system structure consisted of relationships between directories in the application, as depicted in this diagram. Three main directories comprised the created directory: res, Java, and the Machine Learning Model. The res directory consisted of assets to support the visuals of the application that were used in the Java directory later. The directory included the logic that regulated the application path. Machine learning models were applied to the applications in this directory. The deployment diagram in Figure 5 explains the physical settings of the application being created. The Java classes were components that form the flow and logic of an application; resources were the material that was shown to the user. Tflite was the model used for the machine learning prediction process. This GoogleNet-installed mobile application requires an Android operating system, a minimum of 2 GB RAM memory, and a minimum of 32 GB internal storage. The camera's resolution must be at least 4 megapixels for direct image capture. When run, this application has a maximum size of 10 MB.

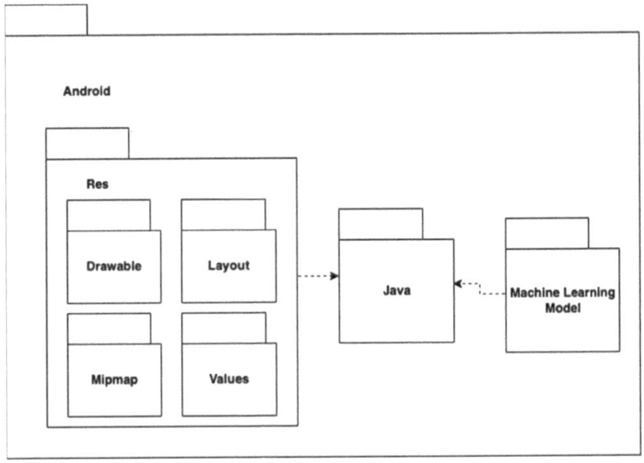

Figure 4. Package diagram.

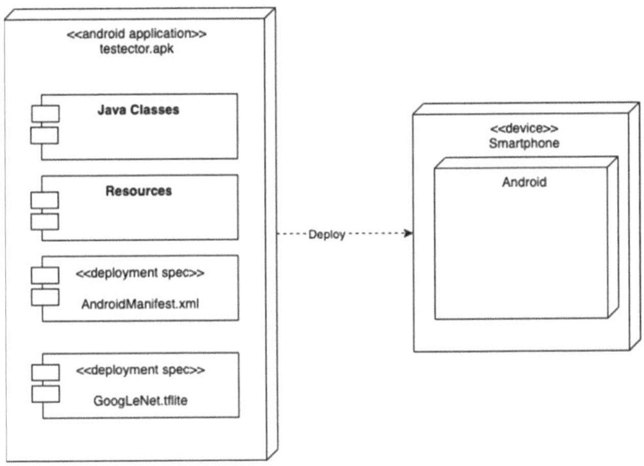

Figure 5. Deployment diagram.

3. Results

At this point, an evaluation of the hyperparameters that were employed is carried out, and its findings are analyzed. The specified hyperparameters were utilized in a total of 12 separate experiments. The metrics taken into consideration were the accuracy score from the testing to ensure that there was no overfitting or underfitting. In the testing process, the model with the best accuracy value was selected from the k-fold process that was previously carried out. This was done because K-fold produced three models and only used one of the models later to be implemented into the prototype to be made. The hyperparameters to be tested were the number of dense layers and the dropout value. The hyperparameters of dense layers had a total of 0, 1, 3, and 5, with some dropout values of 0.3, 0.4, and 0.5.

Table 1 shows that the experiment that produced the best model was the one that had the addition of five dense layers and a dropout of 0.5. Even though it had the same accuracy during testing as a model that did not experience the addition of a dense layer, and even though the dropout value was 0.5, the accuracy of the model with the addition of five dense layers during validation was 1% superior. The worst model from this experiment was the model with the addition of three dense layers, with a dropout value of 0.3% and an accuracy of 89.33%. The best confusion matrix model is in Figure 6.

Table 1. Experimental results.

Dropout	Number of Dense Layer	Number of Neurons	Accuracy
0.3	0	0	92.22%
0.3	1	256	91.33%
0.3	3	512; 256; 128	89.33%
0.3	5	512; 256; 128; 64; 32	93.11%
0.4	0	0	92.88%
0.4	1	256	91.77%
0.4	3	512; 256; 128	92.44%
0.4	5	512; 256; 128; 64; 32	92.22%
0.5	0	0	93.78%
0.5	1	256	93.33%
0.5	3	512; 256; 128	93.55%
0.5	5	512; 256; 128; 64; 32	93.78%

Predicted label

Actual label	A	B	C	D	E	F	G	H	I
A	49	0	0	0	0	0	0	1	0
B	0	49	0	0	0	0	0	1	0
C	0	0	50	0	0	0	0	0	0
D	0	1	0	40	0	1	1	4	3
E	1	0	0	0	48	0	0	1	0
F	0	0	0	4	0	44	1	0	1
G	0	0	0	0	0	2	48	0	0
H	1	1	0	0	1	0	0	45	2
I	0	0	1	0	0	0	0	0	49

A: aphids
B: armyworm
C: beetle
D: bollworm
E: grasshopper
F: mites
G: mosquito
H: sawfly
I: stem borer

Figure 6. Confusion matrix.

In this case, almost all hyperparameters were stable in the range of 91–93% except for the hyperparameter pair number with 3 dense layers and a dropout value of 0.3, which had an accuracy of 89%. Models that experience the addition of five dense layers tend to be stable, with an accuracy in the range of 92–93%. With an increase in the number of dense layers, the model was able to take in deeper features in the image. However, if it took too many features from an image, overfitting occurred. If the model was too focused on a particular training dataset, it made predictions correctly when given another, similar dataset. Furthermore, in this case, the dropout value had no significant effect on producing an optimal model.

The stages of system implementation were carried out by making a prototype application to detect pests on plants. The prototype of this application was based on the Android operating system. In this application prototype, there are four main views: pages from the main menu, pest detection, how to deal with pests, and application description. The application was shown to farmers in Indonesia so that the language of instruction is the national language, namely Indonesian. A detection page is a page containing the primary application features that will be provided to end users, as shown in Figure 7. It exhibits a preview of the detected images and their classification based on nine labels extracted from the dataset. The detection page has two options to select the type of input that will be used for detection. The two categories of input are direct capture with the mobile device's camera and taking images from the gallery. This page employs a pre-built deep-learning model for detection.

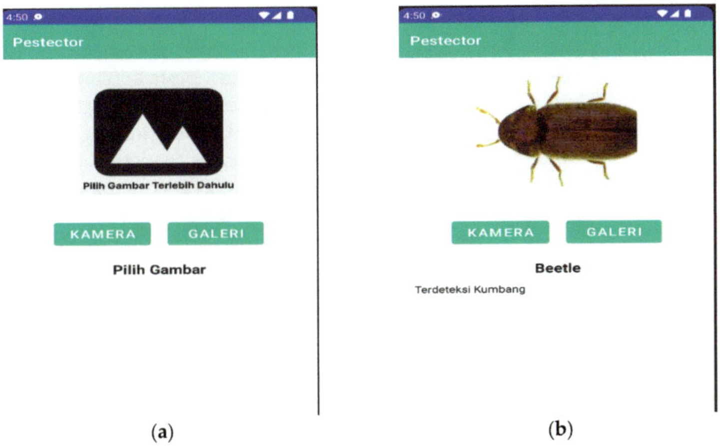

Figure 7. The mobile application is integrated with a pest detection system utilizing GoogleNet. The purpose of this application is to answer the needs of Indonesian farmers, who predominantly communicate in the Indonesian language, by presenting its content in Indonesian. (**a**) Illustrates the user interface for inputting images to be detected. The "Kamera" function refers to the capability of capturing images by utilizing the built-in camera on a mobile device. In contrast, the "Galeri" feature facilitates the inclusion of images from the pre-existing collection into the user's mobile gallery. (**b**) Displays the program interface that exhibits the detected outcomes of the pests that were examined.

This study evaluated the application using a sample size of 36 participants. Table 2 displays the roster of inquiries. The participants in the study included 15 males and 21 females, ranging in age from 16 to 43 years old. The majority of participants were residing in Java and Bali, which are two regions in Indonesia known for their extensive agricultural areas. This survey utilized a Likert scale consisting of 10 assertions, each accompanied by a range of response options that gauged the level of agreement or disagreement with a topic. A rating of 1 signified a strong disagreement, while a rating of 5 signified a strong agreement.

Table 2. Survey results.

No	Questions	Mean	Median	Variation	Min.	Max.
1	Do you think this application will be effective in detecting pests?	4.36	4.00	0.13	3	5
2	Will this application help in identifying pest problems in plants?	4.31	4.00	0.15	3	5
3	Are you satisfied with the performance of this application in detecting pests?	4.08	4.00	0.19	2	5
4	Do you like the features of this app?	4.11	4.00	0.18	2	5
5	Is this application fast and accurate in detecting pests?	3.89	4.00	0.19	2	5
6	Can this app detect pests better than humans?	3.78	4.00	0.19	3	5
7	Will this application increase efficiency or productivity in agriculture?	4.31	4.00	0.17	3	5
8	Will this application make a positive contribution to overcoming pest problems in plants or the surrounding environment?	4.36	4.50	0.16	3	5
9	Is this application easy to use?	4.36	5.00	0.17	3	5
10	Would you recommend this app to others, especially farmers?	4.33	5.00	0.19	3	5

The value 2 appears three times in all responses in the survey. The interviews with the three individuals revealed, respectively, discontentment with the application being available only on Android and a desire for the application to offer many more functionalities to assist farmers in their work; furthermore, one respondent highlighted the lack of accuracy in real-time testing. Nevertheless, the survey findings from the 36 participants indicated a favorable outlook in terms of their contentment with the program. This is evidenced by the fact that the average score falls between the range of 3.78 to 4.36.

4. Conclusions

Using the GoogLeNet architecture, this research tries to apply the CNN method to develop an application that can detect pests in plants. In its implementation, this research showed that the convolution function, which extracted features pixel by pixel in the image, allowed CNN to detect pests on plants. Adding a thick layer improved the accuracy of the model; the number of thick layers made the model extract more features from the dataset, but adding too many thick layers caused the overfitting of the model. The layer dropout value of 0.5 was the most accurate but did not show significant changes between models. The model with an accuracy of 93.78% was the best in this research. By entering the GoogLeNet model into an application, an application can be created to detect pests in plants on Android devices. This application was developed on a smartphone basis so that Indonesian farmers could easily identify pests on their crops. This application is also presented in Indonesian, which is the national language of Indonesia, so it is easy to use. It is the main contribution of this research. It is hoped that this application will be a breakthrough for Indonesian farmers. By identifying pests more easily, farmers can treat pests more appropriately, which will ultimately result in better crop yields.

This research aims to help researchers and practitioners face the challenges that come with real-world implementation by covering topics such as more heterogeneous data, a focus on performance in limited-power devices, integration in end-to-end systems, security, end-user evaluation, the development of custom algorithms, and ongoing evaluation. This study has the potential to be extended further into applications that are more practical by increasing the scale of the experiment. This is also supported by an interview conducted with a respondent who is a staff member from the Plant Quarantine Agency of the Indonesian government. The suggestion is that the impact can be enhanced if the pests listed for detection also encompass the sorts of pests that currently exist in Indonesia but are still prevalent in specific regions. Due to the current process, this task is currently being performed manually by the agency. Therefore, it is possible to ensure that this research offers maximum advantages in a variety of practical applications while ensuring security, privacy, and a good influence on society if a commitment is made to following the relevant rules and standards and taking an ethical approach.

Author Contributions: Conceptualization, I.N.Y.; methodology, I.N.Y., M.F.R.R. and A.S.; software, M.F.R.R.; validation, I.N.Y. and A.S.; formal analysis, I.N.Y.; investigation, I.N.Y. and M.F.R.R.; resources, M.F.R.R.; data curation, I.N.Y. and A.S.; writing—original draft preparation, I.N.Y., M.F.R.R. and A.S.P.; writing—review and editing, I.N.Y. and A.S.P.; visualization, M.F.R.R.; supervision, A.S.P.; project administration, I.N.Y.; funding acquisition, I.N.Y. All authors have read and agreed to the published version of the manuscript.

Funding: The Associate Professor Acceleration Research 2023 initiative at the University of Padjadjaran provided funding for this project, No. Contract: 1549/UN6.3.1/PT.00/2023.

Data Availability Statement: Publicly available datasets were analyzed in this study. This data can be found here: https://www.kaggle.com/simranvolunesia/pest-dataset (accessed on 21 February 2022).

Acknowledgments: The authors would like to express their sincere gratitude to the Universitas Padjadjaran and the Directorate of Research and Community Service (DRPM). In addition, we would like to express our gratitude to the World Class Professor 2023 initiative, which is run by the Indonesian Ministry of Education, Culture, Research, and Technology. We extend our gratitude to the 36 participants who willingly dedicated their time to assessing our application and engaging in interviews about it.

Conflicts of Interest: The authors declare no conflict of interest.

References

1. Rozaki, Z.; Wijaya, O.; Rahmawati, N.; Rahayu, L. Farmers' disaster mitigation strategies in Indonesia. *Rev. Agric. Sci.* **2021**, *9*, 178–194. [CrossRef]
2. Gandharum, L.; Mulyani, M.E.; Hartono, D.M.; Karsidi, A.; Ahmad, M. Remote sensing versus the area sampling frame method in paddy rice acreage estimation in Indramayu regency, West Java province, Indonesia. *Int. J. Remote Sens.* **2021**, *42*, 1738–1767. [CrossRef]
3. Yatoo, A.M.; Ali, M.N.; Baba, Z.A.; Hassan, B. Sustainable management of diseases and pests in crops by vermicompost and vermicompost tea. A review. *Agron. Sustain. Dev.* **2021**, *41*, 7. [CrossRef]
4. Nuryitmawan, T.R. The Impact of Credit on Multidimensional Poverty in Rural Areas: A Case Study of the Indonesian Agricultural Sector. *Agriecobis J. Agric. Socioecon. Bus.* **2021**, *4*, 32–45. [CrossRef]
5. Safitri, K.I.; Abdoellah, O.S.; Gunawan, B.; Suparman, Y.; Mubarak, A.Z.; Pardede, M. The Adaptation of Export-Scale Urban Farmers Amid the COVID-19 Pandemic in Bandung Metropolitan. *Qual. Rep.* **2022**, *27*, 1169–1196. [CrossRef]
6. Agustina, K.K.; Wirawan, I.M.A.; Sudarmaja, I.M.; Subrata, M.; Dharmawan, N.S. The first report on the prevalence of soil-transmitted helminth infections and associated risk factors among traditional pig farmers in Bali Province, Indonesia. *Vet. World* **2022**, *15*, 1154. [CrossRef]
7. Thao, L.Q.; Cuong, D.D.; Anh, N.T.; Minh, N.; Tam, N.D. Pest Early Detection in Greenhouse Using Machine Learning. *Rev. D'intelligence Artif.* **2022**, *36*, 209–214. [CrossRef]
8. Wang, X.; Liu, J.; Zhu, X. Early real-time detection algorithm of tomato diseases and pests in the natural environment. *Plant Methods* **2021**, *17*, 43. [CrossRef]
9. Gupta, N.; Slawson, D.D.; Moffat, A.J. Using citizen science for early detection of tree pests and diseases: Perceptions of professional and public participants. *Biol. Invasions* **2022**, *24*, 123–138. [CrossRef]
10. Pocock, M.J.; Marzano, M.; Bullas-Appleton, E.; Dyke, A.; De Groot, M.; Shuttleworth, C.M.; White, R. Ethical dilemmas when using citizen science for early detection of invasive tree pests and diseases. *Manag. Biol. Invasions* **2020**, *11*, 720–732. [CrossRef]
11. White, R.; Marzano, M.; Fesenko, E.; Inman, A.; Jones, G.; Agstner, B.; Mumford, R. Technology development for the early detection of plant pests: A framework for assessing Technology Readiness Levels (TRLs) in environmental science. *J. Plant Dis. Prot.* **2022**, *129*, 1249–1261. [CrossRef]
12. Rempelos, L.; Baranski, M.; Wang, J.; Adams, T.N.; Adebusuyi, K.; Beckman, J.J.; Brockbank, C.J.; Douglas, B.S.; Feng, T.; Greenway, J.D.; et al. Integrated soil and crop management in organic agriculture: A logical framework to ensure food quality and human health? *Agronomy* **2021**, *11*, 2494. [CrossRef]
13. Braga, M.P.; Janz, N. Host repertoires and changing insect–plant interactions. *Ecol. Entomol.* **2021**, *46*, 1241–1253. [CrossRef]
14. Yulita, I.N.; Amri, N.A.; Hidayat, A. Mobile Application for Tomato Plant Leaf Disease Detection Using a Dense Convolutional Network Architecture. *Computation* **2023**, *11*, 20. [CrossRef]
15. van der Merwe, D.; Burchfield, D.R.; Witt, T.D.; Price, K.P.; Sharda, A. Drones in agriculture. *Adv. Agron.* **2020**, *162*, 1. [CrossRef]
16. Dutta, G.; Goswami, P. Application of drone in agriculture: A review. *Int. J. Chem. Stud.* **2020**, *8*, 203–218. [CrossRef]
17. Ahirwar, S.; Swarnkar, R.; Bhukya, S.; Namwade, G. Application of Drone in Agriculture. *Int. J. Curr. Microbiol. Appl. Sci.* **2019**, *8*, 2500–2505. [CrossRef]
18. Rejeb, A.; Abdollahi, A.; Rejeb, K.; Treiblmaier, H. Drones in agriculture: A review and bibliometric analysis. *Comput. Electron. Agric.* **2022**, *198*, 107017. [CrossRef]
19. Tannous, M.; Stefanini, C.; Romano, D. A Deep-Learning-Based Detection Approach for the Identification of Insect Species of Economic Importance. *Insects* **2023**, *14*, 148. [CrossRef]
20. Gondal, M.D.; Khan, Y.N. Early pest detection from crop using image processing and computational intelligence. *Int. J. Sci. Res. Eng. Manag.* **2022**, *6*, 59–68. [CrossRef]
21. Hadipour-Rokni, R.; Asli-Ardeh, E.A.; Jahanbakhshi, A.; Esmaili paeen-Afrakoti, I.; Sabzi, S. Intelligent detection of citrus fruit pests using machine vision system and convolutional neural network through transfer learning technique. *Comput. Biol. Med.* **2023**, *155*, 106611. [CrossRef]
22. Kasinathan, T.; Singaraju, D.; Uyyala, S.R. Insect classification and detection in field crops using modern machine learning techniques. *Inf. Process. Agric.* **2021**, *8*, 446–457. [CrossRef]
23. Barbedo, J.G.A. Detecting and Classifying Pests in Crops Using Proximal Images and Machine Learning: A Review. *AI* **2020**, *1*, 21. [CrossRef]
24. Sangeetha, T.; Mohanapriya, M. A Novel Exploration of Plant Disease and Pest Detection Using Machine Learning and Deep Learning Algorithms. *Publ. Issue* **2022**, *71*, 1399–1418.
25. Chithambarathanu, M.; Jeyakumar, M.K. Survey on crop pest detection using deep learning and machine learning approaches. *Multimed. Tools Appl.* **2023**, *11*, 1–34. [CrossRef]
26. Du, L.; Sun, Y.; Chen, S.; Feng, J.; Zhao, Y.; Yan, Z.; Zhang, X.; Bian, Y. A Novel Object Detection Model Based on Faster R-CNN for Spodoptera frugiperda According to Feeding Trace of Corn Leaves. *Agriculture* **2022**, *12*, 248. [CrossRef]
27. Jiao, L.; Xie, C.; Chen, P.; Du, J.; Li, R.; Zhang, J. Adaptive feature fusion pyramid network for multi-classes agricultural pest detection. *Comput. Electron. Agric.* **2022**, *195*, 106827. [CrossRef]

28. Prabha, R.; Kennedy, J.S.; Vanitha, G.; Sathiah, N.; Priya, M.B. Android application development for identifying maize infested with fall armyworms with Tamil Nadu Agricultural University Integrated proposed pest management (TNAU IPM) capsules. *J. Appl. Nat. Sci.* **2022**, *14*, 138–144. [CrossRef]
29. Chen, J.W.; Lin, W.J.; Cheng, H.J.; Hung, C.L.; Lin, C.Y.; Chen, S.P. A smartphone-based application for scale pest detection using multiple-object detection methods. *Electronics* **2021**, *10*, 372. [CrossRef]
30. Panggabean, H.; Tampubolon, S.; Sembiring, M. Indonesia's Ambivalent Language Policy on English: Cause and Effect. *Int. J. Innov. Creat. Change* **2020**, *14*, 588–605. Available online: www.ijicc.net (accessed on 16 August 2023).
31. Hasan, K.; Masriadi; Muchlis; Husna, A. Digital Farming and Smart Farming from the Perspective of Agricultural Students at Malikussaleh University 2022. In Proceedings of the 3rd Malikussaleh International Conference on Multidisciplinary Studies (MICoMS), Virtual, 30 November–1 December 2022. [CrossRef]
32. Yuesheng, F.; Jian, S.; Fuxiang, X.; Yang, B.; Xiang, Z.; Peng, G.; Zhengtao, W.; Shengqiao, X. Circular Fruit and Vegetable Classification Based on Optimized GoogLeNet. *IEEE Access* **2021**, *9*, 113599–1135611. [CrossRef]
33. Muhammad, N.A.; Nasir, A.A.; Ibrahim, Z.; Sabri, N. Evaluation of CNN, alexnet and GoogleNet for fruit recognition. *Indones. J. Electr. Eng. Comput. Sci.* **2018**, *12*, 468–475. [CrossRef]
34. Jayalakshmy, S.; Lakshmipriya, B.; Sudha, G.F. Bayesian optimized GoogLeNet based respiratory signal prediction model from empirically decomposed gammatone visualization. *Biomed. Signal Process Control* **2023**, *86*, 105239. [CrossRef]
35. Loo, M.C.; Logeswaran, R.; Salam, Z.A.A. CNN Aided Surface Inspection for SMT Manufacturing. In Proceedings of the 2023 15th International Conference on Developments in eSystems Engineering (DeSE), Baghdad/Anbar, Iraq, 9–12 January 2023. [CrossRef]
36. Panchbhaiyye, V.; Ogunfunmi, T. A Fifo Based Accelerator for Convolutional Neural Networks. In Proceedings of the ICASSP 2020—2020 IEEE International Conference on Acoustics, Speech and Signal Processing (ICASSP), Barcelona, Spain, 4–8 May 2020. [CrossRef]
37. Xie, Y.; Lu, H.; Yan, J.; Yang, X.; Tomizuka, M.; Zhan, W. Active Finetuning: Exploiting Annotation Budget in the Pretraining-Finetuning Paradigm. In Proceedings of the IEEE/CVF Conference on Computer Vision and Pattern Recognition (CVPR), Vancouver, BC, Canada, 17–24 June 2023. [CrossRef]
38. Yapıcı, M.M.; Topaloğlu, N. Performance comparison of deep learning frameworks. *Comput. Inform.* **2021**, *1*, 1–11.
39. Saleh, A.M.; Hamoud, T. Analysis and best parameters selection for person recognition based on gait model using CNN algorithm and image augmentation. *J. Big Data* **2021**, *8*, 1. [CrossRef]
40. Shorten, C.; Khoshgoftaar, T.M. A survey on Image Data Augmentation for Deep Learning. *J. Big Data* **2019**, *6*, 60. [CrossRef]
41. Khalifa, N.E.; Loey, M.; Mirjalili, S. A comprehensive survey of recent trends in deep learning for digital images augmentation. *Artif. Intell. Rev.* **2022**, *55*, 2351–2377. [CrossRef] [PubMed]
42. Xu, M.; Yoon, S.; Fuentes, A.; Park, D.S. A Comprehensive Survey of Image Augmentation Techniques for Deep Learning. *Pattern Recognit.* **2023**, *137*, 109347. [CrossRef]
43. Kattenborn, T.; Leitloff, J.; Schiefer, F.; Hinz, S. Review on Convolutional Neural Networks (CNN) in vegetation remote sensing. *ISPRS J. Photogramm. Remote Sens.* **2021**, *173*, 24–49. [CrossRef]
44. Lei, X.; Pan, H.; Huang, X. A dilated cnn model for image classification. *IEEE Access* **2019**, *7*, 124087–124095. [CrossRef]
45. Tuggener, L.; Schmidhuber, J.; Stadelmann, T. Is it enough to optimize CNN architectures on ImageNet? *Front. Comput. Sci.* **2022**, *4*, 1703. [CrossRef]
46. Lu, T.C. CNN Convolutional layer optimisation based on quantum evolutionary algorithm. *Conn. Sci.* **2021**, *33*, 482–494. [CrossRef]
47. Ardison, A.C.I.; Hutagalung, M.J.R.; Chernando, R.; Cenggoro, T.W. Observing Pre-Trained Convolutional Neural Network (CNN) Layers as Feature Extractor for Detecting Bias in Image Classification Data. *CommIT J.* **2022**, *16*, 149–158. [CrossRef]
48. Hao, K.; Lin, S.; Qiao, J.; Tu, Y. A Generalized Pooling for Brain Tumor Segmentation. *IEEE Access* **2021**, *9*, 159283–159290. [CrossRef]
49. Jang, B.; Kim, M.; Harerimana, G.; Kang, S.U.; Kim, J.W. Bi-LSTM model to increase accuracy in text classification: Combining word2vec CNN and attention mechanism. *Appl. Sci.* **2020**, *10*, 5841. [CrossRef]
50. Wang, S.H.; Tang, C.; Sun, J.; Yang, J.; Huang, C.; Phillips, P.; Zhang, Y.D. Multiple sclerosis identification by 14-layer convolutional neural network with batch normalization, dropout, and stochastic pooling. *Front. Neurosci.* **2018**, *12*, 818. [CrossRef]
51. Kora, P.; Ooi, C.P.; Faust, O.; Raghavendra, U.; Gudigar, A.; Chan, W.Y.; Meenakshi, K.; Swaraja, K.; Plawiak, P.; Acharya, U.R. Transfer learning techniques for medical image analysis: A review. *Biocybern. Biomed. Eng.* **2022**, *42*, 79–107. [CrossRef]
52. Yoo, H.-J. Deep Convolution Neural Networks in Computer Vision: A Review. *IEIE Trans. Smart Process. Comput.* **2015**, *4*, 35–43. [CrossRef]
53. Yu, H.; Yang, L.T.; Zhang, Q.; Armstrong, D.; Deen, M.J. Convolutional neural networks for medical image analysis: State-of-the-art, comparisons, improvement and perspectives. *Neurocomputing* **2021**, *444*, 92–110. [CrossRef]
54. Chen, S.L.; Chen, T.Y.; Huang, Y.C.; Chen, C.A.; Chou, H.S.; Huang, Y.Y.; Lin, W.C.; Li, T.C.; Yuan, J.J.; Abu, P.A.; et al. Missing Teeth and Restoration Detection Using Dental Panoramic Radiography Based on Transfer Learning with CNNs. *IEEE Access* **2022**, *10*, 118654–118664. [CrossRef]
55. Yilmaz, E.; Trocan, M. A modified version of GoogLeNet for melanoma diagnosis. *J. Inf. Telecommun.* **2021**, *5*, 395–405. [CrossRef]

56. Lyu, Z.; Yu, Y.; Samali, B.; Rashidi, M.; Mohammadi, M.; Nguyen, T.N.; Nguyen, A. Back-Propagation Neural Network Optimized by K-Fold Cross-Validation for Prediction of Torsional Strength of Reinforced Concrete Beam. *Materials* **2022**, *15*, 1477. [CrossRef] [PubMed]
57. Fushiki, T. Estimation of prediction error by using K-fold cross-validation. *Stat. Comput.* **2011**, *21*, 137–146. [CrossRef]
58. Wong, T.T.; Yeh, P.Y. Reliable Accuracy Estimates from k-Fold Cross Validation. *IEEE Trans. Knowl. Data Eng.* **2020**, *32*, 1586–1594. [CrossRef]

AgriEngineering

MDPI

Article

Comparative Analysis of YOLO Models for Bean Leaf Disease Detection in Natural Environments

Diana-Carmen Rodríguez-Lira [1], Diana-Margarita Córdova-Esparza [1,*], José M. Álvarez-Alvarado [2], Julio-Alejandro Romero-González [1], Juan Terven [3] and Juvenal Rodríguez-Reséndiz [2]

1 Facultad de Informática, Universidad Autónoma de Querétaro, Av. de las Ciencias S/N, Juriquilla, Querétaro 76230, Mexico; drodriguez129@alumnos.uaq.mx (D.-C.R.-L.); julio.romero@uaq.mx (J.-A.R.-G.)
2 Facultad de Ingeniería, Universidad Autónoma de Querétaro, Querétaro 76010, Mexico; jmalvarez@uaq.edu.mx (J.M.Á.-A.); juvenal@uaq.edu.mx (J.R.-R.)
3 Instituto Politécnico Nacional, CICATA-Unidad Querétaro, Cerro Blanco 141, Col. Colinas del Cimatario, Querétaro 76090, Mexico; jrtervens@ipn.mx
* Correspondence: diana.cordova@uaq.mx

Abstract: This study presents a comparative analysis of YOLO detection models for the accurate identification of bean leaf diseases caused by Coleoptera pests in natural environments. By using a manually collected dataset of healthy and infected bean leaves in natural conditions, we labeled at the leaf level and evaluated the performance of the YOLOv5, YOLOv8, YOLOv9, YOLOv10, and YOLOv11 models. Mean average precision (mAP) was used to assess the performance of the models. Among these, YOLOv9e exhibited the best performance, effectively balancing precision and recall for datasets with limited size and variability. In addition, we integrated the Sophia optimizer and PolyLoss function into YOLOv9e and enhanced it, providing even more accurate detection results. This paper highlights the potential of advanced deep learning models, optimized with second-order optimizers and custom loss functions, in improving pest detection, crop management, and overall agricultural yield.

Keywords: YOLO models; pest detection; image analysis; precision agriculture

Citation: Rodríguez-Lira, D.-C.; Córdova-Esparza, D.-M.; Álvarez-Alvarado, J.M.; Romero-González, J.-A.; Terven, J.; Rodríguez-Reséndiz, J. Comparative Analysis of YOLO Models for Bean Leaf Disease Detection in Natural Environments. *AgriEngineering* **2024**, *6*, 4585–4603. https://doi.org/10.3390/agriengineering6040262

Academic Editors: Ray E. Sheriff and Chiew Foong Kwong

Received: 25 October 2024
Revised: 24 November 2024
Accepted: 28 November 2024
Published: 30 November 2024

1. Introduction

The global agricultural industry is a cornerstone of the world's economy and food security, with certain crops playing an essential role in the livelihoods of millions of farmers. Among these, beans stand out due to their extensive cultivation and high economic value. In 2017, 60.6% of the world's production was concentrated in six countries: India, Myanmar, Brazil, the United States, China, and Mexico [1]. Additionally, the United States, India, Brazil, and Mexico collectively accounted for 40% of the global import volume [2]. However, despite their importance, beans are frequently subjected to a variety of infections and insect infestations, resulting in significant financial losses and presenting a fundamental challenge to food production.

The intensification and increasing frequency of plant pest and disease outbreaks further exacerbate this issue, posing severe threats to agricultural systems and food security [3–5]. These outbreaks not only cause economic instability but also create food safety concerns and, in extreme cases, the potential for widespread events [5–7]. To address these growing challenges, the global community requires better tools, improved capabilities, and better co-ordination to effectively prevent, manage, and eradicate these threats [3,8].

Integrated pest management has long been recognized as a critical approach to mitigating the spread of pests and diseases, reducing yield losses, and improving productivity [9]. However, current methods reveal significant gaps, particularly in early detection, accurate classification, and timely intervention. Emerging automation tools and technologies, particularly those that utilize artificial intelligence (AI), have shown promise in addressing these challenges [8]. Numerous studies have explored AI-based algorithms for the early detection of plant pests and diseases, with deep learning approaches emerging as the predominant focus of recent research. Despite these advancements, several limitations remain, particularly in achieving the precision and speed required for practical real-time applications [10].

One significant challenge is the size and quality of the datasets used to train these models. Limited datasets can lead to overfitting, resulting in poor generalization of new data. Although data augmentation methods offer partial solutions, a substantial amount of high-quality images are still required to achieve robust model accuracy [11]. In addition, symptom variations caused by factors such as plant genotype, environmental conditions, and disease progression stages further complicate the generalization of the model [8].

Another persistent issue is the unbalanced distribution of the training data. Diseases with fewer samples are often under-represented, biasing the models towards more prevalent conditions and reducing their effectiveness in identifying rarer diseases [12]. Moreover, plants are often affected by multiple simultaneous diseases or nutritional deficiencies, a scenario that current single-disease detection models struggle to address [13].

Real-time processing requirements pose yet another obstacle. Deploying these models in practical applications requires algorithms that can process images quickly without compromising accuracy. This often demands a delicate balance between model complexity and computational efficiency. Environmental factors, including lighting variations, background noise, and other external conditions, further challenge the consistency of image-based models in diverse settings [14].

To address these gaps, we conducted a comparative study on the performance of various YOLO (you only look once) [15] object detection models. We hypothesize that using pretrained YOLO models can significantly improve the early diagnosis and intervention of crop diseases, ultimately contributing to better crop management and higher yields in the agricultural domain. By using the effectiveness of these models, our study seeks to enhance the precision and efficiency of pest detection systems. This research not only aims to fill existing gaps in early pest detection but also to provide a robust method that can be widely adopted to improve agricultural practices [16].

We focused on the identification of bean leaf diseases caused by Coleoptera pests under natural field conditions. Unlike previous studies, this work addresses the challenge of using YOLO models in real-world agricultural environments characterized by natural surroundings, varied lighting, and environmental noise. Furthermore, the integration of advanced optimizers such as Sophia and the PolyLoss function into YOLOv9e exhibits measurable performance enhancements, contributing to the reliability and precision of these models in pest detection.

The main contributions of this paper are the following:

1. Our dataset contains bean leaf images captured under natural field conditions, which is less common compared to datasets captured in controlled environments.
2. We demonstrate the effectiveness of integrating the Sophia optimizer and PolyLoss into YOLOv9e, showing measurable performance improvements in mAP50 and mAP50-95.
3. The work focuses on real-world challenges such as complex backgrounds, varied lighting conditions, and natural field variability, which have been less explored in prior studies.

The rest of the paper is divided into six sections:

Section 2 summarizes related work on pest detection using deep learning. Section 3 discusses the importance of Coleoptera pests in beans and the YOLO models. Section 4 details the materials and methods: data collection, annotation, training, and metrics. Section 5 analyzes the YOLO model results. Section 6 covers the main findings, and Section 7 concludes with key insights and future research.

2. Related Work

The use of deep learning for the detection of pests in crop leaves has become a highly effective method to enhance agricultural productivity and improve crop health [17]. In contrast, conventional methods relying on visual inspection by farmers are often time-consuming, costly, and prone to mistakes [18]. Deep learning (DL) techniques, particularly convolutional neural networks (CNNs), have revolutionized this field by enabling the rapid and precise identification of pests and diseases in plant leaves [19,20]. Choudhary et al. [21] proposed a method based on deep convolution neural networks to recognize plant diseases on images. The innovation and advancement of the developed model lie in its simplicity, which used CNN to distinguish between damaged and healthy leaves. Shrestha et al. [22] developed a CNN-based method to detect plant diseases, emphasizing image processing techniques to examine sample images of diseased and healthy plant leaves. These images were used to train the model, resulting in a test precision of 80%. Shobana et al. [23] proposed a technique to determine whether a plant is afflicted by a specific disease by capturing an image of the leaves and entering it into a model to obtain the diagnosis. By using CNN, the plant disease is identified with an accuracy of 86%. The study developed by Survana et al. [24] aimed to improve crop health by precisely identifying leaf diseases in seven plant species (apple, bell pepper, corn, grape, peach, potato, and tomato) using a pretrained convolutional neural network (CNN). The results exceed a 90% accuracy threshold in disease classification, outperforming traditional algorithms such as BPNN, K-Means, MLP, RBF, and SVM. Kim and Kim [25] developed a deep learning model to identify different pest infections in strawberry plants, enhancing pest management in smart farms. EfficientNet [26] models were selected for their superior image processing capabilities and high accuracy in the preprocessing phase, and EfficientNet-B3 demonstrated the highest accuracy when used in conjunction with RegNetY-064 and ensemble learning. The authors of [27] focused on using MobileNet for the classification of diseases in tomato leaves, and their model was able to classify diseases such as early blight, leaf spot, yellow curl, and healthy leaves. Their classification accuracy exceeded 0.980, demonstrating high precision in distinguishing between different types of diseases affecting tomato leaves.

The YOLO family of models has shown significant promise in real-time pest disease detection. The work presented by Mathew and Mahesh [28] focused on the use of YOLOv5 for identifying diseases in bell pepper plants, showcasing its potential in agricultural applications. The study achieved more accurate and efficient disease detection compared to other models, such as YOLOv4. The use of YOLOv5 enabled the rapid and precise identification of disease symptoms in the leaves of the bell pepper plant, allowing early detection and intervention to curb the spread of diseases and improve crop yield. The study developed by the authors of [29] proposed YOLO-Tea in order to improve the detection of tea disease. The proposed method was shown to outperform typical models, such as SSD and Faster R-CNN, by 1.8% and 5.5%, respectively. The enhancements made to the YOLOv5s model, such as the incorporation of the ACmix and CBAM modules, improved the accuracy in identifying small targets such as tea diseases and insect pests. YOLOv7 has also been used for the detection of tea leaf disease, showing superior results with a high detection accuracy of 97.3% and a precision of 96.7% [30]. On the other hand, Trinh et al. [31] presented the development of the Alpha-EIOU-YOLOv8 algorithm for the detection of rice leaf disease. The researchers achieved a high detection accuracy of up to 89.9%, exceeding the performance of existing methods. These advances emphasize the potential of YOLO-based models in improving pest and disease detection in agricultural leaves.

Table 1 presents a detailed analysis of the contributions of this study, comparing them with the results reported in the state of the art.

Table 1. Comparison of key characteristics in different works. This table evaluates studies based on the use of complex backgrounds, multiple classes in the same image, varying brightness conditions, application to beans or soybeans, and data augmentation. Our work addresses all these aspects in comparison to other works.

Work	Use of Complex (Real) Backgrounds	Incidence of Classes in the Same Image	Different Brightness Conditions	Applied to Beans or Soybeans	Use of Data Augmentation
[32]	-	-	-	-	-
[33]	-	-	-	-	✓
[34]	-	-	-	✓	✓
[29]	✓	✓	✓	-	-
[35]	✓	-	-	-	-
[36]	-	-	-	✓	-
[37]	-	-	-	✓	✓
Our work	✓	✓	✓	✓	✓

3. Background

In this study, we focus on the detection of pests, specifically the Coleoptera species, that are known to cause significant damage to crops and stored products. To accurately identify and monitor these pests, we employ object detection models, particularly the YOLO models. YOLO models are renowned for their high accuracy and real-time detection capabilities, making them ideal for the rapid identification of Coleoptera in various environments. This section provides a background on the pests targeted in our study and the object detection methodologies utilized.

Coleoptera pests, such as the bean leaf beetle (Epilachna varivestis) and banded cucumber beetle (Diabrotica balteata), are among the most destructive pests for bean crops. These pests cause significant defoliation during the flowering and pod formation stages, resulting in substantial yield losses. This study prioritizes Coleoptera due to its economic significance in Zacatecas, Mexico. However, the methodology and dataset can be adapted to other pests and diseases in future research.

3.1. Coleoptera

Coleoptera are the most species-rich order on the planet. Some of the families included are Cerambycidae, Chrysomelidae, Scarabaeidae, Cicindelidae, Paussidae, Curculionidae, Erotylidae, Languriidae, Endomychidae, Carabidae, Staphylinidae, Lucanidae, and Passalidae, among others [38]. Figure 1 shows examples of bean leaves affected by Coleoptera pests, specifically from the Coccinellidae or Chrysomalidae families, displaying visible damage such as holes and discoloration.

Figure 1. Examples of bean leaves showing damage caused by Coleoptera pests, specifically from the Coccinellidae and Chrysomalidae families. The damage is characterized by holes and discoloration on the leaf surface.

3.1.1. Bean Leaf Beetle

Epilachna varivestis Mulsant (Coleoptera: Coccinellidae), commonly known as the bean leaf beetle, Mexican bean beetle, bean beetle, or bean weevil, belongs to the order Coleoptera and family Coccinellidae. It is classified as a nonquarantine pest [39]. These beetles feed on the surface of leaf tissues, leaving only the veins and part of the epidermis intact. The remaining tissue quickly dies and turns brown. This defoliation during the flowering and pod formation and filling stages leads to significant yield losses in the crop [40].

3.1.2. Banded Cucumber Beetle

Diabrotica balteata LeConte (Coleoptera: Chrysomelidae) feeds on the leaves, creating oval or irregular perforations throughout the leaf blade. When small plants are affected, their primary leaves show reduced growth, leading to wilting and eventual death. Damage has also been observed in flowers and pods when populations are high [41].

3.2. *YOLO Model*

In 2016, YOLO [15] was introduced as an approach to object detection. This approach addresses object detection as a regression task by simultaneously predicting bounding boxes and class probabilities. YOLO employs a single neural network that processes complete images in one pass, directly predicting bounding boxes and class probabilities. Although YOLO may produce more localization errors than cutting-edge systems, it is less prone to generating false positives on background elements.

The original YOLO model consisted of 24 convolutional layers followed by two fully connected layers. It used 1×1 reduction layers followed by 3×3 convolutional layers. The Fast YOLO model employs a nine-layer neural network. The final output of the model was a $7 \times 7 \times 30$ prediction tensor, as illustrated in Figure 2.

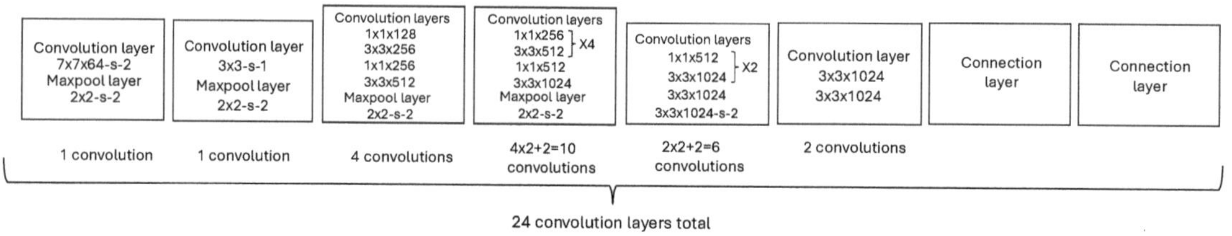

Figure 2. YOLO network's architecture includes 24 convolutional layers, two subsequent fully connected layers, reduction layers of dimensions 1×1, and convolutional layers of 3×3.

3.2.1. YOLOv5

YOLOv5 is characterized by its compact model size and rapid detection speed. Its network structure comprises four main components [42]:

- Input component: The dataset is diversified by implementing the Mosaic data enhancement. Here, four images are randomly selected and then scaled, cropped, and spliced.
- Backbone network: The architecture of this block is based on the CSPDarkNet architecture. Its 6×6 convolutional layers perform feature extraction. The MaxPool layer is included in the spatial pyram pooling fast (SPPF) for the last version of YOLOv5 [43].
- Neck network: The function of this block is to process the extracted features from the backbone network. This task is performed using the feature pyramid network (FPN) or the path aggregate network (PAN) to enhance the feature extraction performance.
- Classification layer: The final stage filters confidence bounding boxes and object classes to yield the highest confidence and category results for detected objects.

3.2.2. YOLOv8

The backbone of the YOLOv8 model retains the YOLOv5 CSP module. The C2f module is used for visual feature extraction, replacing the C3 module. In YOLOv8, the CBS 1 × 1 convolution structure in the PAN-FPN up-sampling stage is removed. The model employs a decoupled head with two convolutions for classification and regression, incorporating distribution focal loss (DFL) [44].

A major update of YOLOv8 includes an anchor-free approach and task alignment learning for classification and regression. The anchor alignment metric is derived by multiplying the classification score with the IOU of predicted and actual frames and is used in the sample allocation and the loss function to optimize the anchor predictions [45] dynamically.

The YOLOv8 model uses variational focal loss (VFL) for classification loss and DFL + CIOU (complete IOU) loss for regression loss. DFL changes the single value of the co-ordinate regression to output n + 1, representing the probability of the corresponding regression distances, and calculates the integral to obtain the final regression distance. This approach enables the network to focus more accurately on nearby target values, increasing their probability [46].

3.2.3. YOLOv9

Launched in February 2024, YOLOv9 [47] incorporates advanced techniques such as programmable gradient information (PGI) and the generalized efficient layer aggregate network (GELAN) [48]. GELAN enhances feature aggregation, which is particularly beneficial for small datasets such as the one used in this work. It optimally utilizes parameters, reducing overfitting while retaining critical features. PGI tackles the information bottleneck issue by retaining critical data across deep network layers, ensuring the production of reliable gradients for precise model updates and improved detection performance. PGI improves gradient flow across layers, making it better suited for extracting fine-grained features in our dataset. GELAN represents a significant architectural enhancement that promotes optimal parameter utilization and computational efficiency. This design facilitates the smooth integration of various computational blocks, making YOLOv9 versatile for a range of applications without sacrificing speed or accuracy. In particular, the medium to large models, YOLOv9m and YOLOv9e, demonstrate significant advances in balancing model complexity and detection performance, achieving notable reductions in parameters and computations while improving accuracy.

3.2.4. YOLOv10

YOLOv10, released in May 2024 [49], was designed to enhance real-time object detection by addressing the limitations of previous models, such as reliance on nonmaximum suppression (NMS) and architectural inefficiencies. It introduces NMS-free training through consistent dual assignments and adopts a holistic model design that balances efficiency and accuracy. The architecture includes an enhanced CSPNet backbone for improved gradient flow, a PAN neck for multiscale feature fusion, and dual heads for training, providing multiple predictions per object and another for inference, offering a single prediction to reduce latency. YOLOv10 also incorporates large-kernel convolutions and self-attention modules [50].

3.2.5. YOLOv11

Launched in October 2024, YOLOv11 [51] represents the latest innovation in the Ultralytics YOLO series at the time of this writing. It offers improvements in accuracy, speed, and efficiency in the COCO dataset but not in our dataset, as we show in the results. YOLOv11 features a modified backbone and neck architecture that improves feature extraction for more precise detection and task performance. With refined designs and optimized training pipelines, YOLOv11 achieves faster processing without sacrificing accuracy. YOLOv11 is adaptable to various environments, including edge devices and cloud platforms, and supports a wide range of tasks such as object detection, segmentation, classification, and pose estimation.

3.2.6. Sophia Optimizer

Sophia is a scalable second-order optimizer specifically developed for pretraining large language models (LLMs). It enhances the optimization process by integrating curvature information through an efficient approximation of the Hessian diagonal, eliminating the need to calculate it at each iteration. This feature makes Sophia particularly beneficial in federated learning (FL) scenarios. Compared to first-order adaptive optimizers such as Adam [52], which is the default optimizer for YOLO models, Sophia achieves faster convergence, requiring fewer steps while keeping the computational cost per step almost the same. Tests on GPT models for language tasks showed that Sophia delivers an improvement of 2× in steps, total computation, and wall clock time [53].

3.2.7. PolyLoss

PolyLoss aims to enhance the performance of commonly used classification loss functions, such as cross-entropy and focal loss, by interpreting them as polynomial expansions. This approach decomposes the loss functions into a series of polynomial terms, with coefficients that can be adjusted to better align with specific tasks and datasets. By providing this flexibility, PolyLoss enables the fine-tuning of the importance assigned to various polynomial terms, leading to improved performance in a wide range of applications, including image classification, segmentation, and object detection [54].

4. Materials and Methods

The methodology of this study is structured into five distinct phases (Figure 3) to ensure the accurate identification of bean leaves using deep learning techniques. A comprehensive image dataset was collected and carefully cleaned to remove any images that did not feature bean leaves. This preprocessing step was crucial to ensure that only relevant images were included in the subsequent stages. The remaining images, which exclusively depicted bean leaves, were then annotated with two classes: healthy and infected. Following annotation, we trained 14 variants of the YOLO model, particularly YOLOv5, YOLOv8, YOLOv9, YOLOv10 and YOLOv11. Finally, the performance of these models was evaluated and compared using standard metrics to determine the most effective variant for detecting and classifying the health status of the bean leaves.

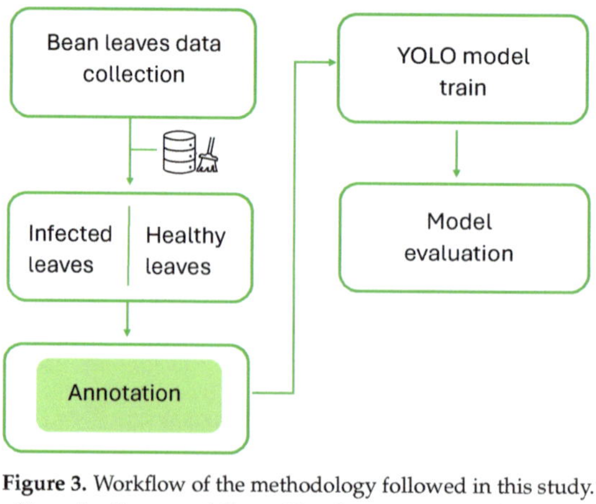

Figure 3. Workflow of the methodology followed in this study. The process begins with the collection of bean leaf images, followed by the annotation of infected and healthy leaves. These annotated images are then used to train YOLO models, and the performance of these models is compared using various metrics.

4.1. Data Acquisition

The dataset applied in this research originates from the prior study conducted by [55]. It consists of images depicting healthy and Coleoptera pest-affected bean leaves. The dataset encompasses 531 images, each with a resolution of 1920 × 1080 pixels and an approximate size of 519 KB in JPEG format. These images were taken in bean fields distributed across different areas of Zacatecas, Mexico. The images were captured with a Logitech C920 camera in midday natural light for even illumination. Photos were taken 170 cm above the ground, with the camera at a 90° angle to ensure uniformity.

4.2. Data Selection

During data selection, the dataset was refined to include only images with bean leaves. Images with weeds or irrelevant objects were removed. As a result, the dataset decreased from 531 to 508 images, ensuring relevance for annotation and training, as illustrated by the removed image in Figure 4.

Figure 4. Example of a deleted image containing an unknown weed, which was removed during the data selection phase to ensure that only images of bean leaves were included in the dataset.

4.3. Annotation

In the annotation phase, the relevant features of the bean leaves were analyzed and extracted using image processing techniques. Based on the literature on the manifestations of Coleoptera pests in bean leaves, two classes were designated for labeling: "healthy" and "infected". To create image labels according to these defined categories, we use a semi-automatic annotation tool called Auto-Seg-annotation [56] that is based on the segment anything model (SAM) [57].

4.4. Models Training

To effectively train and evaluate the models, we selected 508 images and divided them into training, testing, and validation sets. The training set consisted of 308 images, which is approximately 60% of the total; the validation set included 100 images (about 20% of the total), used to fine-tune the model's hyperparameters; and the test set comprised 100 images, used exclusively to evaluate the final performance of the model. This allocation facilitated precision-metric calculations based on a percentage of the total images, allowing for the direct interpretation of results on a scale of 0 to 100.

The hyperparameters and online augmentations selected in this study were consistent for each model to facilitate performance comparison, as shown in Tables 2 and 3.

Table 2. Hyperparameters employed during model training.

Hyperparameter	Value
Epochs	200
Batch Size	16
Patience	20
Image Size	640

Table 3. Data augmentation parameters used during training.

Name	Probability
hsv_hue_augmentation	0.015
hsv_saturation_augmentation	0.7
hsv_value_augmentation	0.4
image_translation	0.1
image_scale	0.5
flip_left_right	0.5
mosaic_augmentation	1.0
random_erasing	0.4
image_crop_fraction	1.0

4.5. Metrics Comparison

When comparing different YOLO versions, various key performance metrics were incorporated to thoroughly evaluate the effectiveness of each model. The precision-recall curve (Figure 5) provides valuable insights into the trade-offs between precision and recall, which is essential to understanding how well the model manages diverse class distributions. Meanwhile, the F1 confidence curve (Figure 6) helps to identify the ideal confidence threshold where the harmony between precision and recall reaches its peak, offering a better assessment of the performance of the model.

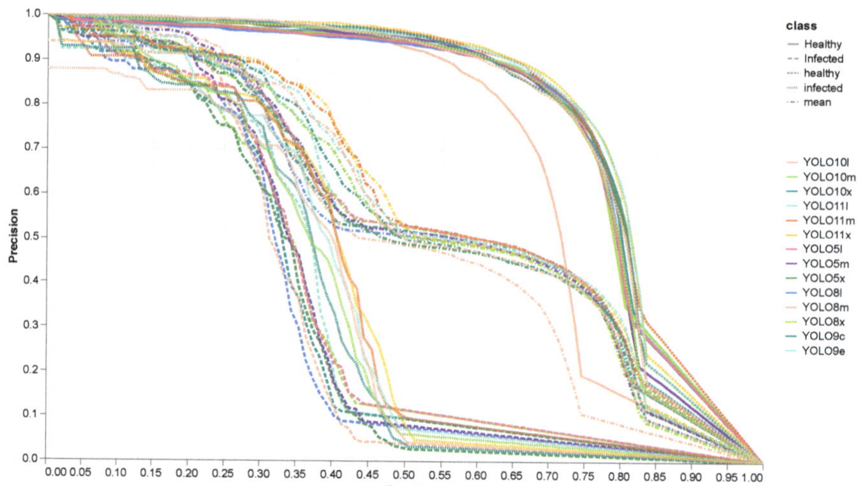

Figure 5. Precision-recall curve for 14 versions of YOLO models, illustrating the trade-off between precision and recall across different thresholds.

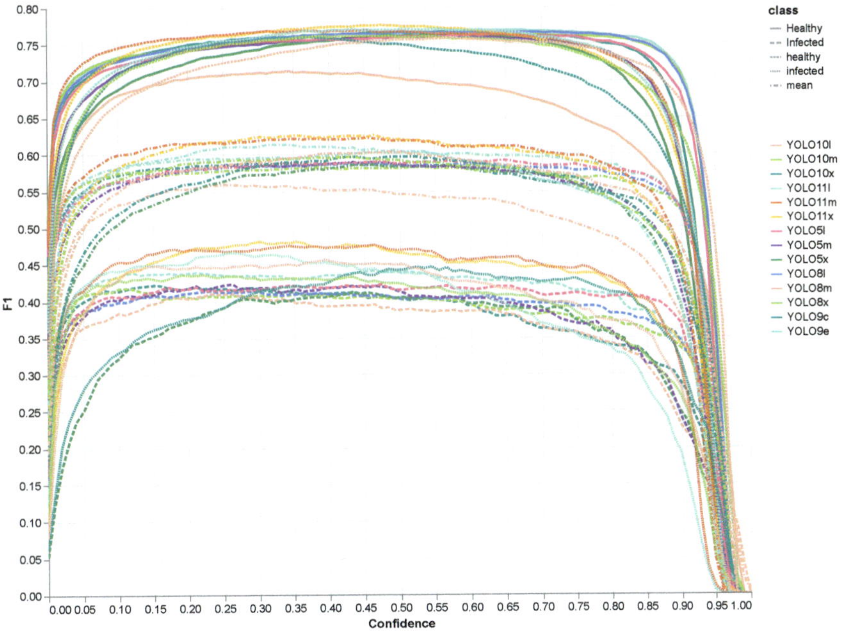

Figure 6. F1 confidence curve for 14 versions of YOLO models, showing the variation in F1 score across different confidence thresholds to identify the optimal balance between precision and recall.

We evaluated the performance of the models using mean average precision, which is based on precision and recall.

Precision measures the ratio of correctly predicted positive observations to the total predicted positives. It is calculated as (1).

$$\text{Precision} = \frac{TP}{TP + FP}, \tag{1}$$

where TP represents true positives, and FP represents false positives. This metric indicates how many of the detected objects are correctly identified.

Recall measures the ratio of correctly predicted positive observations to all observations in the actual class. It is defined as (2).

$$\text{Recall} = \frac{TP}{TP + FN}, \tag{2}$$

with FN being false negatives. Recall highlights the model's ability to identify all relevant instances within the dataset.

Mean average precision at IoU = 0.50 (mAP50) averages the precision across all classes at an intersection over union (IoU) threshold of 0.50. This metric provides insight into the model's performance in detecting objects with a moderate overlap criterion.

For a more detailed evaluation, the mean average precision at IoU thresholds from 0.50 to 0.95 (mAP50-95) is used. This metric averages the precision across multiple IoU thresholds, ranging from 0.50 to 0.95 with a step size of 0.05, offering a more complete view of the model's ability to accurately predict bounding boxes across varying levels of overlap.

5. Results

This section presents the detection results obtained using various YOLO models to identify healthy and infected bean leaves. The comparison focuses on YOLOv5, YOLOv8, YOLOv9, YOLOv10, and YOLOv11 models, highlighting their performance across different metrics.

Tables 4–8 display visual comparisons of the object detection results for the YOLOv5, YOLOv8, YOLOv9, YOLOv10, and YOLOv11 models, respectively. Each row in the table corresponds to a different variant of the model, and each column presents the detection output for a specific image. The blue bounding boxes indicate healthy detected leaves, while the red bounding boxes represent infected leaves.

Table 4. Visual comparison of object detection results using different YOLOv5 models (YOLOv5m, YOLOv5l, and YOLOv5x) on three distinct images. Blue boxes indicate healthy detected leaves, while red boxes represent detections of infected leaves.

Model	Image 1	Image 2	Image 3
YOLOv5m			
YOLOv5l			
YOLOv5x		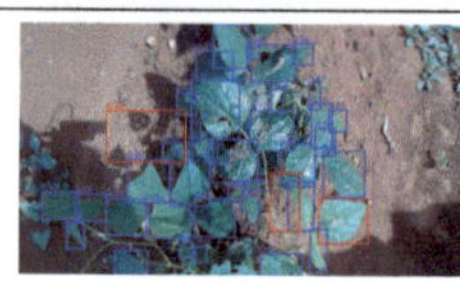	

Table 5. Visual comparison of object detection results using different YOLOv8 models (YOLOv8m, YOLOv8l, and YOLOv8x) on three distinct images. Blue boxes indicate healthy detected leaves, while red boxes represent detections of infected leaves.

Model	Image 1	Image 2	Image 3
YOLOv8m			
YOLOv8l		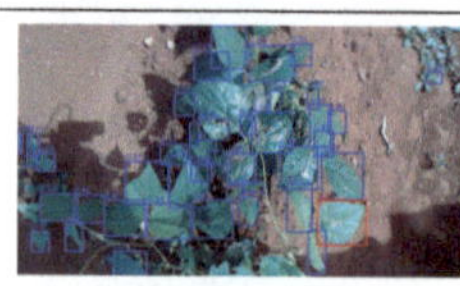	
YOLOv8x			

Table 6. Visual comparison of object detection results using two YOLOv9 models (YOLOv9c and YOLOv9e) on three distinct images. Blue boxes indicate healthy detected leaves, while red boxes represent detections of infected leaves.

Model	Image 1	Image 2	Image 3
YOLOv9c			
YOLOv9e			

The evaluation of the performance of the YOLO models is presented using precision-recall (PR) curves in Figure 5, the F1 confidence curves in Figure 6, and a detailed comparison of performance metrics in Table 9.

The PR curves (Figure 5) illustrate how precision and recall vary between different confidence thresholds. The ideal performance of a model is represented by a curve that approaches the top-right corner, indicating high precision and high recall. Among the models tested, YOLOv9e and YOLOv9c consistently maintained higher precision over a wide range of recall values, demonstrating their effectiveness in accurately identifying healthy and infected bean leaves. In comparison, YOLOv5x and YOLOv8x performed better at higher recall values, but their overall balance between precision and recall was slightly less robust than the YOLOv9 models.

Figure 6 illustrates the F1 confidence curves, highlighting the performance of the model at different thresholds. The F1 score, which reflects precision and recall, helps to pinpoint the optimal confidence threshold. YOLOv9e and YOLOv9c maintained high F1 scores at various levels, proving to be reliable for leaf detection. In contrast, YOLOv5 and YOLOv8 showed moderate F1 performance, with YOLOv5x and YOLOv8x excelling at higher thresholds.

Table 7. Visual comparison of object detection results using two YOLOv10 models (YOLOv10m, YOLOv10l, and YOLOv10x) on three distinct images. Blue boxes indicate healthy detected leaves, while red boxes represent detections of infected leaves.

Model	Image 1	Image 2	Image 3
YOLOv10m			
YOLOv10l			
YOLOv10x			

Table 8. Visual comparison of object detection results using two YOLOv11 models (YOLOv11m, YOLOv11l, and YOLOv11x) on three distinct images. Blue boxes indicate healthy detected leaves, while red boxes represent detections of infected leaves.

Model	Image 1	Image 2	Image 3
YOLOv11m			
YOLOv11l			
YOLOv11x		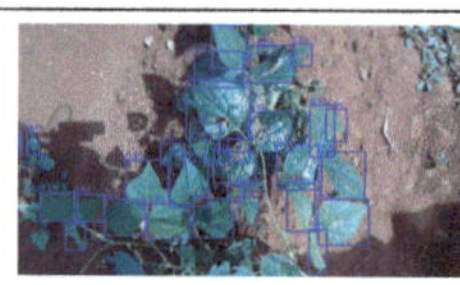	

Figure 6 presents the F1 confidence curves for the 14 tested versions of the YOLO models. The F1 score combines precision and recall, offering a balanced measure of a model's performance. This figure illustrates how the F1 score changes across various confidence thresholds to detect healthy and infected bean leaves.

Table 9 provides a comprehensive comparison of performance metrics for various YOLO models in the detection of healthy and infected bean leaves. The models evaluated include different versions of YOLOv5 (m, l, x), YOLOv8 (m, l, x), YOLOv9 (c, e), YOLOv10 (m, l, x), and YOLOv11 (m, l, x). Each row in the table presents the metrics for a specific model variant, and the results are categorized by the class of leaves (all, healthy, and infected).

The breakdown by class (healthy and infected) shows that the models generally perform better in detecting healthy leaves because the healthy category is more represented in the data.

Table 9 demonstrates the slight superiority of YOLOv9e, with the highest score for mAP50-95 of 0.514, making it the most balanced model crucial for practical applications, as it outperforms in accuracy and reliability at different overlap levels.

The YOLOv10 and YOLOv11 models showed similar strengths. YOLOv10x was the top variant with good precision and recall balance. YOLOv11x outperformed previous versions with higher F1 scores and consistent precision-recall balance, ranking as one of the top models in this study.

The data augmentation parameters and the standardized hyperparameters, as outlined in Tables 2 and 3, respectively, ensured that the evaluation was conducted under consistent and controlled conditions. This standardization facilitates a fair comparison and highlights the real-world performance capabilities of each model.

While YOLOv10 and YOLOv11 introduce innovations such as NMS-free training, large-kernel convolutions, and enhanced feature fusion designed to improve scalability and efficiency in real-time object detection, these features offer advantages for large-scale datasets and multi-class detection tasks, and may not fully capitalize on the focused, single-class nature of our dataset. The architectural design of YOLOv9, specifically its GELAN and PGI components, is more effective at balancing precision and recall for datasets with limited size and variability. This explains the relatively lower performance of YOLOv10 and YOLOv11 compared to YOLOv9e in our experiments.

Table 9. Performance metrics for various YOLO models in detecting healthy and infected bean leaves. The bold numbers indicate the best results for each column and each class.

Model	Class	mAP50	mAP50-95
yolov5m	all	0.56	0.478
	Healthy	0.774	0.673
	Infected	0.345	0.283
yolov5l	all	0.583	0.513
	Healthy	0.783	0.694
	Infected	0.383	**0.333**
yolov5x	all	0.567	0.49
	Healthy	0.783	0.68
	Infected	0.352	0.3
yolov8m	all	0.566	0.473
	Healthy	0.767	0.647
	Infected	0.365	0.3
yolov8l	all	0.565	0.499
	Healthy	0.781	0.696
	Infected	0.348	0.302
yolov8x	all	0.571	0.503
	Healthy	0.778	0.691
	Infected	0.364	0.316
yolov9c	all	0.575	0.505
	Healthy	0.782	0.689
	Infected	0.369	0.322
yolov9e	all	0.578	**0.514**
	Healthy	0.785	**0.706**
	Infected	0.371	0.322
yolov10m	all	0.565	0.481
	Healthy	0.768	0.655
	Infected	0.361	0.307
yolov10l	all	0.56	0.476
	Healthy	0.773	0.664
	Infected	0.346	0.289
yolov10x	all	0.561	0.475
	Healthy	0.777	0.664
	Infected	0.344	0.286
yolov11m	all	**0.59**	0.51
	Healthy	0.792	0.692
	Infected	**0.388**	0.328
yolov11l	all	0.583	0.499
	Healthy	0.787	0.685
	Infected	0.379	0.314
yolov11x	all	0.587	0.507
	Healthy	**0.794**	0.694
	Infected	0.379	0.32

Model Improvement

After determining that YOLOv9e was the model that performed the best in the initial analysis, we performed further experiments by integrating the Sophia optimizer and the PolyLoss function, both separately and in combination. We retrained YOLOv9e with these modifications to see how they would enhance the detection performance of the model. Table 10 shows the numerical results, and Table 11 visually compares the object detection results in three different images, with blue boxes representing healthy leaves and red boxes indicating infected ones. The introduction of Sophia and PolyLoss improved detection accuracy, resulting in more precise and consistent bounding boxes for both healthy and infected leaves.

Table 10. Performance metrics for YOLOv9 improved models in detecting healthy and infected bean leaves. The bold numbers indicate the best results for each column and each class.

Model	Class	mAP50	mAP50-95
yolov9e	all	0.578	0.514
	Healthy	0.785	0.706
	Infected	0.371	0.322
yolov9e with Sophia optimizer	all	0.59	**0.522**
	Healthy	**0.795**	**0.709**
	Infected	0.386	**0.335**
yolov9e with Polyloss	all	**0.592**	511
	Healthy	0.79	0.69
	Infected	**0.394**	0.332
yolov9e with Sophia and Polyloss	all	0.589	0.519
	Healthy	0.793	0.702
	Infected	0.385	**0.335**

Table 11. Visual comparison of object detection results using the YOLOv9e model with Sophia optimizer, Polyloss loss function, and both on three distinct images. Blue boxes indicate healthy detected leaves, while red boxes represent detections of infected leaves.

Model	Image 1	Image 2	Image 3
YOLOv9e with Sophia			
YOLOv9e with Polyloss			
YOLOv9e with Sophia and Polyloss			

6. Discussion

Tables 4–8 reveal that YOLOv9e excels in precision and accuracy for the bounding box for infected leaf areas, which are often smaller and more difficult to detect. This is evident in the superior metrics found in Table 9. Occasionally, YOLOv10 and YOLOv11 incorrectly classified overlapping entities or missed some leaf edges because their architectural design targets larger datasets. As further evidenced in Table 9, the mAP50 and mAP50-95 scores underline the robustness of YOLOv9e. It incorporates architectural features such as GELAN and PGI, enabling better feature aggregation and gradient optimization, which are well-suited for smaller datasets. The precision-recall curves depicted in Figure 5 demonstrate the capacity of YOLOv9e to maintain high precision across different recall rates. Similarly, the F1 confidence curves in Figure 6 show the consistent performance of YOLOv9e across a spectrum of confidence levels. These findings are consistent with the quantitative results presented in Table 9. In Table 10, the incorporation of the Sophia optimizer and the PolyLoss function into YOLOv9e improved both precision and recall, thus improving overall mAP scores. This suggests that utilizing optimization techniques can effectively tailor YOLO architectures to specific datasets and conditions.

7. Conclusions

This work demonstrated the effectiveness of YOLO models in real-world pest detection scenarios, with YOLOv9e emerging as the most reliable model for detecting bean leaf diseases. Despite the limited size of the dataset, the methodological rigor of the study ensured robust model training and evaluation.

We evaluated the performance of various YOLO models in detecting healthy and infected bean leaves. YOLOv9 models showed better overall performance, particularly YOLOv9e, which excelled with a mAP50 of 0.578 and a mAP50-95 of 0.514, making it the most effective model for the detection of bean leaf disease in this study.

Further improvements were explored by incorporating state-of-the-art architectures such as YOLOv10 and YOLOv11, which introduced innovations such as NMS-free training and enhanced feature fusion, yielding a better precision-recall balance. Although YOLOv9e stood out as the most robust model for pest detection, YOLOv10 and YOLOv11 also demonstrated significant potential for precision agriculture applications.

Moreover, the integration of the Sophia optimizer and PolyLoss into YOLOv9e further improved its performance. The combined use of these optimizations yielded the best overall results, improving precision and recall across confidence thresholds and demonstrating their value in the refinement of deep learning models for agricultural tasks. These findings emphasize the potential of YOLOv9e, particularly when enhanced with Sophia and PolyLoss, as a reliable solution to detect pests in bean leaves. Future research will focus on extending the dataset to include additional pests and diseases, enabling broader applications of these models in precision agriculture.

Author Contributions: Conceptualization, D.-C.R.-L., D.-M.C.-E., and J.M.Á.-A.; methodology, D.-C.R.-L., D.-M.C.-E., J.M.Á.-A., J.-A.R.-G., J.T., and J.R.-R.; formal analysis, D.-C.R.-L., D.-M.C.-E., J.M.Á.-A., and J.T.; investigation, D.-C.R.-L., D.-M.C.-E., J.M.Á.-A., and J.T.; resources, D.-C.R.-L., D.-M.C.-E., J.M.Á.-A., J.-A.R.-G., J.T., and J.R.-R.; writing—original draft preparation, D.-C.R.-L., D.-M.C.-E., and J.M.Á.-A.; writing—review and editing; D.-C.R.-L., D.-M.C.-E., J.M.Á.-A., J.-A.R.-G., J.T., and J.R.-R.; supervision, D.-M.C.-E. and J.M.Á.-A.; project administration, D.-C.R.-L., D.-M.C.-E., and J.M.Á.-A.; funding acquisition, D.-C.R.-L., D.-M.C.-E., J.M.Á.-A., J.-A.R.-G., J.T., and J.R.-R. All authors have read and agreed to the published version of the manuscript.

Funding: This research was funded in part by Instituto Politecnico Nacional research grant SIP 20240760.

Data Availability Statement: The raw data supporting the conclusions of this article will be made available by the authors on request.

Acknowledgments: We thank the National Council of Humanities, Sciences, and Technologies (CONAHCYT) for this support through the Postgraduate Scholarship Program in Mexico.

Conflicts of Interest: The authors declare no conflicts of interest.

References

1. Ayala, A.; Acosta, J.; Reyes, L. *El Cultivo del Frijol; Presente y Futuro para México*; Libro Técnico; Instituto Nacional de Investigaciones Forestales, Agrícolas y Pecuarias: Celaya Gto, Mexico, 2021; p. 232.
2. Dirección de Investigación y Evaluación Económica y Sectorial, *Panorama Agroalimentario: Frijol 2016*, 2016. Available online: https://www.gob.mx/cms/uploads/attachment/file/200638/Panorama_Agroalimentario_Frijol_2016.pdf (accessed on 27 November 2024).

3. Naveesh, Y.; Raaga, R.; Sagar, N.; Suresh, K.; Ashwini, M.; Jayashree, A.; Sushma, R.; Krishnamoorthy, P.; Siju, S.; Patil, S. Comprehending farmer's views on climate change and agricultural/livestock adaptation in Halavarthi Village, Koppala District, Karnataka: A case study. *J. Agric. Ecol.* **2024**, *18*, 7–13. [CrossRef]
4. Nitta, A.; Natarajan, V.; Reddy, A.J.; Rakesh, T. Impact of Climate Change on Pest Biology, Behaviour and Their Distributions. *Int. J. Environ. Clim. Change* **2024**, *14*, 46–56. [CrossRef]
5. Ali, S.; Tyagi, A.; Bae, H. Plant microbiome: An ocean of possibilities for improving disease resistance in plants. *Microorganisms* **2023**, *11*, 392. [CrossRef] [PubMed]
6. Tatineni, S.; Hein, G.L. Plant viruses of agricultural importance: Current and future perspectives of virus disease management strategies. *Phytopathology* **2023**, *113*, 117–141. [CrossRef] [PubMed]
7. Reddy, C.A.; Oraon, S.; Bharti, S.D.; Yadav, A.K.; Hazarika, S. Advancing disease management in agriculture: A review of plant pathology techniques. *Plant Sci. Arch.* **2024**, *9*, 16–18. [CrossRef]
8. Elsayed, M.Z.; Hasoon, A.; Zidan, M.K.; Ayyad, S.M. Role of AI for Plant Disease Detection and Pest Detection. In Proceedings of the 2024 International Telecommunications Conference (ITC-Egypt), Cairo, Egypt, 22–25 July 2024 ; IEEE: Piscataway, NJ, USA, 2024; pp. 824–829.
9. Calicioglu, O.; Flammini, A.; Bracco, S.; Bellù, L.; Sims, R. The future challenges of food and agriculture: An integrated analysis of trends and solutions. *Sustainability* **2019**, *11*, 222. [CrossRef]
10. Altalak, M.; Uddin, M.A.; Alajmi, A.; Rizg, A. A hybrid approach for the detection and classification of tomato leaf diseases. *Appl. Sci.* **2022**, *12*, 8182. [CrossRef]
11. Zhang, Z.; Gao, Q.; Liu, L.; He, Y. A high-quality rice leaf disease image data augmentation method based on a dual GAN. *IEEE Access* **2023**, *11*, 21176–21191. [CrossRef]
12. Yadav, R.; Pandey, M.; Sahu, S.K. Cassava plant disease detection with imbalanced dataset using transfer learning. In Proceedings of the 2022 IEEE World Conference on Applied Intelligence and Computing (AIC), Sonbhadra, India, 17–19 June 2022; IEEE: Piscataway, NJ, USA, 2022; pp. 220–225.
13. Kethineni, K.K.; Mohanty, S.P.; Kougianos, E. Stimator: A Method in Agriculture CPS Framework to Estimate Severity of Plant Diseases using Graph Neural Network. In Proceedings of the 2023 OITS International Conference on Information Technology (OCIT), Raipur, India, 13–15 December 2023; IEEE: Piscataway, NJ, USA, 2023; pp. 462–467.
14. Wani, J.A.; Sharma, S.; Muzamil, M.; Ahmed, S.; Sharma, S.; Singh, S. Machine learning and deep learning based computational techniques in automatic agricultural diseases detection: Methodologies, applications, and challenges. *Arch. Comput. Methods Eng.* **2022**, *29*, 641–677. [CrossRef]
15. Redmon, J.; Divvala, S.; Girshick, R.; Farhadi, A. You only look once: Unified, real-time object detection. In Proceedings of the IEEE Conference on Computer Vision and Pattern Recognition, Las Vegas, NV, USA, 27–30 June 2016; pp. 779–788.
16. Boudaa, B.; Abada, K.; Aichouche, W.A.; Nabil Belakermi, A. Advancing Plant Diseases Detection with Pre-trained YOLO Models. In Proceedings of the 2024 6th International Conference on Pattern Analysis and Intelligent Systems (PAIS), El Oued, Algeria, 24–25 April 2024; pp. 1–6. [CrossRef]
17. Saleem, M.H.; Potgieter, J.; Arif, K.M. Plant Disease Detection and Classification by Deep Learning. *Plants* **2019**, *8*, 468. [CrossRef]
18. Pandian, J.A.; Kumar, V.D.; Geman, O.; Hnatiuc, M.; Arif, M.; Kanchanadevi, K. Plant disease detection using deep convolutional neural network. *Appl. Sci.* **2022**, *12*, 6982. [CrossRef]
19. Li, L.; Zhang, S.; Wang, B. Plant disease detection and classification by deep learning—A review. *IEEE Access* **2021**, *9*, 56683–56698. [CrossRef]
20. Shoaib, M.; Shah, B.; Ei-Sappagh, S.; Ali, A.; Ullah, A.; Alenezi, F.; Gechev, T.; Hussain, T.; Ali, F. An advanced deep learning models-based plant disease detection: A review of recent research. *Front. Plant Sci.* **2023**, *14*, 1158933.
21. Choudhary, S.; Swathi, G.; Kumar, S.; Srinivas, K.; Deepika, M. Automatic Leaf Disease Detection Using Convolution Neural Network. In Proceedings of the 2023 International Conference on Computational Intelligence and Sustainable Engineering Solutions (CISES), Greater Noida, India, 28–30 April 2023; IEEE: Piscataway, NJ, USA, 2023; pp. 271–276.
22. Shrestha, G.; Deepsikha; Das, M.; Dey, N. Plant disease detection using CNN. In Proceedings of the 2020 IEEE Applied Signal Processing Conference (ASPCON), Kolkata, India, 7–9 October 2020; IEEE: Piscataway, NJ, USA, 2020; pp. 109–113.
23. Shobana, M.; S, V.; C, G.P.; SP, P.K.; P, M.K.; C, N.; N, K. Plant Disease Detection Using Convolution Neural Network. In Proceedings of the 2022 International Conference on Computer Communication and Informatics (ICCCI), Coimbatore, India, 25–27 January 2022; pp. 1–5. [CrossRef]
24. Suvarna, T.S.L.; Sai, N.N.; Rao, S.D.; Gowtham, G.; Anoosha, C. Enhancing Crop Health by Identifying Leaf Disease with High Accuracy Using CNN. In Proceedings of the 2023 Third International Conference on Secure Cyber Computing and Communication (ICSCCC), Jalandhar, India, 26–28 May 2023; pp. 403–408. [CrossRef]
25. Kim, H.; Kim, D. Deep-Learning-Based Strawberry Leaf Pest Classification for Sustainable Smart Farms. *Sustainability* **2023**, *15*, 7931. [CrossRef]
26. Tan, M.; Le, Q. Efficientnet: Rethinking model scaling for convolutional neural networks. In Proceedings of the International Conference on Machine Learning Research, Long Beach, CA, USA, 9–15 June 2019; PMLR: Seattle, WA, USA, 2019; pp. 6105–6114.

27. Nguyen, H.T.; Luong, H.H.; Huynh, L.B.; Le, B.Q.H.; Doan, N.H.; Le, D.T.D. An improved MobileNet for disease detection on tomato leaves. *Adv. Technol. Innov.* **2023**, *8*, 192–209. [CrossRef]
28. Mathew, M.P.; Mahesh, T.Y. Leaf-based disease detection in bell pepper plant using YOLO v5. *Signal Image Video Process.* **2022**, *16*, 841–847. [CrossRef]
29. Xue, Z.; Xu, R.; Bai, D.; Lin, H. YOLO-tea: A tea disease detection model improved by YOLOv5. *Forests* **2023**, *14*, 415. [CrossRef]
30. Soeb, M.J.A.; Jubayer, M.F.; Tarin, T.A.; Al Mamun, M.R.; Ruhad, F.M.; Parven, A.; Mubarak, N.M.; Karri, S.L.; Meftaul, I.M. Tea leaf disease detection and identification based on YOLOv7 (YOLO-T). *Sci. Rep.* **2023**, *13*, 6078. [CrossRef]
31. Trinh, D.C.; Mac, A.T.; Dang, K.G.; Nguyen, H.T.; Nguyen, H.T.; Bui, T.D. Alpha-EIOU-YOLOv8: An Improved Algorithm for Rice Leaf Disease Detection. *AgriEngineering* **2024**, *6*, 302–317. [CrossRef]
32. Raja, D.; Karthikeyan, M. Red Deer Optimization with Deep Learning Enabled Agricultural Plant Disease Detection and Classification Model. *Int. J. Intell. Eng. Syst.* **2023**, *16*, 21–30. [CrossRef]
33. Chug, A.; Bhatia, A.; Singh, A.P.; Singh, D. A novel framework for image-based plant disease detection using hybrid deep learning approach. *Soft Comput.* **2023**, *27*, 13613–13638. [CrossRef]
34. Towfek, S.; Khodadadi, N. Deep Convolutional Neural Network and Metaheuristic Optimization for Disease Detection in Plant Leaves. *J. Intell. Syst. Internet Things* **2023**, *10*, 66–75. [CrossRef]
35. Khoramshahi, E.; Näsi, R.; Rua, S.; Oliveira, R.A.; Päivänsalo, A.; Niemeläinen, O.; Niskanen, M.; Honkavaara, E. A Novel Deep Multi-Image Object Detection Approach for Detecting Alien Barleys in Oat Fields Using RGB UAV Images. *Remote Sens.* **2023**, *15*, 3582. [CrossRef]
36. Aabidi, M.H.; EL Makrani, A.; Jabir, B.; Zaimi, I. A Model Proposal for Enhancing Leaf Disease Detection Using Convolutional Neural Networks (CNN): Case Study. *Int. J. Online Biomed. Eng.* **2023**, *19*, 127–143. [CrossRef]
37. Sahu, S.K.; Pandey, M. Hybrid Xception transfer learning with crossover optimized kernel extreme learning machine for accurate plant leaf disease detection. *Soft Comput.* **2023**, *27*, 13797–13811. [CrossRef]
38. Chandra, K.; Gupta, D.; Ghosh, J.; Das, P.; Bhunia, D. Insecta: Coleoptera. In *Faunal Diversity of Biogeographic Zones of India: North-East*; Zoological Survey of India: Kolkata, India, 2021; pp. 377–451.
39. de Términos Fitosanitarios, G. *NIMF No. 5*; FAO: Rome, Italy, 2002.
40. Nottingham, L.; Dively, G.; Schultz, P.; Herbert, D.; Kuhar, T. Natural history, ecology, and management of the Mexican bean beetle (Coleoptera: Coccinellidae) in the United States. *J. Integr. Pest Manag.* **2016**, *7*, 2. [CrossRef]
41. Rodríguez-Del-Bosque, L.; Magallanes-Estala, A. Seasonal abundance of Diabrotica balteata and other diabroticina beetles (Coleoptera: Chrysomelidae) in northeastern Mexico. *Environ. Entomol.* **1994**, *23*, 1409–1415. [CrossRef]
42. Hui, Y.; You, S.; Hu, X.; Yang, P.; Zhao, J. SEB-YOLO: An Improved YOLOv5 Model for Remote Sensing Small Target Detection. *Sensors* **2024**, *24*, 2193. [CrossRef]
43. Omer, S.M.; Ghafoor, K.Z.; Askar, S.K. Lightweight improved yolov5 model for cucumber leaf disease and pest detection based on deep learning. *Signal Image Video Process.* **2024**, *18*, 1329–1342. [CrossRef]
44. Terven, J.; Córdova-Esparza, D.M.; Romero-González, J.A. A comprehensive review of yolo architectures in computer vision: From yolov1 to yolov8 and yolo-nas. *Mach. Learn. Knowl. Extr.* **2023**, *5*, 1680–1716. [CrossRef]
45. Xiao, B.; Nguyen, M.; Yan, W.Q. Fruit ripeness identification using YOLOv8 model. *Multimed. Tools Appl.* **2024**, *83*, 28039–28056. [CrossRef]
46. Ma, B.; Hua, Z.; Wen, Y.; Deng, H.; Zhao, Y.; Pu, L.; Song, H. Using an improved lightweight YOLOv8 model for real-time detection of multi-stage apple fruit in complex orchard environments. *Artif. Intell. Agric.* **2024**, *11*, 70–82. [CrossRef]
47. Wang, C.Y.; Yeh, I.H.; Liao, H.Y.M. YOLOv9: Learning What You Want to Learn Using Programmable Gradient Information. *arXiv* **2024**, arXiv:2402.13616.
48. Vo, H.T.; Mui, K.C.; Thien, N.N.; Tien, P.P. Automating Tomato Ripeness Classification and Counting with YOLOv9. *Int. J. Adv. Comput. Sci. Appl.* **2024**, *15* . [CrossRef]
49. Wang, A.; Chen, H.; Liu, L.; Chen, K.; Lin, Z.; Han, J.; Ding, G. Yolov10: Real-time end-to-end object detection. *arXiv* **2024**, arXiv:2405.14458.
50. Hussain, M.; Khanam, R. In-depth review of yolov1 to yolov10 variants for enhanced photovoltaic defect detection. *Solar* **2024**, *4*, 351–386. [CrossRef]
51. Jocher, G.; Qiu, J. Ultralytics YOLO11. Version 11.0.0. 2024. Available online: https://docs.ultralytics.com/models/yolo11/ (accessed on 25 October 2024).
52. Kingma, D.P. Adam: A method for stochastic optimization. *arXiv* **2014**, arXiv:1412.6980.
53. Liu, H.; Li, Z.; Hall, D.; Liang, P.; Ma, T. Sophia: A scalable stochastic second-order optimizer for language model pre-training. *arXiv* **2023**, arXiv:2305.14342.
54. Leng, Z.; Tan, M.; Liu, C.; Cubuk, E.D.; Shi, X.; Cheng, S.; Anguelov, D. Polyloss: A polynomial expansion perspective of classification loss functions. *arXiv* **2022**, arXiv:2204.12511.
55. Sandoval-Alemán, S.A.; Córdova-Esparza, M.d.J.; Gutiérrez-García, J.D. Diseño y Construcción de Robot Agrícola para Detección y Clasificación de Plagas Mediante Visión Artificial. Bachelor's Thesis, Unidad Profesional Interdisciplinaria de Ingeniería Campus Zacatecas (UPIIZ), Instituto Politécnico Nacional, Zacatecas, Mexico, 2021.

56. Terven, J. Auto-Seg-Annotation: Automatic Annotation Segmentation tool based on SAM (Segment Anything Model). 2023. Available online: https://github.com/jrterven/auto-seg-annotation (accessed on 10 June 2024).
57. Kirillov, A.; Mintun, E.; Ravi, N.; Mao, H.; Rolland, C.; Gustafson, L.; Xiao, T.; Whitehead, S.; Berg, A.C.; Lo, W.Y.; et al. Segment anything. In Proceedings of the EEE/CVF International Conference on Computer Vision, Paris, France, 1–6 October 2023; pp. 4015–4026.

AgriEngineering

Article

Climate-Based Prediction of Rice Blast Disease Using Count Time Series and Machine Learning Approaches

Meena Arumugam Gopalakrishnan [1], Gopalakrishnan Chellappan [1,*], Santhosh Ganapati Patil [2], Santosha Rathod [3], Kamalakannan Ayyanar [1], Jagadeeswaran Ramasamy [4], Sathyamoorthy Nagaranai Karuppasamy [5] and Manonmani Swaminathan [6]

1 Department of Plant Pathology, Tamil Nadu Agricultural University, Coimbatore 641003, India; meena.phdpat2021@tnau.ac.in (M.A.G.); kamalakannan.a@tnau.ac.in (K.A.)
2 Agricultural Statistics, Tamil Nadu Agricultural University, Coimbatore 641003, India; patilsgstat@tnau.ac.in
3 Agricultural Statistics, Indian Council of Agricultural Research-Indian Institute of Rice Research, Hyderabad 500030, India; santosh.rathod@icar.gov.in
4 Department of Remote Sensing (RS) and Geographic Information System (GIS), Tamil Nadu Agricultural University, Coimbatore 641003, India; jagawaran@tnau.ac.in
5 Department of Agriculture Climate Research Centre (ACRC), Tamil Nadu Agricultural University, Coimbatore 641003, India; nks6@tnau.ac.in
6 Department of Rice, Tamil Nadu Agricultural University, Coimbatore 641003, India; manonmanis@tnau.ac.in
* Correspondence: pcgopalagri@tnau.ac.in

Abstract: *Magnaporthe oryzae*, the source of the rice blast, is a serious threat to the world's rice supply, particularly in areas like Tamil Nadu, India. In this study, weather-based models were developed based on count time series and machine learning techniques like INGARCHX, Artificial Neural Networks (ANNs), and Support Vector Regression (SVR), to forecast the incidence of rice blast disease. Between 2015 and 2023, information on rice blast occurrence was gathered weekly from three locations (Thanjavur, Tirunelveli, and Coimbatore), together with relevant meteorological data like temperature, humidity, rainfall, sunshine, evaporation, and sun radiation. The associations between the occurrence of rice blast and environmental factors were investigated using stepwise regression analysis, descriptive statistics, and correlation. Mean Squared Error (MSE) and Root Mean Squared Error (RMSE) were used to assess the model's prediction ability. The best prediction accuracy was given by the ANN, which outperformed SVR and INGARCHX in every location, according to the results. The complicated and non-linear relationships between meteorological variables and disease incidence were well-represented by the ANN model. The Diebold–Mariano test further demonstrated that ANNs are more predictive than other models. This work shows how machine learning algorithms can improve the prediction of rice blast, offering vital information for early disease management. The application of these models can help farmers make timely decisions to minimize crop losses. The findings suggest that machine learning models offer promising potential for accurate disease forecasting and improved rice management.

Keywords: rice blast; *Magnaporthe oryzae*; weather parameters; INGARCHX; SVRX; ANNX

Citation: Arumugam Gopalakrishnan, M.; Chellappan, G.; Patil, S.G.; Rathod, S.; Ayyanar, K.; Ramasamy, J.; Nagaranai Karuppasamy, S.; Swaminathan, M. Climate-Based Prediction of Rice Blast Disease Using Count Time Series and Machine Learning Approaches. *AgriEngineering* **2024**, *6*, 4353–4371. https://doi.org/10.3390/agriengineering6040246

Academic Editors: Ray E. Sheriff and Chiew Foong Kwong

Received: 18 September 2024
Revised: 15 October 2024
Accepted: 16 October 2024
Published: 19 November 2024

1. Introduction

More than half of the world's population is fed by rice (*Oryza sativa* L.), which is an essential component of global agriculture. It is grown in over 100 countries, with Asia accounting for the majority of production. With an anticipated production of 11 million metric tonnes in 2022 [1], the USA is another major producer of rice. For many people in India, rice is a staple item and an essential component of the diet [2]. However, India's average rice yield is still relatively low, which makes it difficult to meet both domestic consumption and export requirements despite growing local and worldwide demand [3]. Meeting rice production has become more challenging due to changing climatic conditions [2]. With an estimated total production of 7.98 million tonnes, Tamil Nadu is the fourth-most productive state in India. Tamil Nadu is ranked third in terms of yield per hectare, producing about 3379 kg per hectare. The physiological processes of rice, such as growth, development, and grain production, are impacted by climate change. It also indirectly affects the prevalence of pests and diseases, which has an effect on yields [4]. The crop is susceptible to pathogen infection, which can result in a variety of diseases depending on the weather. This highlights the importance of early detection and suitable treatment to minimize damage and guarantee maximum efficacy [5–7]. A fungus known as *Magnaporthe oryzae* attacks rice crops from seedling to harvesting. When the infection is severe, it can cause the panicles to topple over. It can also occur on the seeds, neck, and branches of the panicles, which show greyish-brown, spindle-shaped lesions when infected (Figure 1) [8]. Blast is considered a major disease of rice because of its wide distribution and extent of destruction under favourable conditions. Count time series modelling is an extensively used statistical methodology in which integer autocorrelated discrete count observations are available as the inputs with a typically assumed Poisson or negative binomial distribution. This method is especially pertinent in the modelling of crop disease, as data for the number of days or weeks it takes until the first individual becomes infected by a particular incidence are correlated. While the count time series models and machine learning methods have been applied extensively to different types of problems, including epidemiological studies [9], these models are under-exploited in modelling and predicting rice blast incidence data streams.

Figure 1. Rice blast incidence documented in study area. The image was captured by the author at Paddy Breeding Station, TNAU, Coimbatore, Tamil Nadu, India.

Count time series models have been applied to a variety of scenarios, including disease prediction, manufacturing industry claims, and stock market data [10–13]. Machine learning models have been successfully used to predict agricultural outcomes like oilseed production, banana and rice yields, susceptibility to pests in rice, and disease severity in crops like tomatoes and sugarcane [14–19]. The adoption of preventive measures can be greatly aided by an accurate weather-based prediction of rice blast. However, multiple regression analysis and conventional time series models were frequently used in previous studies on crop disease forecasting. The accuracy of the prediction model is not improved by normalizing count data that follows non-Gaussian distributions, such as those characterized by Poisson and negative binomial distributions. Although the generalized linear model (INGARCH) is well suited for count data, little to no use has been made of it for disease modelling, including rice blast. In addition, it has been shown that the standard models such as multiple linear regression and ARIMA are less successful for highly variable and nonlinear data [14]. On the other hand, data-driven machine learning models such as SVR and ANN, which do not rely on rigid assumptions, exhibit potential in these kinds of situations. Modelling and forecasting disease incidence like rice blast is crucial for informed decision-making and effective crop management. The goal of this research is to use count time series (INGARCH) and machine learning techniques (ANN, SVR) to create a reliable statistical model for rice blast prediction based on important weather data. In order to minimize yield losses, the models are made to predict when rice blast will occur. This work highlights the innovative use of machine learning techniques like ANN and SVR in the field of rice disease modelling by combining weather variables with the count time series model or INGARCH for the first time. In order to determine the causal linkages between environmental conditions and the incidence of rice blast, this study used stepwise regression analysis, basic descriptive statistics, and correlation analysis.

This study aims to develop a robust statistical model for predicting rice blast based on critical weather parameters using count time series (INGARCH) and machine learning methods (ANN, SVR). The models are designed to forecast the occurrence of rice blast, thereby minimizing yield losses. For the first time, this study applies the count time series model, specifically INGARCH, with weather variables to the area of rice disease modelling, highlighting the novel application of machine learning techniques like ANN and SVR in this field. In this study, the work carried out included conducting basic descriptive statistics, performing correlation analysis, and utilizing stepwise regression analysis to identify the causal relationships between environmental factors and the incidence of rice blast. Utilizing exogenous meteorological information, sophisticated computational methods including INGARCH, ANN, and SVR are used to predict the incidence of rice blast in high-risk areas.

2. Materials and Method

2.1. Data Collection

The rice blast disease data were collected from three different survey points in Tamil Nadu (Thanjavur, Tirunelveli, and Coimbatore) from the year 2015 to 2023 (Figure 2). The disease incidence was calculated using percent disease incidence (PDI) according to IRRI, 1990. The weekly disease data were obtained from the cumulative of the daily data collected. Corresponding meteorological data, including maximum temperature (MaxT), minimum temperature (MinT), relative humidity morning (RHm), relative humidity evening (RHe), rainfall (RF), sunshine (SS), solar radiation (SR), wind speed (WS), leaf wetness (LW), and evaporation (EVP) were also collected from the automatic weather stations situated at the respective locations. Standard Meteorological Week (SMW) wise disease incidence and weekly meteorological observation were considered for this study. A total of 80% of the dataset was used as a training set and the remaining 20% of observations were used as a testing set.

2.2. Pre-Processing of Meteorological Variables

Missing meteorological data were addressed using linear interpolation and mean imputation techniques. These methods were implemented using R software (version 4.x) with the "zoo" and "imputeTS" packages for handling time series data gaps and missing values [20,21].

Meteorological data were normalized using the min-max scaling method to ensure comparability across variables. This was executed in R using the "scales" package [22]. Time lags (1 to 3 weeks) were introduced for the weather variables to account for the delayed effects of meteorological factors on rice blast disease [23]. This step was implemented using the "dplyr" and "zoo" packages in R.

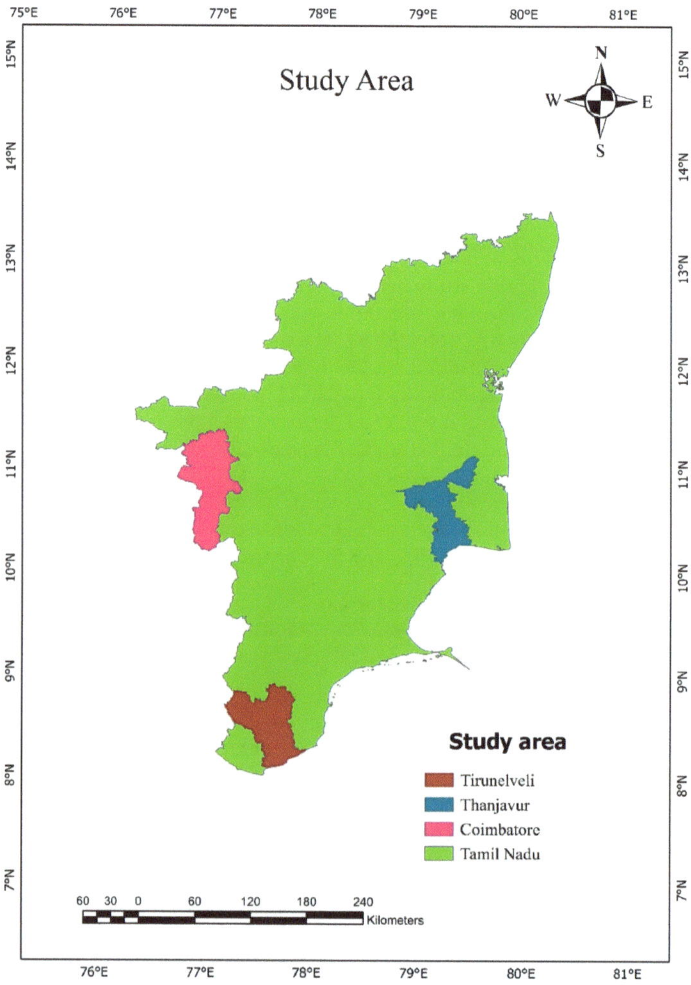

Figure 2. Study area for rice blast incidence in Tamil Nadu.

2.3. Statistical Models

Descriptive statistics were carried out with the parameters including mean, standard error (SE), skewness, kurtosis, minimum and maximum observation, and coefficient of variance (CV) to describe the nature of the data under study. The data were graphically distributed as the time series plots. The interrelationships between the variables were analysed with Pearson product–moment correlation, and stepwise multiple regression was carried out to observe the cause-and-effect relationship between rice blast disease incidence and exogenous weather variables [9]. The regression equation in matrix form was given as follows:

$$Y = X\beta + e \tag{1}$$

where 'Y' is the dependent variable, 'X' is the vector of exogenous/independent variable, 'β' is the regression coefficient and 'e' is the residual error assumed to be normally distributed with $e \sim N\ (0, \sigma^2)$. The three models, INGARCH, ANN, and SVR, as well as time series plots, were completed with R software (version 4.3.1, R Core Team 2018). Correlation and stepwise regression analysis were carried out in SAS (version 9.3).

2.4. Integer-Valued Generalized Autoregressive Conditional Heteroscedastic (INGARCH) Model

The time series analysis is structured within the generalized linear model (GLM) framework, which has been extensively explored by [24]. INGARCH models, a subset of these GLMs, assume that the conditional distribution of the dependent variable follows one of the well-known discrete distributions such as Poisson, negative binomial, generalized Poisson, or double Poisson [11,25,26]. For a count time series $\{Y_t: t \in N\}$ where λt represents the time-varying r-dimensional covariate vector $\{X_t: t \in N\}$ i.e., $X_t = (X_{t,1}, \ldots, X_{t,r,})^T$, the conditional mean of Y_t, denoted as $E(Y_t \mid F_{t-1}) = \lambda_t$, is based on historical data F_t. The generalized form of INGARCH model is expressed as follows

$$g(\lambda_t) = \beta_0 + \sum_{k=1}^{p} \alpha_k \tilde{g}\left(Y_{t-i_k}\right) + \sum_{l=1}^{q} \beta_1 g\left(\lambda_{t-j_1}\right) + \eta^T \quad (2)$$

In the case where the functions g and $\tilde{g}$ are identity functions, that is $g(x) = \tilde{g}(x) = x$, the model simplifies significantly. Under this configuration, the process Y_t is assumed to follow a Poisson distribution conditional on its past values. Specifically, Y_t is Poisson distributed with a conditional mean $\lambda_t = E(Y_t|Y_{t-1}, Y_{t-2}, \ldots)$, which is a linear function of both its past values and the lagged values of λ_t. The model is characterized by the equation:

$$\lambda_t = \beta_0 + \sum_{i=1}^{p} \alpha y_{t-i} + \sum_{j=1}^{q} \beta_j \lambda_{t-j} \quad (3)$$

with $\beta_0 > 0$ and $\alpha_1, \ldots, \alpha_p, \ldots, \beta_1, \ldots, \beta_q \geq 0$ where the coefficients α_k and β_i are non-negative. When the model is simplified in this way, it is referred to as an INGARCH model of order p and q (abbreviated as INGARCH (p, q)), or simply as an INGARCH(p) model if q = 0. These models are also known as autoregressive conditional Poisson (ACP) models [10].

In situations where the conditional variance of Y_t is larger than its mean λ_t, indicating over-dispersion, the negative binomial distribution is employed instead of the Poisson distribution. In this scenario, Y_t is assumed to follow a negative binomial distribution conditional on past information, specifically $Y_t|F_{t-1} \sim \text{NegBinom}(\lambda_t, \varnothing)$, where $\varnothing$ represents the overdispersion parameter [27]. As $\varnothing$ increases, the model converges towards the Poisson distribution. This negative binomial extension accounts for over-dispersion and provides a more flexible model in such cases.

The estimation of parameters in INGARCH models typically uses conditional likelihood methods, with special attention given to the asymptotic properties. The standard INGARCH model relies on past values of the forecast variable, predicting future values based on this historical data. The INGARCHX model is an extension of this framework that incorporates additional external covariates to the INGARCH model, allowing for a more comprehensive analysis by including exogenous factors in the prediction model [28].

2.5. Support Vector Regression (SVR)

The methodology employed in this study involves the use of Support Vector Regression (SVR) to transform the original input data into a higher-dimensional feature space [29]. This transformation enables the construction of a linear regression model in the newly mapped feature space, facilitating more accurate predictions. Given a dataset $Z = \{x_i y_i\}_{i=1}^{N}$, where $x_i \in R^n$ represents the input variables, y_i is the corresponding scalar output and N is the size of the data set, the SVR model is formulated as:

$$f(x) = w^T \phi(x) + b \quad (4)$$

In this equation, w denotes the weight vector, b is the bias term, and $\phi(x)$ is a function that maps the input data x into a higher-dimensional space. The parameters w and b are determined by minimizing a regularized risk function, which combines model complexity and prediction error:

$$R(\theta) = \frac{1}{2}\|w\|^2 + c\left[\frac{1}{N}\Sigma_{i=1}^{N} L_\varepsilon(y_i, f(x_i))\right] \quad (5)$$

The first component of the risk function, $\frac{1}{2}\|w\|^2$ is the regularisation term, which controls the complexity of the model by penalising large weights, ensuring that the model remains as flat as possible. The second component $\frac{1}{N}\Sigma_{i=1}^{N} L_{\varepsilon}(y_i, f(x_i))$ is the empirical error term, evaluated using the ε-insensitive loss function as defined by:

$$L_{\varepsilon}(y_i, f(x_i)) = f(x) = \begin{cases} |y_i, f(x_i) - \varepsilon|; |y_i - f(x_i)| \geq \varepsilon, \\ 0\ |y_i - f(x_i)| < \varepsilon, \end{cases} \tag{6}$$

This loss function allows the model to ignore small deviations (within ε) from the actual values, thereby focusing only on significant prediction errors. For this study, the Radial Basis Function (RBF) kernel was utilized to map the input data into the higher-dimensional feature space. The RBF kernel is defined as:

$$k(x_i, x_j) = \exp\left\{-\gamma\|x - x_i\|^2\right\} \tag{7}$$

The effectiveness of the RBF kernel depends on the proper selection of two hyper-parameters: the regularization parameter C, which balances the trade-off between the model's complexity and its prediction accuracy, and the kernel parameter γ, which defines the influence of a single training example. These parameters were optimized during the training process to enhance the model's performance. Additionally, SVR, in conjunction with the radial basis function, was applied to model the relationship between input variables and outcomes, similar to the approach used in the INGARCHX model for time series forecasting. The structural design of the SVR architecture is typically illustrated visually to provide a clearer understanding of the kernel's composition and data flow (Figure 3).

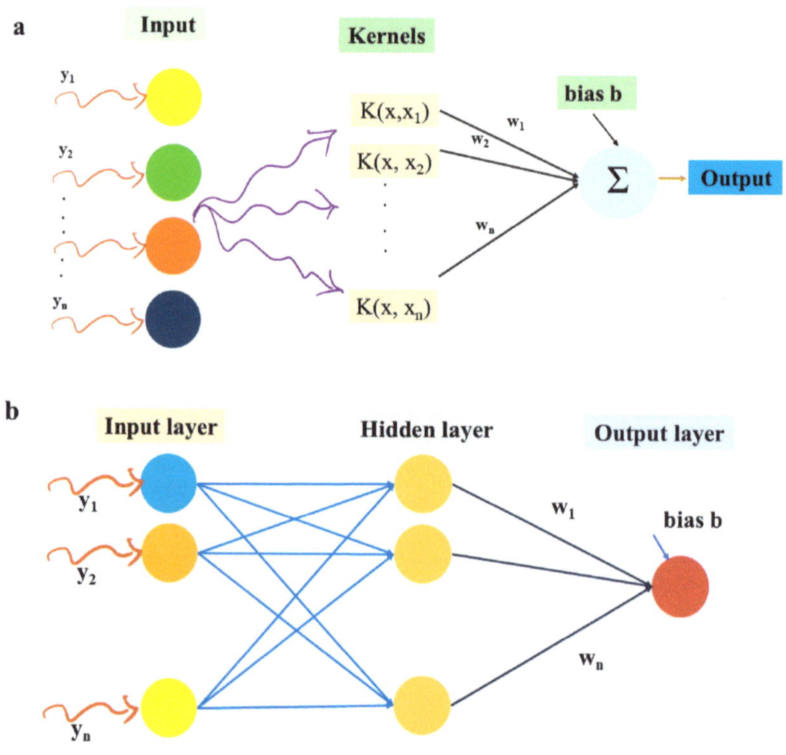

Figure 3. Architecture of the SVR (**a**) and ANN (**b**) models.

2.6. Artificial Neural Network

Over the past few decades, Artificial Neural Networks (ANNs) have emerged as a dominant machine learning methodology. In the realm of time series analysis, a specialized form of ANN has gained traction, known as the autoregressive neural network. This variant is particularly adept at handling temporal data by incorporating time lags as input features.

The mathematical framework underpinning ANNs in time series contexts can be conceptualized as a neural network with an implicit functional representation of time. For multi-layer feed-forward autoregressive neural networks, the final output Y_t can be expressed through the following equation

$$Y_t = \alpha_0 + \sum_{j=1}^{q} \alpha_j g\left(\beta_{0j} + \sum_{i=1}^{p} \beta_{ij} Y_{t-p}\right) + \varepsilon_t \tag{8}$$

In this formulation the α_j (j = 0, 1, 2, . . ., q) and β_{ij} (i = 0, 1, 2, . . ., p; j = 0, 1, 2, . . ., q) represent the model parameters also referred to as synapsis weights, p signifies the number of input nodes, q denotes the number of hidden nodes, and g represents the activation function. The training process for ANNs in this context focuses on minimizing the disparity between actual and predicted values. This is achieved through an error function specific to autoregressive ANNs, which can be expressed as

$$E = \frac{1}{N}\Sigma_{t=1}^{N}(e_t)^2 = \frac{1}{N}\Sigma_{t=1}^{N}\{X_t - \left(w_0 + \left(\sum_{j=1}^{q} w_J g\left(w_{0_j} + \Sigma_{i=1}^{p} w_{ij} X_{t-i}\right)\right)\right)\}^2 \tag{9}$$

where N represents the total number of error terms. The parameters of the neural network (w_{ij}) are adjusted based on the magnitude of changes in Δw_{ij} as $\Delta w_{ij} = -\eta \frac{\partial E}{\partial w_{ij}}$, where η is the learning rate [14,30]. This approach shares similarities with INGARCHX and SVRX models in that exogenous variables can be incorporated to account for external factors. When extended in this manner, the model becomes known as an ANNX model. The structural design of the ANN architecture is typically illustrated visually to provide a clearer understanding of the network's composition and data flow (Figure 4).

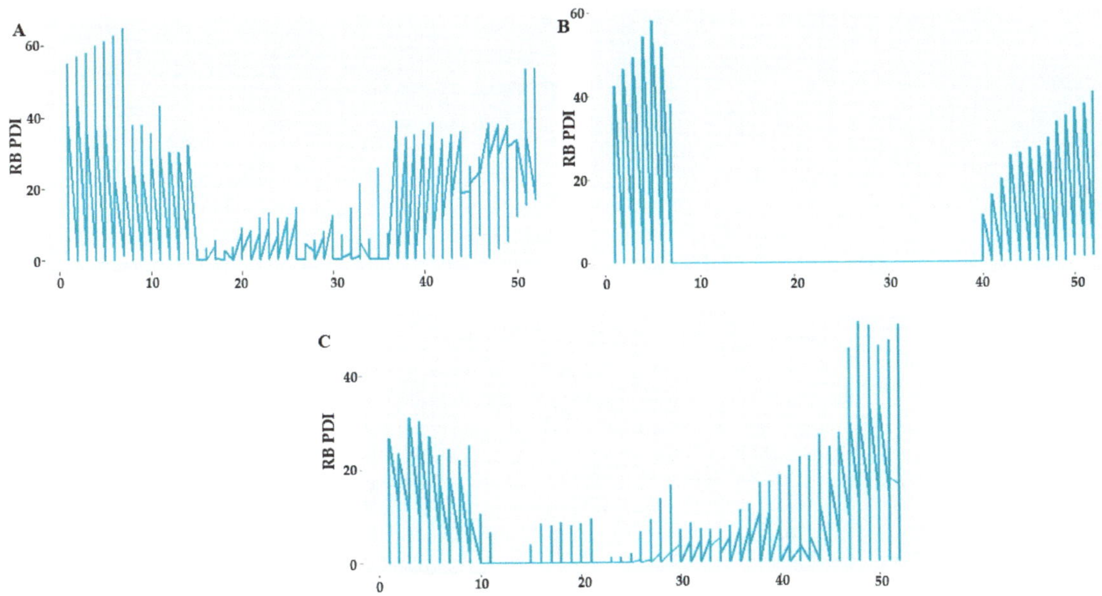

Figure 4. Time series plots of rice blast occurrence at Tirunelveli (**A**), Thanjavur (**B**), and Coimbatore (**C**).

2.7. Model Evaluation Metrics

To assess the accuracy and reliability of predictive models, two key performance indicators were employed: the mean square error (MSE) and its square root, the root mean square error (RMSE).

The Mean Square Error (MSE) is defined as the average of the squared differences between predicted and actual values. It is calculated by summing the squared differences between each predicted value and its corresponding actual value, then dividing by the total number of observations. The formula for MSE is presented as

$$\text{MSE} = \frac{\sum_{i=1}^{N}(y_i - \hat{y}_i)^2}{N} \tag{10}$$

Root Mean Square Error (RMSE) is described as the standard error of estimate in regression analysis. It is calculated by taking the square root of the MSE. The formula for RMSE is given as

$$\text{RMSE} = \sqrt{\frac{\sum_{i=1}^{N}(y_i - \hat{Y}_i)}{N}^{2}} \tag{11}$$

where y_i is the actual value, $\hat{y}_i$ is the predicted value, and N is the number of observations The sum is taken from i = 1 to N.

2.8. Diebold–Mariano Test

The Diebold–Mariano (DM) test, is used to assess whether there is a statistically significant difference between two forecasting models based on their residuals. The test considers the residuals of two models (r_1 and r_2) and d_i was the absolute difference between these residuals:

$d_i = |r_1| - |r_2|$. An autocovariance function γ_k is calculated as

$$\gamma_k = \frac{1}{n}\sum_{i=k+1}^{n}\left(d_1 - \bar{d}\right)\left(d_{i-k} - \bar{d}\right) \tag{12}$$

where $\bar{d}$ is the mean of d_i, and the sum is from i = k + 1 to n.

The DM test statistic is then given by

$$\text{DM} = \frac{\bar{d}}{\sqrt{\left[\gamma_0 + 2\sum_{k=1}^{h-1}\gamma_k\right]/n}} \tag{13}$$

where the sum in the denominator goes from k = 1 to h − 1, and $h = n^{1/3} + 1$. For hypothesis testing: Null hypothesis (H_0): E(d) = 0, meaning the forecast accuracy is similar for both models. Alternative hypothesis (H_1): $E(d) \neq 0$, indicating different forecast accuracies between the models. This test helps researchers determine if one forecasting model significantly outperforms another in terms of accuracy.

3. Results

The time series plot of rice blast disease incidence from 2015 to 2023 across three locations in Tamil Nadu is presented in Figure 4. The plot reveals that, in all the locations studied, the disease was most severe between the 40th and 15th Standard Meteorological Weeks (SMWs), with minor occurrences during other periods as well. However, an exception was observed at the Thanjavur centre, where severe incidence was confined to the period between the 40th and 6th SMWs.

The summary statistics of the dependent variable, i.e., the rice blast, and the exogenous weather variable given in Table 1 are self-explanatory. For instance, rice blast incidence at Tirunelveli, Thanjavur and Coimbatore were 11.54, 3.88 and 3.30 per cent, respectively. The data under consideration were highly heterogeneous in nature, with rice blast incidence varying in a wide range (0 to 65), leading to a high standard deviation (SD) and abnormality in data distribution as indicated by skewness and kurtosis values out of the normal range.

Table 1. Summary statistics of rice blast incidence at study locations.

Location	Statistics	RB	MaxT	MinT	RHm	RHe	WS	RF	SR	LW
Tirunelveli	Mean	11.54	33.08	25.63	66.22	72.92	5.21	1.71	496.07	5.93
	Std. Error	0.73	0.15	0.09	0.85	0.68	0.14	0.3	4.99	0.12
	Skewness	1.21	0.189	−0.086	−1.01	−1.13	0.14	0.3	4.99	0.12
	Kurtosis	0.59	−0.224	−0.472	1.93	1.17	−0.07	51.26	0.34	−0.88
	Minimum	0	24.77	20.5	42.8	53.72	1.75	0	128.7	0.02
	Maximum	65	40.97	29.81	92.18	95.38	15.64	63.14	693.75	10.82
	SD	14.97	3.24	1.98	17.43	17.58	2.91	6.19	101.92	2.55
	Statistics	RB	MaxT	MinT	RHm	RHe	WS	RF	SS	EVP
Thanjavur	Mean	3.88	33.21	23.27	89.60	59.12	4.89	2.69	6.05	4.41
	Std. Error	0.36	0.10	0.09	0.24	0.64	0.10	0.20	0.08	0.06
	Skewness	2.96	−0.32	1.39	−0.94	−1.08	0.96	4.16	−0.51	−0.12
	Kurtosis	8.60	−0.97	18.19	0.17	2.43	1.37	25.18	−0.25	−0.24
	Minimum	0.00	26.37	16.44	68.43	52.78	0.00	0.00	0.00	0.09
	Maximum	58.23	38.89	15.29	99.71	98.43	15.74	55.29	10.11	9.76
	SD	9.78	2.78	2.56	6.59	17.56	2.74	5.66	2.18	1.74
	Statistics	RB	MaxT	MinT	RHm	RHe	WS	RF	SS	EVP
Coimbatore	Mean	3.30	31.9	22.62	84.3	52.64	6.36	2.17	6.46	97.16
	Std. Error	0.25	0.12	0.09	0.18	0.53	0.13	0.21	0.12	8.32
	Skewness	2.26	0.88	−1.06	−0.61	−0.38	1.7	3.29	0.94	1.06
	Kurtosis	5.57	−0.72	1.18	0.77	−0.16	3.17	13.2	7.15	−0.71
	Minimum	0	26.64	16.1	69.05	22.71	2.35	0	1.07	3.77
	Maximum	25.60	37.5	25.85	94	75	17.68	28.02	23.74	9.36
	SD	4.94	2.33	1.74	3.48	10.23	2.49	4.03	2.32	6

RB—Rice Blast; MaxT—Maximum Temperature; MinT—Minimum Temperature; RHm—Relative Humidity Morning; RHe—Relative Humidity Evening; WS—Wind Speed; RF—Rainfall; SR—SolarRadiation; SS—Sunshine; LW—Leaf wetness; EVP—Evaporation.

3.1. Correlation Analysis

The Pearson correlation coefficients between rice blast incidence and weather parameters are presented in Table 2. A strong positive correlation was observed between rice blast incidence and evening relative humidity (RHe), morning relative humidity (RHm), and leaf wetness (LW). Conversely, a strong negative correlation was noted with minimum temperature (MinT) at the Tirunelveli centre. In Thanjavur, the morning and evening relative humidity (RHm and RHe) showed strong positive correlations, while maximum temperature (MaxT) and minimum temperature (MinT) exhibited strong negative correlations. Similarly, in Coimbatore, RHm and RHe had strong positive correlations with disease incidence, whereas MinT displayed a strong negative correlation. Overall, the correlation analysis indicates that rice blast incidence is associated with MaxT, MinT, RHm, RHe, and LW, highlighting the influence of weather heterogeneity on the disease's variability.

3.2. Stepwise Regression Analysis

To determine the weather factors influencing rice blast incidence, stepwise regression analysis was conducted, and the results are presented in Table 3. The findings indicate that weather parameters such as MaxT, RHe, EVP, and RF in Thanjavur; MinT, RHm, WS, LW, and RF in Tirunelveli; and MaxT, RHm, RHe, and EVP in Coimbatore have a significant impact on rice blast incidence. Although these variables exhibit a notable influence, the model's R^2 value suggests moderate explanatory power for Tirunelveli and Coimbatore. However, the R^2 value for Thanjavur is lower, indicating a weaker fit, likely due to non-linearity and high heterogeneity in the dependent variables.

Table 2. Pearson correlation coefficients between rice blast incidence and meteorological variables at study locations.

Location		RB	MaxT	MinT	RHm	RHe	WS	RF	SR	LW
Tirunelveli	RB	1.00								
	MaxT	0.46	1.00							
	MinT	−0.58	−0.07	1.00						
	RHm	0.59	0.04	−0.44	1.00					
	Rhe	0.72	−0.04	0.51	−0.14	1.00				
	WS	−0.20	−0.04	0.56	−0.19	0.78	1.00			
	RF	0.03	−0.01	−0.14	0.19	−0.09	−0.16	1.00		
	SR	−0.32	−0.02	0.38	−0.45	0.07	0.16	−0.30	1.00	
	LW	0.59	0.05	−0.29	0.22	−0.16	−0.19	0.08	−0.25	1.00
		RB	MaxT	MinT	RHm	RHe	WS	RF	SS	EVP
Thanjavur	RB	1.00								
	MaxT	−0.48	1.00							
	MinT	−0.33	0.63	1.00						
	RHm	0.82	−0.60	−0.57	1.00					
	RHe	0.72	−0.52	−0.25	0.37	1.00				
	WS	−0.17	0.27	0.24	−0.68	−0.36	1.00			
	RF	0.16	−0.33	0.00	0.23	0.31	−0.14	1.00		
	SS	−0.21	0.44	−0.04	−0.08	−0.30	−0.04	−0.41	1.00	
	EVP	−0.45	0.70	0.45	−0.60	−0.55	0.36	−0.36	0.56	1.00
		RB	MaxT	MinT	RHm	Rhe	WS	RF	SS	EVP
Coimbatore	RB	1								
	MaxT	−0.43	1							
	MinT	−0.87	0.49	1						
	RHm	0.50	−0.25	−0.11	1					
	Rhe	0.62	−0.58	0.29	0.34	1				
	WS	−0.28	−0.01	0.22	−0.52	0.09	1			
	RF	0.25	−0.24	0.09	0.35	0.46	−0.14	1		
	SS	−0.09	0.64	−0.08	−0.23	−0.74	−0.11	−0.28	1	
	EVP	−0.28	0.07	−0.04	−0.15	−0.22	−0.16	−0.03	0.05	1

RB—Rice Blast; MaxT—Maximum Temperature; MinT—Minimum Temperature; RHm—Relative Humidity Morning; RHe—Relative Humidity Evening; WS—Wind Speed; RF—Rainfall; SR—SolarRadiation; SS—Sunshine; LW—Leaf wetness; EVP—Evaporation.

Table 3. Stepwise regression analysis of rice blast incidence and meteorological variables at study locations.

Centre	**Variable**	**Estimate**	**S.E.**	**t Value**	**Pr > t**	**R^2**	**Model R^2**	***p*-Value**
Thanjavur	Intercept	47.73	5.32	8.97	<0.0001	0.2783	0.2822	$<2.2 \times 10^{-16}$
	MaxT	−1.18	0.15	−7.42	<0.0001			
	RHe	0.03	0.02	1.62	0.01			
	EVP	−1.34	0.26	−5.14	<0.0001			
	RF	−0.27	0.05	−4.6	<0.0001			
Tirunelveli	Intercept	81.73	10.14	8.06	<0.0001	0.4716	0.4779	$<2.22 \times 10^{-16}$
	MinT	−3.78	0.36	−10.47	<0.0001			
	RHm	0.21	0.03	6.29	<0.0001			
	WS	0.84	0.22	3.8	<0.0001			
	LW	1.4	0.21	6.38	<0.0001			
	RF	−0.18	0.08	−2.1	<0.01			
Coimbatore	Intercept	29.15	7.76	3.75	<0.0001	0.439	0.4483	$<2.22 \times 10^{-16}$
	MaxT	−1.43	0.11	−12.79	<0.0001			
	RHm	0.31	0.07	4.21	<0.0001			
	RHe	−0.16	0.03	−5.23	<0.0001			
	EVP	0.01	0.001	6.37	<0.0001			

3.3. INGARCHX Model

Prior to the development of the INGARCH model, the Box–Pierce non-correlation test was used to assess the presence of autocorrelation in the blast incidence count time series data (Table 4). The test results indicated a highly significant autocorrelation ($p < 0.0001$) across all centres, confirming the presence of autocorrelation in the data. The INGARCH model, incorporating exogenous weather variables, was applied and while the overall model parameters were significant, none of the weather variables were significant across all locations. The over-dispersion parameters obtained for Thanjavur (3.97), Tirunelveli (1.64), and Coimbatore (7.10) suggest that the data are heterogeneous and over-dispersed, fitting a negative binomial distribution. The over-dispersion parameters for Thanjavur (3.97), Tirunelveli (1.64), and Coimbatore (7.10) indicate that the data are heterogeneous and over-dispersed, aligning with a negative binomial distribution. Further diagnostic checks using the Box–Pierce test on the residuals confirmed that they were autocorrelated and non-random ($p < 0.0001$) at all locations. The inability of the INGARCHX model to account for the data's heterogeneity and complexity likely explains the non-significance of the weather variables and the persistence of significant residuals. This suggests that the model may not fully capture the intricate relationships within the data.

Table 4. Parameter estimation of the INGARCHX model for rice blast incidence at study locations.

Center	Parameters	Estimation	S.E.	Z Value	Probability	Box–Pierce Non-Correlation Test	
						Original	Residual
Thanjavur	Intercept	5.07×10^{-7}	2.513	0	1	$\chi^2 = 573.03$ ($p < 0.0001$)	$\chi^2 = 464.61$ ($p < 0.0001$)
	beta_1	1.28×10^{-1}	0.046	2.74	<0.001		
	MaxT	2.99×10^{-10}	0.056	0	1		
	MinT	2.48×10^{-9}	0.049	0	1		
	RHm	1.72×10^{-2}	0.025	1.67	0.501		
	RHe	1.49×10^{-9}	0.034	0	1		
	WS	9.06×10^{-6}	0.014	0	0.999		
	RF	3.52×10^{-9}	0.06	0	1		
	SS	1.44×10^{-2}	0.002	5.32	<0.0001		
	EVP	8.28×10^{-9}	0.093	0	1		
	Overdispersion Parameter (Ø)	3.97					
Tirunelveli	Intercept	1.17	354.38	0.0330	0.9737	$\chi^2 = 308.88$ ($p < 0.0001$)	$\chi^2 = 277.66$ ($p < 0.0001$)
	beta_1	5.83×10^{-5}	0.0709	0.0008	0.9993		
	MinT	1.84×10^{-8}	0.3848	0	1		
	MaxT	2.34×10^{-6}	0.7412	0	1		
	RHm	2.66×10^{-5}	0.1380	0.0002	0.9998		
	RHe	4.77×10^{-6}	0.4840	0	1		
	WS	1.69×10^{-5}	0.0575	0.0003	0.9998		
	RF	3.19×10^{-9}	0.3349	0	1		
	SR	3.79×10^{-10}	0.3457	0	1		
	LW	3.34×10^{-13}	0.0118	0	1		
	Overdispersion Parameter (Ø)	1.64					
Coimbatore	Intercept	3.27×10^{-4}	4.1765	0	1	$\chi^2 = 301.73$ ($p < 0.0001$)	$\chi^2 = 200$ ($p < 0.0001$)
	beta_1	3.25×10^{-1}	0.0561	5.789	<0.0001		
	MaxT	1.32×10^{-13}	0.1339	0	1		
	MinT	2.49×10^{-9}	0.1313	0	1		
	RHm	6.19×10^{-7}	0.0316	0	1		
	RHe	4.79×10^{-4}	0.0464	0.0103	0.9918		
	WS	5.28×10^{-8}	0.0533	0	1		
	RF	1.44×10^{-2}	0.0264	0.543	0.5869		
	SS	6.01×10^{-9}	0.0660	0	1		
	EVP	7.43×10^{-9}	0.0049	0.510	0.1310		
	Overdispersion Parameter (Ø)	7.10					

3.4. SVRX Model

The parameter specifications for the SVRX model applied to rice blast incidence at the study locations are outlined in Table 5. The model used a Radial Basis Function (RBF) kernel across all three centres (Thanjavur, Tirunelveli, and Coimbatore). The number of support vectors ranged from 223 to 269, with a cost of 1. The gamma values were 0.125 for Thanjavur, 0.09 for Tirunelveli, and 0.111 for Coimbatore. The epsilon value was set to 0.1 across all locations, with cross-validation errors of 0.019 for Thanjavur, 0.66 for Tirunelveli, and 0.370 for Coimbatore. The Box–Pierce non-correlation test for residuals showed significant autocorrelation ($p < 0.001$) for all locations, indicating that the residuals were not random.

Table 5. Parameter specifications of SVRX and ANNX models for rice blast incidence at study locations.

	Thanjavur	**Tirunelveli**	**Coimbatore**
		SVRX Model	
Kernel function	RBF	RBF	RBF
No. of Support Vectors	223	269	231
Cost	1	1	1
Gamma	0.125	0.09	0.111
Epsilon	0.1	0.1	0.1
Cross-validation error	0.019	0.66	0.370
Box–Pierce non-correlation test for residuals	370.11 ($p < 0.0001$)	110.67 ($p < 0.0001$)	152.5 ($p < 0.0001$)
		ANNX	
Input lag	9	2	2
Dependent/output variable	1	1	1
Hidden layer	1	1	1
Hidden nodes	1	2	2
Exogenous variables	8	8	8
Model	9:1S:1L	13:2S:1L	11:2S:1L
Total number of parameters	12	31	27
Network type		Feed Forward	
Activation function I:H		Sigmoidal	
Activation function H:O		Identity	
Box–Pierce non-correlation test for residuals	0.203 ($p = 0.65$)	0.313 ($p = 0.575$)	0.921 ($p = 0.3372$)

3.5. ANNX Model

The ANNX model parameters presented in Table 5 show that for Thanjavur, Tirunelveli, and Coimbatore, the model configurations included input lags of 9, 2, and 2, respectively. Each location had one dependent variable and one hidden layer, with the number of hidden nodes set to 1, 2 and 2 for each location, respectively. Eight exogenous variables were used across all locations. The model architecture was a feed-forward network with a total of 12, 31, and 27 parameters (or synaptic weights) for Thanjavur, Tirunelveli, and Coimbatore, respectively. The activation function used for the input-to-hidden layer was sigmoidal, and an identity function was applied from the hidden layer to the output layer. Diagnostic checking using the Box–Pierce non-correlation test indicated that the residuals were uncorrelated and random for Thanjavur ($p = 0.065$), and highly uncorrelated for Tirunelveli ($p = 0.575$) and Coimbatore ($p = 0.337$), suggesting a good fit for these locations.

3.6. Model Performance Evaluation Across Different Centres

Comparisons of model performance for predicting rice blast incidence in Thanjavur, Tirunelveli, and Coimbatore using three different models—INGARCHX, ANNX, and SVRX—are shown in Table 6. The performance is evaluated based on Mean Squared Error (MSE) and Root Mean Squared Error (RMSE) for both the training and testing datasets. Table 6 highlights that the ANNX model consistently demonstrated superior performance across all locations and datasets, particularly excelling in the training sets with the lowest MSE and RMSE values.

Table 6. Comparison criteria for different models for rice blast incidence in training and testing datasets.

		Criteria	INGARCHX	ANNX	SVRX
Thanjavur	Training	MSE	50.941	12.12	36.54
		RMSE	7.137	3.481	6.045
	Testing	MSE	472.58	407.89	387.44
		RMSE	21.738	20.196	19.683
Tirunelveli	Training	MSE	228.86	31.38	56.832
		RMSE	15.12	5.601	7.538
	Testing	MSE	175.49	14.974	44.44
		RMSE	13.24	3.869	6.66
Coimbatore	Training	MSE	11.774	2.542	7.784
		RMSE	3.431	1.594	2.789
	Testing	MSE	39.746	10.890	13.757
		RMSE	6.304	3.30	3.709

4. Discussion

Based on MSE and RMSE in both training and testing datasets, the results of modelling and forecasting rice blast incidence at the research areas using various models were evaluated, as indicated in Table 6. The investigation revealed that the stepwise regression model's fitness was low (low R^2) as a result of the dependent variable's high heterogeneity and non-linearity. However, ref. [23] demonstrated a strong association between the onset of rice blast illness and the morning, afternoon, and dew point temperatures, as well as evaporation and dew explained using stepwise regression. The reduced MSE and RMSE values of the ANNX model among the evaluated methods show that it performs better than other machine learning models and classical statistical models, including SVR and INGARCH, in both training and testing datasets in forecasting blast incidence. Additionally, in both datasets, the SVRX model fared better than the INGARCHX model. ANNX > SVRX > INGARCHX was the model's performance rating in both the training and validation stages at all three locations.

The ANN model outperformed the conventional autoregressive integrated moving average and SVR models, according to similar findings published by [31,32]. Table 5 illustrates how candidate hyperparameters for the SVR model were chosen from a variety of user-defined combinations. The robustness of the models was confirmed by 10-fold cross-validation. This method allowed the models to be trained on 80% of the data and tested on the remaining 20%, thereby providing reliable performance estimates and reducing the risk of overfitting. In addition to cross-validation, an independent test set comprising 20% of the data was held out for further validation. Using a feed-forward network and the "Levenberg-Marquardt back-propagation algorithm", the ANN model was improved via repeated testing. The model was modified over 25 iterations and a maximum of 1000 cycles, with the learning rate and momentum terms set at 0.03 and 0.01, respectively. Following testing of various input lag and hidden node combinations, the final model parameters were selected based on the lowest possible training error. Figure 5 shows how well the ANNX model predicted the incidence of rice blasts, exceeding both the INGARCHX and SVRX models in terms of accuracy. The model's prediction disparities were emphasized by the MSE and RMSE comparison criterion (Figure 6). The ANN model's consistent performance across both cross-validation and independent test datasets confirmed its ability to generalize well to unseen data, reinforcing its suitability for practical applications. The Diebold–Mariano test (Table 7) further validated the ANN model's performance, confirming the statistical significance of its superiority over traditional models such as

INGARCHX and SVR. These findings highlight the practical importance of the ANN model in capturing complex, non-linear relationships between weather variables and rice blast incidence. In both datasets, the ANN model outperformed the INGARCHX and SVRX models. The superior performance of the ANN model across different datasets and statistical validation through tests like Diebold–Mariano strongly supports its applicability for real-time disease prediction in varied environmental conditions.

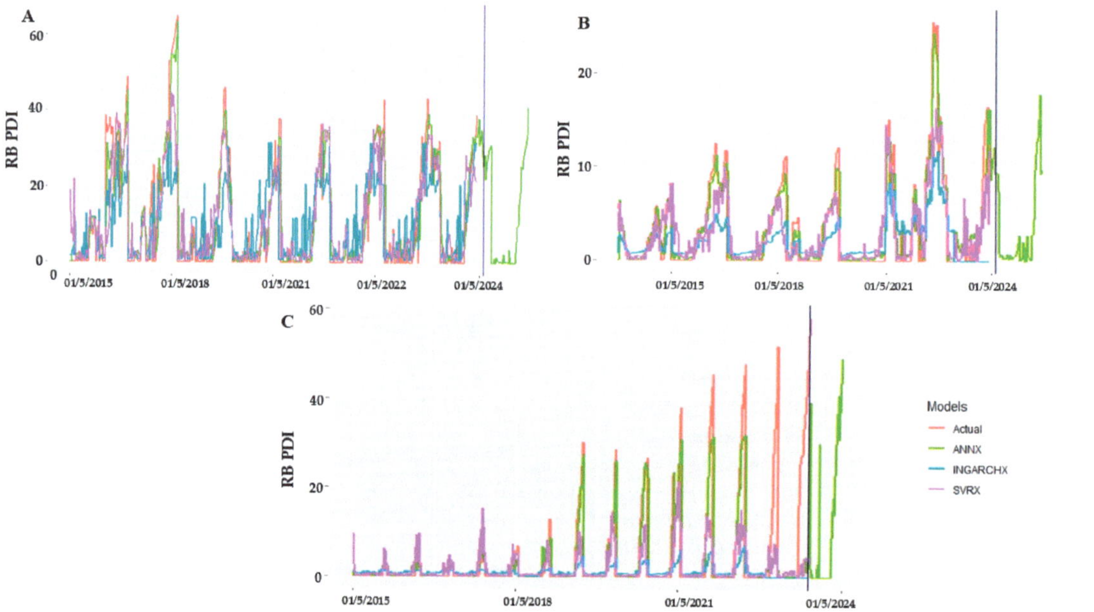

Figure 5. Actual vs. fitted plots of rice blast occurrence, Tirunelveli (**A**), Thanjavur (**B**), and Coimbatore (**C**).

Table 7. Diebold–Mariano test for comparison of performance of different models with training and testing data sets at study locations.

	Data Type	M1, M2	M1, M3	M2, M3
Thanjavur	Training	5.054 (<0.0001)	5.757 (<0.0001)	−3.848 (<0.0001)
	Testing	−6.161 (<0.0001)	−5.303 (<0.0001)	−3.152 (0.0023)
Tirunelveli	Training	10.253 (<0.0001)	11.01 (<0.0001)	−3.154 (0.001741)
	Testing	2.4021 (0.021)	−1.8683 (0.069)	1.7255 (0.0921)
Coimbatore	Training	5.7704 (<0.0001)	4.0486 (<0.0001)	−4.3679 (<0.0001)
	Testing	−3.1281 (0.003535)	−3.4715 (0.001395)	2.2673 (0.0296)

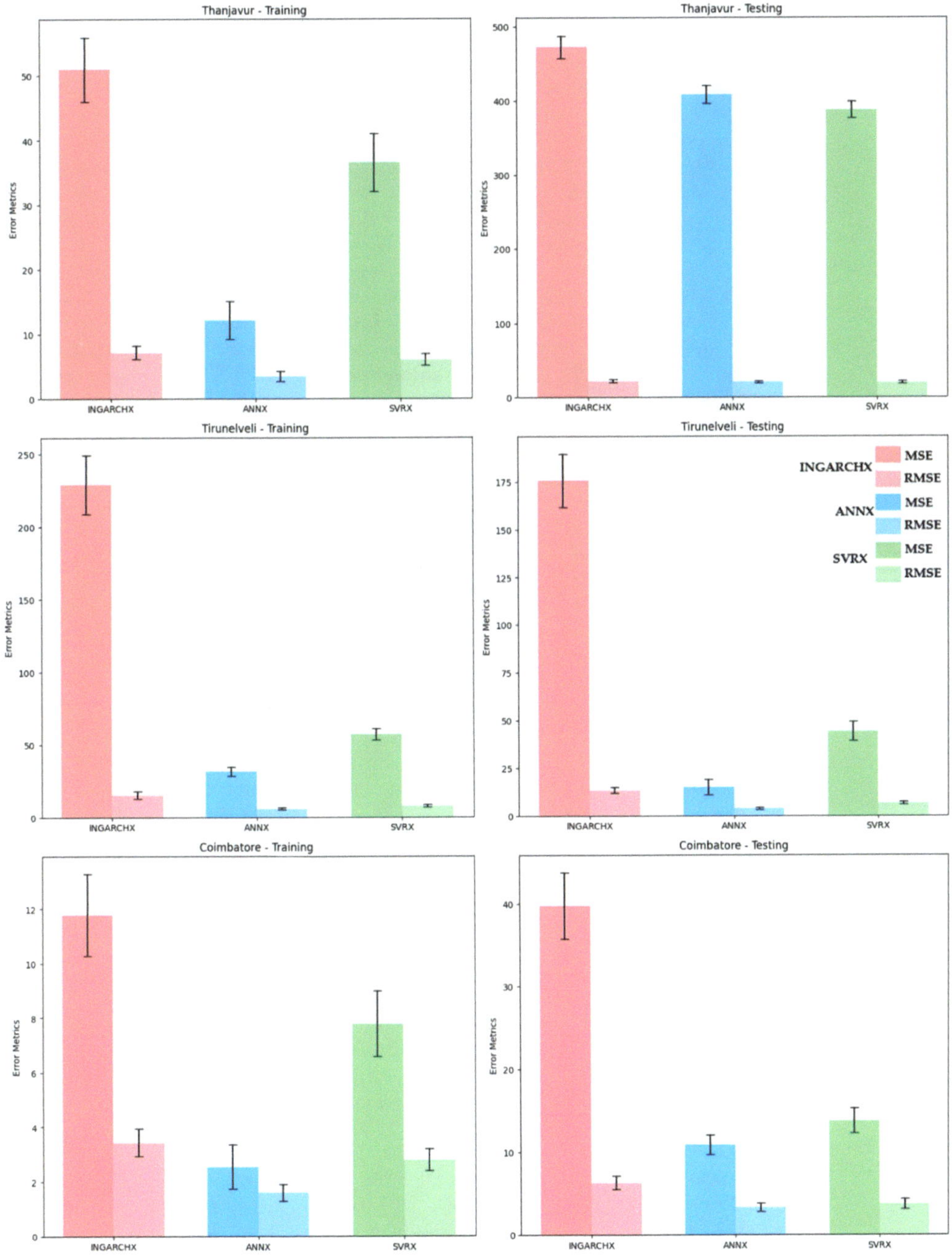

Figure 6. Comparison of performance of each model based on error metrics MSE and RMSE of both testing and training set for Tirunelveli, Thanjavur, and Coimbatore.

The RBF function in SVR approached a Gaussian distribution with an increasing gamma value, whereas the ANN model used a sigmoidal activation function to map inputs to the hidden layer. Since count time series data are produced from non-Gaussian distributions, this may help to explain why INGARCH has trouble identifying patterns in these kinds of data. Similar outcomes were noted by [14], who found that when it came to modelling and forecasting rice gall midge, ANN performed better than INGARCH and SVR. Further evidence that the ANNX model fits the data better than the other two comes from diagnostic checks that showed the residuals from the SVRX and INGARCHX models were non-random and correlated, while those from the ANNX model were random and uncorrelated. Table 7 presents the inter-model significance differences in a simple manner.

Machine learning algorithms provide better predictive performance, as shown by similar studies by [33] for early potato yield prediction, [34] for rice blast, and [35] for wheat yield in Pakistan. Earlier models, such as those developed in Japan, Korea, and India, often relied on linear regression models and statistical methods, which limited their ability to capture complex non-linear interactions between environmental factors and disease incidence. Many models, including EPIBLAST in Korea and PYRICULARIA in Japan, struggled to integrate important variables like nitrogen fertilization and canopy-level leaf wetness (LW) [36,37]. The limited incorporation of these factors reduced the reliability of predictions, as excessive nitrogen or microclimatic variability can alter disease severity significantly. Another shortcoming in these earlier models was the lack of real-time data integration. Many models, such as those developed in Taiwan and Iran, relied heavily on historical weather data collected from stations located far from rice fields [37]. This spatial disconnect resulted in inaccurate predictions, especially during periods of rapid environmental changes. Our machine learning models (ANN and SVR) address these limitations by incorporating a wider range of inputs—such as solar radiation, wind speed, and localized meteorological data—collected from in-field sensors. This reduces the spatial and temporal uncertainties that limit older models. Furthermore, statistical validation through the Diebold–Mariano (DM) test confirmed the superiority of ANN over traditional approaches, ensuring more robust predictions for Tamil Nadu's climate. Unlike simpler regression-based models, the developed ANN can detect non-linear relationships, making it better suited for conditions where multiple variables interact simultaneously, such as high humidity combined with fluctuating temperatures. This allows for early disease detection and timely interventions. Recent models, like EPIRICE, and modifications to older frameworks, such as BLASTSIM, have shown improvements by using climate change projections and geographic information systems (GIS) to predict disease risk under different future scenarios [36]. The model developed in this study aligns with these advancements by integrating real-time environmental inputs with localized weather data, thus improving prediction accuracy under variable climatic conditions.

The ANNX model has been shown to be more accurate than other models in predicting outbreaks of rice blast disease through field-level application. Farmers can gain useful insights into how weather variations may impact disease risk by utilizing the model, which incorporates essential meteorological characteristics including sunshine, minimum temperature and relative humidity. This model is a particularly useful tool for site-specific disease management because it provided localized predictions after being trained on data from certain regions of Tamil Nadu, namely Thanjavur, Tirunelveli, and Coimbatore. This makes it possible for farmers to prevent rice blast disease by adopting preventative measures, including modifying irrigation schedules, using fungicides at the required times and choosing more resilient rice types which were identified as critical but underutilized in previous models. These actions greatly lower the frequency and intensity of rice blast disease. Agricultural extension services can assist in converting the complex predictions of the ANNX model into practical insights for farmers. Based on the model's predictions, these services can offer frequent updates that help farmers decide when to take preventive measures. The predictive models developed in this study have significant practical applications for real-time rice blast disease management. Farmers and agricultural extension services can leverage these models through mobile apps or web-based platforms. By inputting current meteorological data, the models can provide real-time alerts on the likelihood of disease outbreaks, allowing farmers to take preventive actions such as applying fungicides or adjusting irrigation schedules. Additionally, the models can be integrated into Decision Support Systems (DSS) used by agricultural officers to recommend the optimal timing for interventions based on weather forecasts. Governments and cooperatives could also use these models to develop early warning systems, notifying farmers of potential risks during critical growth stages. Moreover, combining the models with smart farming technologies and IoT devices would enable automatic weather data input, offering real-time recommendations for precise disease management, reducing crop losses, and improving overall yield sustainability.

5. Conclusions

This study demonstrates the effectiveness of using machine learning models, particularly the Artificial Neural Network (ANN) model, for predicting rice blast disease based on meteorological data. The models developed in this research can serve as vital tools for farmers and agricultural extension services in Tamil Nadu, India, allowing for timely interventions to minimize crop losses and optimize fungicidal sprays. There is significant scope to scale up these models to other rice-growing regions beyond Tamil Nadu. The developed models not only improve upon the predictive accuracy of traditional approaches but also offer a practical solution for early disease management. By addressing the limitations identified in previous studies, such as inadequate data collection, lack of non-linear modelling, and accounting for regional climatic variations, the proposed model is expected to forecast rice blast disease more accurately. By integrating these models into national agricultural strategies and facilitating the development of real-time disease monitoring systems and decision support tools, this research paves the way for more reliable and accessible forecasting systems for sustainable agriculture.

Author Contributions: Conceptualization, M.A.G., G.C., S.G.P. and S.R.; methodology, M.A.G., G.C., S.G.P. and S.R.; software, M.A.G., S.G.P. and S.R.; validation M.A.G., G.C., S.G.P. and S.R.; formal analysis, M.A.G., G.C., S.G.P. and S.R.; investigation, G.C., K.A., S.G.P. and S.R.; resources G.C. and M.S.; data curation, M.A.G.; writing—original draft preparation, M.A.G. and G.C.; writing—review and editing, G.C., J.R., S.N.K. and K.A.; visualization, G.C., S.G.P. and S.R.; supervision, G.C., S.G.P. and S.R.; project administration, G.C. and K.A. All authors have read and agreed to the published version of the manuscript.

Funding: This research received no external funding.

Data Availability Statement: The data that support the findings of this study are available from the corresponding author upon reasonable request.

Acknowledgments: The authors gratefully acknowledge the support provided by the Department of Plant Pathology and Department of Rice, TNAU, Coimbatore, Tamil Nadu, India.

Conflicts of Interest: The authors declare no conflicts of interest.

References

1. Mahajan, G.; Kumar, V.; Chauhan, B.S. Rice Production in India. In *Rice Production Worldwide*; Springer: Berlin/Heidelberg, Germany, 2017; pp. 53–91.
2. Ramakrishna, B.; Degaonkar, C. Rice Export from India: Trends, Problems and Prospects. *Int. J. Res. Granthaalayah* **2016**, *4*, 122–136. [CrossRef]
3. Datta, P.; Debnath, U.; Panda, C. Climate Change Implications for Rice Cultivation. *Asian J. Agric. Hortic. Res.* **2019**, *4*, 1–4. [CrossRef]
4. Huluka, W.; Kumsa, L. Analysis of Rice (*Oryza sativa* L. Ssp. Japonica) Wall Associated Receptor-like Protein Kinase Gene's Promoter Region and Regulatory Elements. *Curr. Plant Biol.* **2022**, *31*, 100254. [CrossRef]
5. Kumar, A.; Bhowmik, B. Rice Cultivation and Its Disease Classification in Precision Agriculture. In Proceedings of the 2023 International Conference on Artificial Intelligence and Smart Communication (AISC), Greater Noida, India, 27–29 January 2023; pp. 200–205.
6. Windi, W.; Novriyenni; Khair, H. Application of Case Based Reasoning Method to Diagnose Rice Plant Diseases. *J. Artif. Intell. Eng. Appl.* **2023**, *3*, 498–503. [CrossRef]
7. Simkhada, K.; Thapa, R. Rice Blast, a Major Threat to the Rice Production and Its Various Management Techniques. *Turk. J. Agric.-Food Sci. Technol.* **2022**, *10*, 147–157. [CrossRef]

8. Rathod, S.; Yerram, S.; Arya, P.; Katti, G.; Rani, J.; Padmakumari, A.P.; Somasekhar, N.; Padmavathi, C.; Ondrasek, G.; Amudan, S. Climate-Based Modeling and Prediction of Rice Gall Midge Populations Using Count Time Series and Machine Learning Approaches. *Agronomy* **2021**, *12*, 22. [CrossRef]
9. Fokianos, K.; Rahbek, A.; Tjøstheim, D. Poisson Autoregression. *J. Am. Stat. Assoc.* **2009**, *104*, 1430–1439. [CrossRef]
10. Zhu, F. Modeling Time Series of Counts with COM-Poisson INGARCH Models. *Math. Comput. Model.* **2012**, *56*, 191–203. [CrossRef]
11. Weiß, C.H. Modelling Time Series of Counts with Overdispersion. *Stat. Methods Appl.* **2009**, *18*, 507–519. [CrossRef]
12. Tanawi, I.N.; Vito, V.; Sarwinda, D.; Tasman, H.; Hertono, G.F. Support Vector Regression for Predicting the Number of Dengue Incidents in DKI Jakarta. *Procedia Comput. Sci.* **2021**, *179*, 747–753. [CrossRef]
13. Rathod, S.; Singh, K.; Patil, S.; Naik, R.; Ray, M.; Meena, V.S. Modeling and Forecasting of Oilseed Production of India through Artificial Intelligence Techniques. *Indian J. Agric. Sci.* **2018**, *88*, 22–27. [CrossRef]
14. Amaratunga, V.; Wickramasinghe, L.; Perera, A.; Jayasinghe, J.; Rathnayake, U. Artificial Neural Network to Estimate the Paddy Yield Prediction Using Climatic Data. *Math. Probl. Eng.* **2020**, *2020*, 8627824. [CrossRef]
15. Su, Y.; Xu, H.; Yan, L. Support Vector Machine-Based Open Crop Model (SBOCM): Case of Rice Production in China. *Saudi J. Biol. Sci.* **2017**, *24*, 537–547. [CrossRef] [PubMed]
16. Ma, C.; Liang, Y.; Lyu, X. Weather Analysis to Predict Rice Pest Using Neural Network and D-S Evidential Theory. In Proceedings of the 2019 International Conference on Cyber-Enabled Distributed Computing and Knowledge Discovery (CyberC), Guilin, China, 17–19 October 2019; pp. 277–283.
17. Paul, R.; Vennila, S.; BHAT, M.; Yadav, S.; Sharma, V.; Nisar, S.; Panwar, S. Prediction of Early Blight Severity in Tomato (*Solanum lycopersicum*) by Machine Learning Technique. *Indian J. Agric. Sci.* **2019**, *89*, 169–175. [CrossRef]
18. Huang, T.; Yang, R.; Huang, W.; Huang, Y.; Qiao, X. Detecting Sugarcane Borer Diseases Using Support Vector Machine. *Inf. Process. Agric.* **2018**, *5*, 74–82. [CrossRef]
19. Moritz, S.; Sardá, A.; Bartz-Beielstein, T.; Zaefferer, M.; Stork, J. Comparison of Different Methods for Univariate Time Series Imputation in R. *arXiv* **2015**, arXiv:1510.03924.
20. Chhabra, G. Comparison of Imputation Methods for Univariate Time Series. *Int. J. Recent Innov. Trends Comput. Commun.* **2023**, *11*, 286–292. [CrossRef]
21. Sinsomboonthong, S. Performance Comparison of New Adjusted Min-Max with Decimal Scaling and Statistical Column Normalization Methods for Artificial Neural Network Classification. *Int. J. Math. Math. Sci.* **2022**, *2022*, 3584406. [CrossRef]
22. Amith, G.; Geethalakshmi, V.; Dheebakaran, G.; Karthikeyan, G. Influence of Weather Variables on the Incidence and Development of Rice Leaf Blast (Magnapor the Oryzae). *Res. Crops* **2022**, *23*, 682–688.
23. Kedem, B.; Fokianos, K. *Regression Models for Time Series Analysis*; John Wiley & Sons: Hoboken, NJ, USA, 2005; ISBN 0-471-46168-7.
24. Heinen, A. Modelling Time Series Count Data: An Autoregressive Conditional Poisson Model. 2003. Available online: https://ssrn.com/abstract=1117187 (accessed on 1 July 2003).
25. Ferland, R.; Latour, A.; Oraichi, D. Integer-valued GARCH Process. *J. Time Ser. Anal.* **2006**, *27*, 923–942. [CrossRef]
26. Christou, V.; Fokianos, K. Quasi-Likelihood Inference for Negative Binomial Time Series Models. *J. Time Ser. Anal.* **2014**, *35*, 55–78. [CrossRef]
27. Liboschik, T.; Fokianos, K.; Fried, R. Tscount: An R Package for Analysis of Count Time Series Following Generalized Linear Models. *J. Stat. Soft.* **2017**, *82*, 1–51. [CrossRef]
28. Vapnik, V. *The Nature of Statistical Learning Theory*; Springer Science & Business Media: Berlin/Heidelberg, Germany, 2013; ISBN 1-4757-3264-3.
29. Zhang, G.P. Time Series Forecasting Using a Hybrid ARIMA and Neural Network Model. *Neurocomputing* **2003**, *50*, 159–175. [CrossRef]
30. Kumari, P.; Mishra, G.; Srivastava, C. Time Series Forecasting of Losses Due to Pod Borer, Pod Fly and Productivity of Pigeonpea (*Cajanus cajan*) for North West Plain Zone (NWPZ) by Using Artificial Neural Network (ANN). *Int. J. Agric. Stat. Sci.* **2014**, *10*, 15–21.
31. Chitikela, G.; Admala, M.; Ramalingareddy, V.K.; Bandumula, N.; Ondrasek, G.; Sundaram, R.M.; Rathod, S. Artificial-Intelligence-Based Time-Series Intervention Models to Assess the Impact of the COVID-19 Pandemic on Tomato Supply and Prices in Hyderabad, India. *Agronomy* **2021**, *11*, 1878. [CrossRef]
32. Piekutowska, M.; Niedbała, G.; Piskier, T.; Lenartowicz, T.; Pilarski, K.; Wojciechowski, T.; Pilarska, A.A.; Czechowska-Kosacka, A. The Application of Multiple Linear Regression and Artificial Neural Network Models for Yield Prediction of Very Early Potato Cultivars before Harvest. *Agronomy* **2021**, *11*, 885. [CrossRef]
33. Liu, L.-W.; Hsieh, S.-H.; Lin, S.-J.; Wang, Y.-M.; Lin, W.-S. Rice Blast (*Magnaporthe oryzae*) Occurrence Prediction and the Key Factor Sensitivity Analysis by Machine Learning. *Agronomy* **2021**, *11*, 771. [CrossRef]
34. Haider, S.A.; Naqvi, S.R.; Akram, T.; Umar, G.A.; Shahzad, A.; Sial, M.R.; Khaliq, S.; Kamran, M. LSTM Neural Network Based Forecasting Model for Wheat Production in Pakistan. *Agronomy* **2019**, *9*, 72. [CrossRef]
35. Katsantonis, D.; Kadoglidou, K.; Dramalis, C.; Puigdollers, P. Rice Blast Forecasting Models and Their Practical Value: A Review. *Phytopathol. Mediterr.* **2017**, *56*, 187–216.

36. Kim, C.K.; Kim, C.H. *The Rice Leaf Blast Simulation Model EPIBLAST*; Springer: Berlin/Heidelberg, Germany, 1993; pp. 309–321.
37. Izadyar, M.; Baradaran, P. A New Method for Rice Leaf Blast Forecasting. 1990. Available online: https://www.cabidigitallibrary.org/doi/full/10.5555/19922317492 (accessed on 11 April 1992).

AgriEngineering MDPI

Article

Alpha-EIOU-YOLOv8: An Improved Algorithm for Rice Leaf Disease Detection

Dong Cong Trinh [1], Anh Tuan Mac [2], Khanh Giap Dang [2], Huong Thanh Nguyen [2], Hoc Thai Nguyen [3] and Thanh Dang Bui [1,2,*]

[1] Institute for Control Engineering and Automation, Hanoi University of Science and Technology, Dai Co Viet Street, Hanoi 100000, Vietnam; dong.trinhcong@hust.edu.vn
[2] School of Electrical and Electronic Engineering, Hanoi University of Science and Technology, Dai Co Viet Street, Hanoi 100000, Vietnam; anh.mt192182@sis.hust.edu.vn (A.T.M.); khanh.gd192219@sis.hust.edu.vn (K.G.D.); huong.nguyenthanh3@hust.edu.vn (H.T.N.)
[3] Faculty of Engineering, Vietnam National University of Agriculture, Trauquy, Gialam, Hanoi 131000, Vietnam; nguyenthaihoc@vnua.edu.vn
* Correspondence: thanh.buidang@hust.edu.vn; Tel.: +84-9158-97699

Abstract: Early detection of plant leaf diseases is a major necessity for controlling the spread of infections and enhancing the quality of food crops. Recently, plant disease detection based on deep learning approaches has achieved better performance than current state-of-the-art methods. Hence, this paper utilized a convolutional neural network (CNN) to improve rice leaf disease detection efficiency. We present a modified YOLOv8, which replaces the original Box Loss function by our proposed combination of EIoU loss and α-IoU loss in order to improve the performance of the rice leaf disease detection system. A two-stage approach is proposed to achieve a high accuracy of rice leaf disease identification based on AI (artificial intelligence) algorithms. In the first stage, the images of rice leaf diseases in the field are automatically collected. Afterward, these image data are separated into *blast leaf*, *leaf folder*, and *brown spot* sets, respectively. In the second stage, after training the YOLOv8 model on our proposed image dataset, the trained model is deployed on IoT devices to detect and identify rice leaf diseases. In order to assess the performance of the proposed approach, a comparative study between our proposed method and the methods using YOLOv7 and YOLOv5 is conducted. The experimental results demonstrate that the accuracy of our proposed model in this research has reached up to 89.9% on the dataset of 3175 images with 2608 images for training, 326 images for validation, and 241 images for testing. It demonstrates that our proposed approach achieves a higher accuracy rate than existing approaches.

Keywords: rice leaf diseases; blast leaf; leaf folder; brown spot; YOLOv8

Citation: Trinh, D.C.; Mac, A.T.; Dang, K.G.; Nguyen, H.T.; Nguyen, H.T.; Bui, T.D. Alpha-EIOU-YOLOv8: An Improved Algorithm for Rice Leaf Disease Detection. *AgriEngineering* **2024**, *6*, 302–317. https://doi.org/10.3390/agriengineering6010018

Academic Editors: Chiew Foong Kwong and Ray E. Sheriff

Received: 24 November 2023
Revised: 19 January 2024
Accepted: 22 January 2024
Published: 4 February 2024

1. Introduction

Rice (*Oryza sativa*) is one of the most important staple crops in the world, not only providing a rich source of carbohydrates but also plenty of vitamins and minerals for human healthy growth and development. This adaptable grain is an essential element of diets and Asian cuisine [1]. Unfortunately, the recent successful cultivation of rice faces numerous challenges, and one of the most significant threats to its yield and quality comes in the form of rice leaf diseases. These diseases, caused by various pathogens, can severely impact the health of rice plants, resulting in reduced crop yields and low-quality harvests. Therefore, accurate and timely detection of rice leaf diseases is of paramount importance and has become a primary agricultural concern.

Nowadays, science and technology are growing rapidly with many breakthrough inventions that make people's lives easier and simpler. The adoption of modern technologies, including machine learning and deep learning, has opened up new avenues for tackling these challenges in the agriculture sector. One of the most popular technologies that can be used to solve this problem is the use of deep learning technology on plants to identify and classify diseases. In recent years, more and more studies have focused on disease detection and crop management based on these technologies with more and more improvement in accurate detection rate [2–5].

Deep learning is a subset of machine learning that seeks to mimic the data processing and decision-making mechanisms of the human brain. Within deep learning, CNN stands as a vital element in various fields such as speech recognition, image recognition, natural language processing, and the field of genomics [6]. Its utility extends across diverse real applications such as health care, finance, and even education as in the E-Learning Modeling Technique and CNN as proposed in [7]. Notably, there have been encouraging outcomes in scientific investigations that employ deep learning to identify diseases in specific plant species. In this research area, several noteworthy studies have explored disease detection in plants. For instance, study [8] focused on disease detection in cucumber plants using the YOLOv4 network for leaf image analysis. The authors achieved impressive accuracy rates exceeding 80% with a dataset comprising over 7000 images. However, since this study relied solely on a convolutional neural network (CNN) model, there were limitations to its accuracy. Another study [9] delved into deep learning methods for surveillance and the detection of plant foliar diseases. It examined images at various infection levels and proposed a model to assess disease progression over the plant growth cycle. This study also investigated factors influencing image quality, such as dataset size, learning rate, illumination, and more. In a different context, research [10] utilized YOLOv3 to detect *brown spot* and *leaf blast* diseases. Nevertheless, the practicality of this approach was hindered because it exclusively used images with white backgrounds, limiting its flexibility. Meanwhile, studies [11,12] leveraged ResNet and YOLOv3 to detect diseases in tomato leaves. In [13], the authors introduced their CNN model for classifying *leaf blight* diseases. Furthermore, research [14] conducted a comprehensive comparison of CNN models, including DenseNet-121, ResNet-50, ResNeXt-50, SE-ResNet-50, and ResNeSt-50. They even combined these models to improve detection results. Another study [15] employed a CNN model to detect *brown spot* and *leaf blast* diseases in rice, achieving an impressive overall accuracy of 0.91. Vimal K. Shrivastava et al. in [16], utilized AlexNet and SVM to classify three rice leaf diseases, achieving an accuracy rate of 91.37%. Additionally, study [17] performed a survey and comparative analysis of different models for detecting diseases in rice leaves and seedlings. Research [18] used VGG16 and InceptionV3 to classify five distinct rice diseases. Lastly, in research [19], SVM was combined with a DCNN model to identify and classify nine different types of rice diseases, achieving an impressive accuracy rate of 97.5%. Article [20] presents a method for recognizing and classifying paddy leaf diseases, such as *leaf blast*, *bacterial blight*, *sheath rot*, and *brown spot*, using an optimized deep neural network with the Jaya algorithm. The classification of diseases is carried out using an optimized deep neural network with the Jaya optimization algorithm. The experimental results show high accuracy for different types of diseases. In [21], the author used the hue value to extract the non-diseased and diseased parts during the pre-processing process and used CNN for feature extraction to be able to identify the diseases including rice *blast*, *sheath rot*, *bacterial leaf blight*, *brown spot*, *rymv*, and *rice tungro*. Studies [22,23] used YOLOv5 to detect various typical rice leaf diseases.

In this research, we use the YOLOv8 model to detect the three most common rice leaf diseases in Vietnam, namely *leaf blast* (Section 1.1), *leaf folder* (Section 1.2), and *brown spot* (Section 1.3).

1.1. Leaf Blast

The primary culprit behind *leaf blast* disease is the fungus *Magnaporthe oryzae*, which can afflict all components of the rice plant, including the leaves, collar, node, neck, portions of the panicle, and occasionally the leaf sheath. The occurrence of blast disease is contingent upon the presence of blast spores [24]. The *leaf blast* disease is depicted in Figure 1a.

(**a**) *Leaf blast* disease (**b**) *Leaf folder* disease (**c**) *Brown spot* disease

Figure 1. Common rice leaf diseases in Vietnam.

To identify *leaf blast* disease, a thorough examination of the leaf and collar is necessary. Initial signs manifest as pale to grayish-green lesions or spots, characterized by dark green borders. As these lesions mature, they assume an elliptical or spindle-shaped configuration, featuring whitish-to-gray centers surrounded by reddish-to-brownish or necrotic boundaries. Another indicative symptom resembles a diamond shape, broader at the center and tapering toward both ends [24].

1.2. Leaf Folder

The *leaf folder* is formed by *leaf folder* caterpillars, which encircle the rice leaf and secure the leaf edges with threads of silk. They consume the inner part of the folded leaves, leading to the development of whitish and see-through streaks along the surface of the leaf [25]. *Leaf folder* disease is shown in Figure 1b.

Infected leaves typically exhibit vertical and translucent white streaks, along with tubular leaf structures. Occasionally, the leaf tips are affixed to the leaf base. Fields heavily affected by this condition may present an appearance of scorching due to the prevalence of numerous folded leaves [25].

1.3. Brown Spot

Brown spot is a fungal disease that infects cotyledons, leaves, sheaths, panicles, and shoots. Its most noticeable impact is the formation of numerous prominent spots on the leaves, which have the potential to lead to the demise of the entire leaf. In cases of seed infection, it results in unfilled seeds or seeds displaying spots or discoloration [24]. The *brown spot* disease is shown in Figure 1c.

Infected seedlings display small, circular lesions that are yellowish-brown or brown in color and may encircle the cotyledons while distorting the primary and secondary leaves. Starting from the tillering stage, lesions begin to appear on the leaves. Initially, these lesions are small and circular and range from dark brown to purplish-brown. As they fully mature, they adopt a round-to-oval shape with a light brown-to-gray center, encircled by a reddish-brown border produced by *mycotoxins*. On susceptible rice varieties, lesions measuring between 5 and 14 mm in length can result in leaf wilting. Conversely, on resistant varieties, the lesions are brown and approximately the size of a fingertip [24].

The rest of the paper is planned as follows. Section 2 presents the methodology with the system overview, a hardware design, and data preparation. It also demonstrates the YOLOv8 architecture, along with evaluation metrics and a loss calculation. The experimental results and the discussion are given in Section 3. Finally, conclusions, acknowledgments, and references are made in Section 4.

2. Materials and Methods

2.1. System Overview

The procedure can be outlined in the following steps: First, the images of rice leaf diseases are collected and added to dataset. Then, the YOLOv8 model is employed to swiftly and accurately detect diseases in rice leaves. Finally, the symptoms of unhealthy leaves of rice are used to alert farmers via email and text message. The structure of the system is depicted in Figure 2.

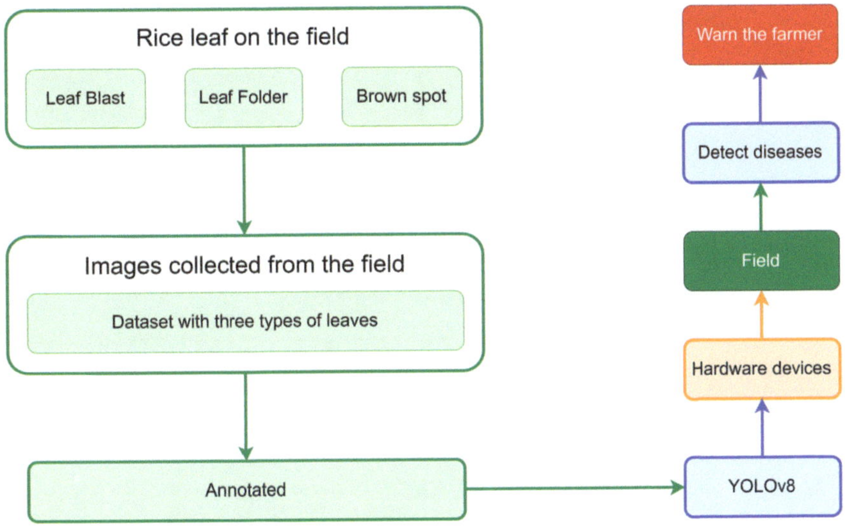

Figure 2. The system of detecting rice leaf diseases.

2.2. System Hardware Design

Figure 3 illustrates the configuration of the hardware devices responsible for the identification and alerting of *leaf folders*, *leaf blasts*, and *brown spots* on rice leaves. The system includes 1 micro-controller, 1 USB camera, and 1 GSM module (Table 1). First, the camera is connected to a micro-controller to capture real-time images of the rice plant on the field. The AI algorithm (implemented in micro-controller) detects and identifies plant diseases. The detection outcomes are then transmitted to the GSM module, which subsequently sends warning messages to the farmers.

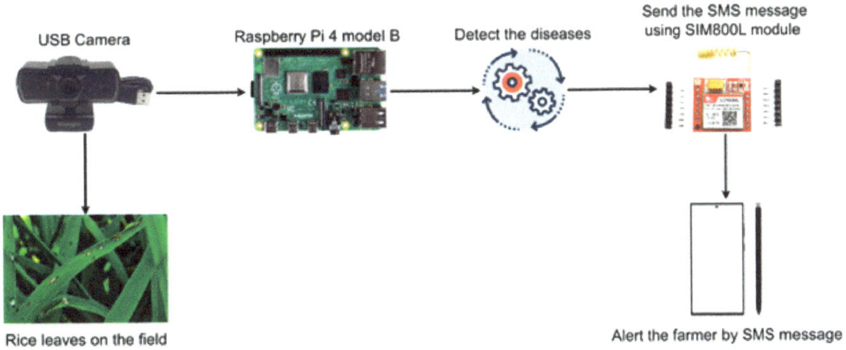

Figure 3. Auto-detecting rice disease system with Raspberry Pi 4 model B.

Table 1. Main features of parts in system

No.	Name of Main Part	Specifications
1	USB Camera Essager C3	• Resolution: 1920 × 1080; • Frame rate: 30 fps; • Pixel: 2 mp.
2	Raspberry Pi 4 model B	• Chipset: Broadcom BCM2711, Quad core Cortex-A72 (ARM v8) 64-bit SoC @ 1.8 GHz; • 8 GB LPDDR4-3200 SDRAM; • 2.4 GHz and 5.0 GHz IEEE 802.11ac wireless, Bluetooth 5.0, BLE Gigabit Ethernet; • 2 USB 3.0 ports; 2 USB 2.0 ports; 40 pin GPIO header; • 5 V DC via USB-C connector (minimum 3 A).
3	SIM 800 L	• Input voltage: 3.7–4.2 V; • GSM antenna connector: IPX and ordinary wire; • SIM type: Micro SIM; • Allows calls, SMS/MMS sending; • Integrated protocols for internet connection (TCP, UDP, HTTP, IP).

2.3. Data Preparation

2.3.1. Image Collection

The experiments are conducted on the rice field at the Vietnam National University of Agriculture (latitude: 21.001792, longitude: −254.068270), Hanoi, Vietnam. A total of 1634 images of rice plants were collected in the experimental rice field with some type of rice variance, such as *DCG66*, *DCG72*, *DH15*, or *Hat ngoc 9*. The dataset contains images of three common diseases in rice plants namely *leaf folder*, *leaf blast*, and *brown spot*. The images were captured under various conditions: different weather conditions (sunny, cloudy, rainy); different environment light. All these images of the rice plant diseases were confirmed manually by agricultural experts.

2.3.2. Dataset Splitting

Our final dataset includes 3175 images and is divided into three parts: a training set, validation set, and test set. The training set includes 2608 images, the validation set includes 326 images, and the test set includes 241 images. The number of samples in each class are 1231 for the leaf folder, 1377 for the blast, and 1237 for the brown spot. One important note is that the images from our test set do not overlap with the images from the validation and training sets; the same is true of the validation set. Our dataset contains small images with the sizes of 150 pixels × 150 pixels and 106 pixels × 200 pixels to large images with the sizes of 3024 pixels × 4032 pixels and 4312 pixels × 5760 pixels.

2.3.3. Data Augmentation

In YOLOv8, different augmentation methods are implemented for the training process; the value of these methods can be adjusted in the default.yaml file. Some of these methods include horizontal and vertical flipping, translation, mosaic, etc. One thing that must be noted is that the augmentation process does not create additional images. Instead, it is applied straightforwardly during each epoch to the dataset. This approach can generate a greater number of unique examples rather than a fixed number of images. Consequently, the model can learn from a significantly larger amount of data. As a result, the augmentation methods can lead to better performance of the model and help the model adapt better to various conditions. Table 2 gives the augmentation hyperparameters in our model.

Table 2. Augmentation hyperparameters table.

No.	Name	Function
1	Hsv_h	0.015
2	Hsv_s	0.7
3	Hsv_v	0.4
4	Translate	0.1
5	Scale	0.5
6	Flipud	0.5
7	Fliplr	0.5
8	Mosaic	1.0

2.4. YOLOv8 Architecture

2.4.1. Overview

YOLOv8 is one of the newest YOLO versions along with YOLO-NAS. Some honorable mentions of earlier versions of YOLO are YOLOv1, YOLO9000v2 [26], YOLOv3 [27], YOLOv4 [28], YOLOv5, YOLOv6 [29], YOLOv7 [30], etc. YOLOv8 itself contains different versions of architecture including YOLOv8n, YOLOv8s, YOLOv8m, YOLOv8l, and YOLOv8x. The lightest model is YOLOv8n, and the heaviest model is YOLOv8x. The differences that make one version heavier than another are the convolutional kernel and the feature extraction number. It is important to note that the YOLOv8 version and the YOLOv5 version come from the same developer, Ultralytics. This is the main reason for the similarity of the YOLOv8 architecture and the YOLOv5 architecture. In this project, YOLOv8n is investigated for the rice plant disease detection problem because it has satisfied our requirements for accuracy, interference speed, and lightness. Figure 4 depicts the YOLOv8 architecture diagram.

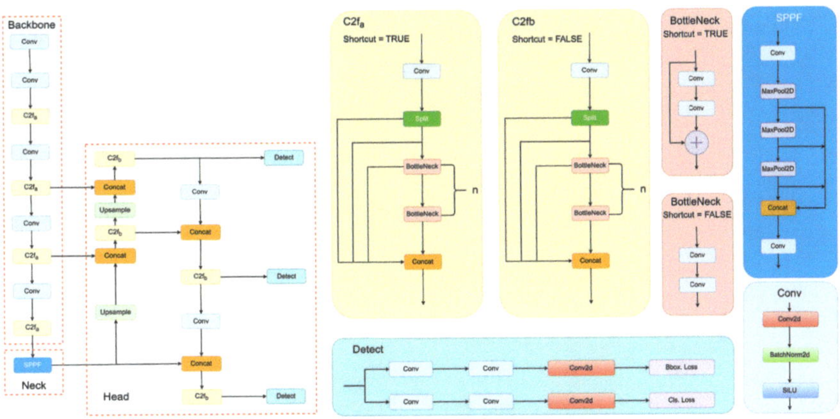

Figure 4. YOLOv8 architecture diagram.

2.4.2. Backbone

YOLOv8 has many similarities to YOLOv5. The C3 module in YOLOv5 is replaced in YOLOv8 with C2f. The C2f idea is the fusion between the C3 module of YOLOv5 and the ELAN model idea. This idea of the ELAN block [31] was influenced by the CSP strategy and VoVNet [32] The two main blocks of the backbone are C2f and Conv. The C2f block contains a split block, bottleneck, a Conv Block, and a concatenat layer. The number of bottleneck layers is based on the position of the C2f block in the backbone and the depth of the architecture. In the first C2f block, the number of bottleneck layers will be $n = 3 \times d$; for the second and the third, it will be $n = 6 \times d$; and, finally, in the last C2f, the value will be $n = 3 \times d$. One important thing to note is that the feature map of the second and third C2f blocks will be concatenated with the head part. The Conv Block contains Conv2d, BatchNorm2d, and SiLU.

2.4.3. Neck

Same as in YOLOv5, YOLOv8 uses SPPF in the neck part. The neck part's mission is to perform feature fusion. Feature fusion combines features from the backbone and forwards them into the head part. SPPF is an improved version of SPP. The advantage of SPP is that it allows the model to effectively capture and encode information from objects of various sizes within an image, regardless of their spatial locations. In SPP, max-pooling with different kernel sizes will be applied to each channel in the feature maps of the backbone to create three feature maps with different sizes. This feature map will then be produced into a fixed-length vector. One of the two main differences between SPP and SPPF is the kernel size of the max-pooling layer; while SPP uses different kernel sizes, as mentioned above, SPPF uses the same kernel size for each layer. The second difference is that, instead of applying three max-pooling layers in parallel, SPPF positions these max-pooling layers in series. These differences have made SPPF much faster than SPP.

2.4.4. Head

The main blocks in the head structure are Conv Block, C2f Block, Upsample, and Concat. Note that the C2f of the head part will not contain shortcuts for the bottleneck block. One of the changes in YOLOv8, compared to other versions, is the shift to Anchor-Free instead of Anchor-Based, as seen in the older version. The second change is the use of Decouple-Head. In the older versions of Object Detection, such as Faster R-CNN and YOLO, the Localization and Classification mission is processed on the same branch of the head. This may lead to a problem because Classification needs discriminative features, while Localization needs features containing information on boundary regions. The difference in the features of these tasks is referred to as task conflict. To avoid this, the author of YOLOv8 has separated these tasks into two different branches. However, this led to a task misalignment problem. To solve this, the authors of YOLOv8 have applied TAL in the TOOD paper [33], which is a Label Assignment strategy. YOLOv8 uses TAL to measure the alignment level of Anchor, and it is given in Equation (1):

$$t = s^{\alpha} \times u^{\beta} \tag{1}$$

where s and u are the classification score and IoU score, respectively. Alpha and beta are the numbers used to adjust the impact of the two tasks in the Anchor alignment metric, and t is the alignment metric. Depending on the value of t, positive samples and negative samples will be separated for training through the loss function.

2.5. Metrics

In this study, evaluation metrics including Precision (2), Recall (3), mAP (mean Average Precision) (4), and F1-score (5) were used. They are given in Equations (2)–(5), respectively.

$$precision = \frac{TP}{TP + FP} \times 100\% \tag{2}$$

$$recall = \frac{TP}{TP + FN} \times 100\% \tag{3}$$

$$mAP = \frac{\sum_{k=1}^{N} P(k)R(k)}{C} \tag{4}$$

$$F_{1\text{-}score} = 2 \times \frac{precision \times recall}{precision + recall} \tag{5}$$

where TP is the number of exactly identified rice leaves, FP is the number of backgrounds misidentified as the target leaves, FN represents an unspecified number of leaf targets, C represents the number of target categories of rice leaves, N means the number of IOU thresholds, k is the IOU threshold, $P(k)$ is the precision, and $R(k)$ is the recall.

2.6. Loss

2.6.1. Classification Loss

The loss function of YOLOv5 is divided into 3 subfunctions: class loss (BCE with logits loss), objectness loss (BCE with logits loss), and location loss (CIoU loss). The loss function is expressed in Equation (6).

$$Loss = l_1 L_{cls} + l_2 L_{box} + l_3 L_{dfl} \tag{6}$$

2.6.2. Loss Calculation

With BCE with logits loss, this loss function combines a sigmoid layer and BCE Loss into a single layer. This version is more numerically stable than using sigmoid alone followed by BCE Loss because, by combining operations into one class, the logsum-exp trick for numerical stability is used. The formula of BCE Loss with logits in the multi-label classification case can be demonstrated in Equations (7) and (8).

$$\ell_c(x, y) = L_c = \{l_{1,c}, \ldots, l_{N,c}\}^T \tag{7}$$

$$L_{n,c} = -w_{n,c} \begin{bmatrix} p_c y_{n,c} \cdot \log \sigma(x_{n,c}) + \\ +(1 - y_{n,c}) \log(1 - \sigma(x_{n,c})) \end{bmatrix} \tag{8}$$

where:

- c: class number, in this case, $c > 1$, otherwise $c = 1$;
- pc: weight of the positive answer for class c;
- N: the batch size;
- n: number of samples in a batch;
- $y_{n,c}$: positive class relates to that logit, while $1 - y_{n,c}$ represents that for a negative class.

2.6.3. Box Loss

CIoU or Complete IoU loss [34] is a loss function that was created to address the limitations of earlier versions of IoU loss functions, such as GIoU and IoU, and aims to provide a more accurate bounding box regression. CIoU is computed in Equation (9).

$$L_{CIoU} = 1 - IoU + \frac{p^2(b, b^{gt})}{c^2} + \alpha v \tag{9}$$

While p indicates the Euclidean distance, α and v are used to calculate the discrepancy in the width-to-height ratio.

$$\alpha = \frac{v}{(1 - IoU) + v} \tag{10}$$

$$v = \frac{4}{\pi^2} \left(\arctan \frac{w^{gt}}{h^{gt}} - \arctan \frac{w}{h} \right)^2 \tag{11}$$

2.6.4. DFL Loss

According to the authors in [35], inside each image, the target object could suffer from various conditions, such as shadow, blur, occlusion, or the boundary being ambiguous (being covered by another object). So, for these reasons, the authors of [35] pointed out that the ground-truth labels are sometimes not trustworthy. To address this problem, DFL was created to force the network to quickly concentrate on learning the probabilities associated with values in the surrounding continuous regions of the target bounding boxes. The DFL is computed in Equation (12).

$$\mathrm{DFL}(S_i, S_{i+1}) = -((y_{i+1} - y)\log(S_i) + (y - y_i)\log(S_{i+1})) \quad (12)$$

where y_{i+1} and y_{i-1} are the two closest values to y or, in other words, the ground-truth bounding box. S_i and S_{i+1} can be described as the probability for y_{i+1} and y_{i-1}.

2.7. Loss

2.7.1. Efficient IoU Loss

In [36], the experimental result has pointed out the improvement in accuracy and convergence speed compared to the other versions. However, the authors of [36] have spotted one big disadvantage of CIoU [34], which is the parameter v, in which:

- v does not represent the relation between w and w^{gt} or h and h^{gt}; instead, it represents the difference between their aspect ratios. In cases where $w = kw^{gt}$, we will have $v = 0$, which is inconsistent with reality. Because of this, the loss function will only try to increase the similarity of the aspect ratio, rather than decrease the discrepancy between (w, h) and (w^{gt}, h^{gt});
- The gradient of v with respect to w and h can be demonstrated as:

$$\begin{aligned} \frac{\partial v}{\partial w} &= \frac{8}{\pi^2}\left(\arctan\frac{w^{gt}}{h^{gt}} - \arctan\frac{w}{h}\right) * \frac{h}{w^2+h^2}, \\ \frac{\partial v}{\partial h} &= -\frac{8}{\pi^2}\left(\arctan\frac{w^{gt}}{h^{gt}} - \arctan\frac{w}{h}\right) * \frac{w}{w^2+h^2}; \end{aligned} \quad (13)$$

$$\frac{\partial v}{\partial w} = -\frac{h}{w}\frac{\partial v}{\partial h} \quad (14)$$

With the second equation, it is easy to see that if w is increased, then h will decrease and vice versa. According to the author, it is unreasonable when $w < w^{gt}$ and $h < h^{gt}$ or $w > w^{gt}$ and $h > h^{gt}$. To solve the above problem, the authors have suggested a new version of IoU loss, which is EIoU. The loss function is computed in Equation (15).

$$L_{EIoU} = 1 - IoU + \frac{\rho^2(b,b^{gt})}{(w^c)^2+(h^c)^2} + \frac{\rho^2(w,w^{gt})}{(w^c)^2} + \frac{\rho^2(h,h^{gt})}{(h^c)^2} \quad (15)$$

where w^c and h^c are the width and height of the smallest enclosing box covering two boxes.

2.7.2. Alpha-IoU Loss

Alpha-IoU loss created by [37] is a new family of losses based on the fusion of IoU loss functions and the power parameter α. There are two important properties of alpha-IoU that should be mentioned:

- The loss re-weighting: $w_{l_r} = L_{\alpha-IoU}/L_{IoU} = 1 + (IoU - IoU^{\alpha})/(1 - IoU)$ with $w_{l_r}(IoU = 0) = 1$ and $\lim_{IoU \to 1} w_{l_r} = \alpha.$;
- $L_{\alpha-IoU}$ will adaptively down-weight and up-weight the relative loss of all objects according to their IoUs when $0 < \alpha < 1$ and $\alpha > 1$, respectively. When $\alpha > 1$, the reweighting factor increases monotonically with the increase in IoU (w_{l_r} decays from 1 to α). In other words, with $\alpha > 1$, the model will focus more on one high-IoU object;
- The loss gradient reweighting: $w_{\nabla_r} = |\nabla_{IoU} L_{\alpha-IoU}| / |\nabla_{IoU} L_{IoU}| = \alpha IoU^{\alpha-1}$ with the turning point at $IoU = \alpha^{\frac{1}{1-\alpha}} \in \left(0, \frac{1}{e}\right)$ when $0 < \alpha < 1$ and $IoU = \alpha^{\frac{1}{1-\alpha}} \in \left(\frac{1}{e}, 1\right)$ if $\alpha > 1$. When $\alpha > 1$, the above reweighting factor $w_{\Delta r}$ increases monotonically with the increase in IoU, while decreasing monotonically with the increase in IoU when $0 < \alpha < 1$. In other words, $L\alpha - IoU$ with $\alpha > 1$ helps detectors learn faster on high-IoU objects.

2.7.3. Alpha-EIoU Loss

In this paper, we propose a method that replaces the original Box Loss function of YOLOv8 with the combination of the EIoU loss and alpha-IoU loss. The alpha-EIoU loss function is given in Equation (16).

$$L_{\alpha-EIoU} = 1 - IoU^{\alpha} + \frac{\rho^2\left(b,b^{gt}\right)}{(w^c)^{2\alpha}+(h^c)^{2\alpha}} + \frac{\rho^{2\alpha}\left(w,w^{gt}\right)}{(w^c)^{2\alpha}} + \frac{\rho^{2\alpha}\left(h,h^{gt}\right)}{(h^c)^{2\alpha}} \tag{16}$$

$\alpha = 3$ as recommended by the author in [37].

3. Results and Discussion

3.1. Parameter Setting

The input size of the image is changed to 640 pixels × 640 pixels, in terms of optimizers, we choose the Stochastic Gradient Descent, the momentum for the Stochastic Gradient Descent is set to 0.937, the initial learning rate and the final learning rate is set by default and equal to 0.01 and 0.0001, respectively. Finally, the dataset is trained within 300 epochs with batch size equal to 16, the figures for the parameter can be seen in the Table 3.

Table 3. Parameter Settings.

Batch size	16
Optimizer	SGD
Epoch	300
Initial learning rate	0.01
Final learning rate	0.0001
Momentum	0.937

3.2. Evaluation of the Proposed Method

The final result of rice disease detection using YOLOv8 with the loss function change is shown in Table 4 and the PR-Curve. This result will be compared with the YOLOv8 model without the change in loss function. This result was achieved in the test set with a total number of 95 leaves containing *folder* disease, 95 leaves containing *blast* disease, and 84 leaves containing *brown spot* disease. This result was achieved by using the nano version of YOLOv8, which is the YOLOv8n model.

The PR-Curve of YOLOv8 is given in Figure 5.

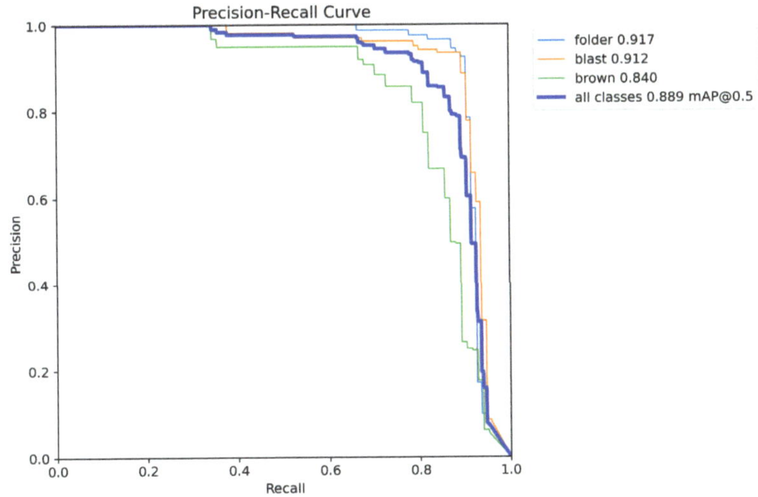

Figure 5. PR-Curve of YOLOv8.

Table 4. The final result of rice disease detection using YOLOv8 with loss function change.

YOLOv8	PRECISION	RECALL	F1	mAP@50
All	89.6	83.5	86.4	88.9
Leaf Folder	93.9	88.4	91.1	91.7
Leaf Blast	86.7	90.5	88.6	91.2
Brown Spot	88.2	71.5	79.0	84.0

The PR-Curve of YOLOv8 with alpha-EIoU loss is given in Figure 6.

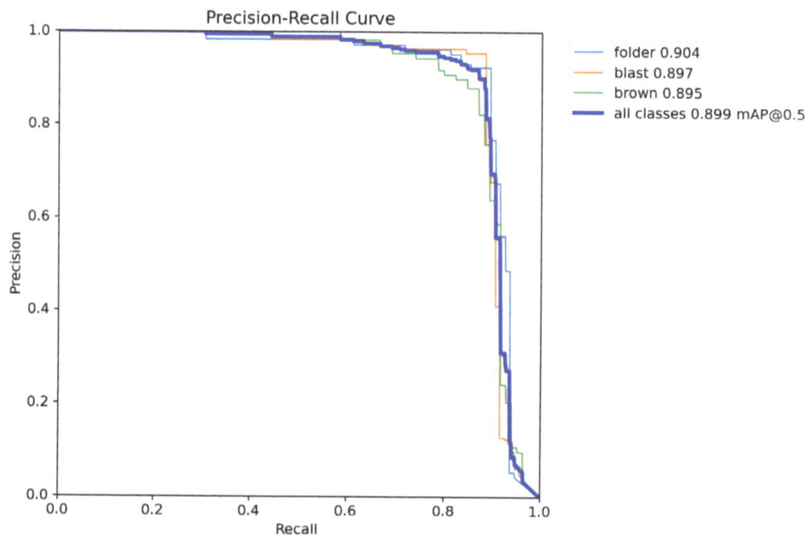

Figure 6. PR-Curve of YOLOv8.

Figure 7 depicts the inference results of our proposed model. Take note that the red, pink, and orange bounding boxes represent the folder, blast, and brown, respectively. Based on the given results of both YOLOv8 and YOLOv8 with the alpha-EIoU loss, the accuracy of the modified loss function has increased the model accuracy up to 1%, which is significant. In terms of other metrics, the proposed method of using alpha-EIoU also yields a better result. In precision and recall, the proposed method achieved 90% and 84.4%, respectively, while YOLOv8 without the modification to the Box Loss function achieved 89.6% and 83.5%. It can be seen that the improvements in these metrics for the proposed method increase from 0.4% to 0.9%. In order to consider both economic factors and the precision factors, the F1-score is known as an important parameter in detecting diseases. Our proposed method achieves a higher F1-score (87.1%) than that of the original method (86.4%). However, if we look at the specific accuracy value of each class, there are still some improvements that need to be made in order to improve our proposed model. The accuracy of the folder disease is 91.7% when predicted by the original method; however, the proposed method only achieved 90.4%. The same can be seen in the blast disease, where the value of mAP@0.5 is 91.2% for the conventional way and only 89.7% for the proposed method. The proposed method greatly increased the accuracy of brown spot disease, in which, the alpha-EIoU method provides an accuracy of 89.5% compared to 84% for the CIoU method, which is more than a 5% increase in accuracy. Compared to the algorithm of YOLOv5 in [22,23], the accuracy of our proposed model on *leaf blast* has reached 89.7% compared to 69.4% in the study in [22] and 80.3% proposed in [23]. The efficient disease detection of brown leaf spot of our proposed model is also given in Table 5. It is shown that the accuracy of our proposed model on the *brown spot* has reached 89.5%. This is much better than that of

YOLOv5, used in [23], which gives only a 55.4% accuracy. As depicted in the pictures of bacterial blight in [22], the authors of that study have mistaken the *leaf folder* disease for bacterial blight; therefore, this shows that our model's accuracy on the *leaf folder*, which is 90.4%, is much better than the 65% from [22]. In the comparison of the proposed method with one of the state-of-the-art models, YOLOv7, in the paper [38], written by the same author as that of [18], our method increases the accuracy of the leaf blast class by more than 19%. In terms of leaf folder, which the author mistook for bacterial blight, our model achieves 90.4% compared to only 69.9%. For the cases of precision and recall, our method only achieves 90% and 84.4%, respectively, and does not surpass the values of 1 and 0.92 for YOLOv5 [18] and YOLOv7 [38]. However, our proposed method reaches 87.1% compared to 77% and 76% of the two other methods for the F1-score.

(a) (b) (c)

Figure 7. Inference results on rice leaf diseases: (**a**) inference result on brown spot disease, (**b**) inference result on leaf folder disease, and (**c**) inference result on leaf blast disease.

Table 5. YOLOv8 with alpha-EIoU loss results.

YOLOv8	Precision	Recall	F1	mAP@50
All	90.0	84.4	87.1	89.9
Leaf Folder	93.1	85.2	89.0	90.4
Leaf Blast	86.2	88.4	87.3	89.7
Brown Spot	90.6	79.8	84.9	89.5

The complexity and the FPS of our proposed method for YOLOv8n, YOLOv5n, and the YOLOv7-tiny method provided by [39–41] can be seen in Table 6:

Table 6. Computional complexity of YOLOv8n, YOLOv7-tiny, and YOLOv5n.

	Size (pixels)	Params (M)	FLOPs (B)	NVIDIA Jetson AGX Orin (FPS)
YOLOv8n	640	3.2	8.7	383
YOLOv5n	640	1.9	4.5	370
YOLOv7-tiny	640	6.2	13.8	290

As can be seen in the table, YOLOv8n and YOLOv5 are clearly more suitable to implement on low-cost devices. While YOLOv8n's params and flops are higher than those of YOLOv5n, YOLOv8 is still able to yield a better performance. Because of these reasons, our proposed method is a suitable option for low-cost devices. This will further benefit the economic aspect, especially as it requires a large number of devices when being utilized on large-scale rice fields.

For the application of our model, we implement it on Raspberry Pi 4 Model B, as a Single-Board Computer (SBC), supporting a variety of packages and software libraries across different programming languages. In this study, we use Python packages, such as Serial and Ultralytics. The Serial library was used to enable control of the GPIO pins with the SIM800L module on Raspberry Pi. The Ultralytics library was used to run the YOLOv8 model. Raspberry Pi is a low-cost device, so it took more time to process the AI model. Therefore, in this project, we used YOLOv8n (the lightest version of the YOLOv8 model) so that the Rpi could handle an average processing time of about 1500 ms per frame.

In this research, the prediction results of our approach have been evaluated by the agricultural experts at the Vietnam National University of Agriculture, Hanoi, Vietnam. Despite some mistakes in the prediction process, our proposed model still has great potential to warn farmers of infected regions in their rice fields (Figure 8). The performance of our hardware system, according to experts, is sufficient to apply to real-world scenarios.

Figure 8. The SMS message sent to farmers.

3.3. Discussion

As mentioned above, while our method can provide better accuracy for all of the combined classes, our method seems to have drawbacks in detecting folder disease and leaf blast disease. In terms of the model complexity, although our method has been applied to the lightest version of YOLOv8, which is YOLOv8 nano, the change in our method did not further reduce the complexity of the original model or increase it. In the case of images captured in different weather conditions, our dataset does not contain images captured in heavily windy conditions, a case that can make the images extremely blurry. Adding images in windy weather can greatly enhance our model performance in real-world scenarios. For the folder class, in extreme lighting conditions, a healthy leaf can be easily mistaken for a diseased leaf, which can lead to a reduced accuracy of the model.

In future research works, we intend to improve our methods in three terms, namely, (1) deep learning techniques, (2) practical application techniques, and (3) the scale of the dataset. For the first term, we would like to test replacing the classification loss function of the original YOLOv8 with the Varifocal Loss Function [42] and train the model on different learning rate schedulers, as inspired by [43], to further increase the accuracy of the proposed model. In the practical aspect, our proposed system may be implemented on a real agricultural robot that can automatically move in the field for early detection and prevention of rice leaf diseases in Vietnam. With high accuracy detection of rice diseases, our proposed method may improve the effectiveness of the rice disease management systems. Our proposed system reduces the human resources and the time needed for rice disease detection; therefore, it can easily diagnose the diseases at an early stage. Moreover, this approach reduces the pesticide residue in the final product of rice and decreases the environmental impact of rice cultivation. Although we only apply the proposed method to three different types of disease, e.g., leaf folder, leaf blast, and brown spot, the number of disease classes and the number of images in our dataset can be further extended. In terms of large datasets, this can easily be performed because our modification to the model does not affect the learning process on either large-scale or small-scale datasets. In addition, in this research, we do not change the optimizer of the original YOLOv8 version, which is the Stochastic Gradient Descent. This is an optimization algorithm that is commonly used in the case of large-scale datasets. In further extending the number of classes, the new diseases might contain relatively small disease marks. This could be a challenge, however, in this research, our prediction accuracies on the brown spot disease and the leaf blast disease, characterized by the smallest disease marks, are relatively good, which bodes well for future improvements.

4. Conclusions

Our paper introduces a new approach using the YOLOv8 architecture with the YOLOv8n model and the alpha-EIoU loss function to identify the disease of rice plants with an accuracy is up to 89.9%, which is a better result compared to 62% in [23], 80.7% in [22], and 82.8% in [38]. This result was achieved on 241 photos, and this is a relatively good result for further applications in the real world.The experiment system worked well and provided warnings immediately after detection to farmers with knowledge equivalent to an expert; this will help farmers have quick responses to deal with the diseases. With this result, we implement the achieved model into the low-cost hardware that we designed to provide farmers with a solution to detect diseases in a timely manner in the rice field. Our research also provides a completely new rice leaf disease image dataset of the three most common diseases, (*leaf blast*, *leaf folder*, and *brown spot*), in Vietnam. Therefore, it may help researchers reduce the time and money recquired to obtain such data in future research. It is proved that our proposed modification of the YOLOv8 framework achieved better performance than some other state-of-the-art methods.

Author Contributions: Conceptualization, H.T.N. (Hoc Thai Nguyen) and H.T.N. (Huong Thanh Nguyen); methodology, T.D.B. and H.T.N. (Huong Thanh Nguyen); validation, D.C.T.; formal analysis, H.T.N. (Hoc Thai Nguyen) and T.D.B.; literature review, D.C.T., K.G.D. and A.T.M.; project administration, D.C.T. and T.D.B.; supervision, T.D.B. All authors have read and agreed to the published version of the manuscript.

Funding: This research is funded by the Hanoi University of Science and Technology (HUST) under grant number: T2023-PC-069.

Institutional Review Board Statement: Not applicable.

Informed Consent Statement: Not applicable.

Data Availability Statement: The data presented in this study are available on request from the corresponding author.

Acknowledgments: The authors gratefully acknowledge the support of the Industrial Instrumentation & IoT Laboratory at the School of Electrical and Electronic Engineering, HUST, and other support from experts at the Vietnam National University of Agriculture, Vietnam.

Conflicts of Interest: The authors declare no conflict of interest.

Abbreviations

The following abbreviations are used in this manuscript:

AI	Artificial Intelligence
IoT	Internet of Things
CNN	Convolutional Neural Network
SVM	Support vector machines
GSM	Global system for mobile communications
mAP	Mean Average Precision

References

1. Fairhurst, T.; Dobermann, A. Rice in the global food supply. *World* **2002**, *5*, 454349–511675.
2. Wijayanto, A.K.; Junaedi, A.; Sujaswara, A.A.; Khamid, M.B.; Prasetyo, L.B.; Hongo, C.; Kuze, H. Machine Learning for Precise Rice Variety Classification in Tropical Environments Using UAV-Based Multispectral Sensing. *AgriEngineering* **2023**, *5*, 2000–2019. [CrossRef]

3. de Oliveira Carneiro, L.; Coradi, P.C.; Rodrigues, D.M.; Lima, R.E.; Teodoro, L.P.R.; de Moraes, R.S.; Teodoro, P.E.; Nunes, M.T.; Leal, M.M.; Lopes, L.R.; et al. Characterizing and Predicting the Quality of Milled Rice Grains Using Machine Learning Models. *AgriEngineering* **2023**, *5*, 1196–1215. [CrossRef]
4. Paidipati, K.K.; Chesneau, C.; Nayana, B.M.; Kumar, K.R.; Polisetty, K.; Kurangi, C. Prediction of rice cultivation in India—Support vector regression approach with various kernels for non-linear patterns. *AgriEngineering* **2021**, *3*, 182–198. [CrossRef]
5. Rahman, H.; Sharifee, N.H.; Sultana, N.; Islam, M.I.; Habib, S.A.; Ahammad, T. Integrated Application of Remote Sensing and GIS in Crop Information System—A Case Study on Aman Rice Production Forecasting Using MODIS-NDVI in Bangladesh. *AgriEngineering* **2020**, *2*, 264–279. [CrossRef]
6. Krichen, M. Convolutional Neural Networks: A Survey. *Computers* **2023**, *12*, 151. [CrossRef]
7. Naved, M.; Devi, V.A.; Gaur, L.; Elngar, A.A. (Eds.) E-Learning Modeling Technique and Convolution Neural Networks in Online Education. In *IoT-Enabled Convolutional Neural Networks: Techniques and Applications*; River Publishers: Aalborg, Denmark, 2023.
8. Uoc, N.Q.; Duong, N.T.; Thanh, B.D. A novel automatic detecting system for cucumber disease based on the convolution neural network algorithm. *GMSARN Int. J.* **2022**, *16*, 295–302.
9. Saleem, M.H.; Potgieter, J.; Arif, K.M. Plant disease detection and classification by deep learning. *Plants* **2002**, *8*, 468. [CrossRef] [PubMed]
10. Agbulos, M.K.; Sarmiento, Y.; Villaverde, J. Identification of leaf blast and brown spot diseases on rice leaf with yolo algorithm. In Proceedings of the IEEE 7th International Conference on Control Science and Systems Engineering (ICCSSE), Qingdao, China, 30 July–1 August 2021; pp. 307–312. [CrossRef]
11. Zhang, K.; Wu, Q.; Liu, A.; Meng, X. Can deep learning identify tomato leaf disease? *Adv. Multimed.* **2018**, *8*, 6710865. [CrossRef]
12. Liu, J.; Wang, X. Tomato diseases and pests detection based on improved yolo v3 convolutional neural network. *Front. Plant Sci.* **2020**, *11*, 898. [CrossRef]
13. Sharma, R.; Kukreja, V.; Kaushal, R.K.; Bansal, A.; Kaur, A. Rice leaf blight disease detection using multi-classification deep learning model. In Proceedings of the 10th International Conference on Reliability, Infocom Technologies and Optimization (Trends and Future Directions) (ICRITO), Noida, India, 13–14 October 2022; pp. 1–5. [CrossRef]
14. Deng, R.; Tao, M.; Xing, H.; Yang, X.; Liu, C.; Liao, K.; Qi, L. Automatic diagnosis of rice diseases using deep learning. *Front. Plant Sci.* **2021**, *12*, 701038. [CrossRef]
15. Altinbİlek, H.F.; Kizil, U. Identification of paddy rice diseases using deep convolutional neural networks. *Yuz. YıL Univ. J. Agric. Sci.* **2021**, *32*, 705–713. [CrossRef]
16. Shrivastava, V.K.; Pradhan, M.K.; Minz, S.; Thakur, M.P. Rice plant disease classification using transfer learning of deep convolution neural network. *ISPRS Int. Arch. Photogramm. Remote Sens. Spat. Inf. Sci.* **2019**, *42*, 631–635. [CrossRef]
17. Chaudhari, D.J.; Malathi, K. A survey on rice leaf and seedlings disease detection system. *Indian J. Comput. Sci. Eng. (IJCSE)* **2021**, *12*, 561–568. [CrossRef]
18. Rahman, C.R.; Arko, P.S.; Ali, M.E.; Khan, M.A.I.; Apon, S.H.; Nowrin, F.; Wasif, A. Identification and recognition of rice diseases and pests using convolutional neural networks. *Biosyst. Eng.* **2020**, *194*, 112–120. [CrossRef]
19. Hasan, M.J.; Mahbub, S.; Alom, M.S.; Nasim, M.A. Rice disease identification and classification by integrating support vector machine with deep convolutional neural network. In Proceedings of the 2019 1st International Conference on Advances in Science, Engineering and Robotics Technology, Dhaka, Bangladesh, 3–5 May 2019; pp. 1–6. [CrossRef]
20. Ramesh, S.; Vydeki, D. Recognition and Classification of Paddy Leaf Diseases Using Optimized Deep Neural Network with Jaya Algorithm. *Inf. Process. Agric.* **2020**, *7*, 249–260. [CrossRef]
21. Maheswaran, S.; Sathesh, S.; Rithika, P.; Shafiq, I.M.; Nandita, S.; Gomathi, R.D. Detection and Classification of Paddy Leaf Diseases Using Deep Learning (CNN). In *Computer, Communication, and Signal Processing. ICCCSP 2022. IFIP Advances in Information and Communication Technology*; Neuhold, E.J., Fernando, X., Lu, J., Piramuthu, S., Chandrabose, A., Eds.; Springer: Cham, Switzerland, 2022; Volume 651. [CrossRef]
22. Haque, M.E.; Rahman, A.; Junaeid, I.; Hoque, S.U.; Paul, M. Rice Leaf Disease Classification and Detection Using YOLOv5. *arXiv* **2022**, arXiv:2209.01579.
23. Jhatial, M.J.; Shaikh, R.A.; Shaikh, N.A.; Rajper, S.; Arain, R.H.; Chandio, G.H.; Bhangwar, A.Q.; Shaikh, H.; Shaikh, K.H. Deep learning-based rice leaf diseases detection using yolov5. *Sukkur IBA J. Comput. Math. Sci.* **2022**, *6*, 49–61. [CrossRef]
24. Sparks, N.C.A.; Cruz, C.V. Brown Spot. Available online: http://www.knowledgebank.irri.org/training/fact-sheets/pest-management/diseases/item/brown-spot (accessed on 5 September 2023).
25. Catindig, J. Rice Leaf Folder. Available online: http://www.knowledgebank.irri.org/training/fact-sheets/pest-management/insects/item/rice-leaffolder (accessed on 5 September 2023).
26. Redmon, J.; Farhadi, A. YOLO9000: Better, faster, stronger. In Proceedings of the IEEE Conference on Computer Vision and Pattern Recognition, Honolulu, HI, USA, 21–26 July 2017; pp. 7263–7271.
27. Redmon, J.; Farhadi, A. Yolov3: An incremental improvement. *arXiv* **2018**, arXiv:1804.02767.
28. Bochkovskiy, A.; Wang, C.Y.; Liao, H.Y.M. Yolov4: Optimal speed and accuracy of object detection. *arXiv* **2020**, arXiv:2004.10934.
29. Li, C.; Li, L.; Jiang, H.; Weng, K.; Geng, Y.; Li, L.; Ke, Z.; Li, Q.; Cheng, M.; Nie, W.; et al. YOLOv6: A single-stage object detection framework for industrial applications. *arXiv* **2022**, arXiv:2209.02976.

30. Wang, C.Y.; Bochkovskiy, A.; Liao, H.Y.M. YOLOv7: Trainable bag-of-freebies sets new state-of-the-art for real-time object detectors. In Proceedings of the IEEE/CVF Conference on Computer Vision and Pattern Recognition, Vancouver, BC, Canada, 17–24 June 2023; pp. 7464–7475.
31. Wang, C.Y.; Liao, H.Y.M.; Yeh, I.H. Designing network design strategies through gradient path analysis. *arXiv* **2022**, arXiv:2211.04800.
32. Lee, Y.; Hwang, J.W.; Lee, S.; Bae, Y.; Park, J. An energy and GPU-computation efficient backbone network for real time object detection. In Proceedings of the IEEE/CVF Conference on Computer Vision and Pattern Recognition Workshops, Long Beach, CA, USA, 16–17 June 2019.
33. Feng, C.; Zhong, Y.; Gao, Y.; Scott, M.R.; Huang, W. Tood: Task-aligned one-stage object detection. In *2021 IEEE/CVF International Conference on Computer Vision (ICCV)*; IEEE Computer Society: Washington, DC, USA, 2021; pp. 3490–3499. [CrossRef]
34. Zheng, Z.; Wang, P.; Ren, D.; Liu, W.; Ye, R.; Hu, Q.; Zuo, W. Enhancing geometric factors in model learning and inference for object detection and instance segmentation. *IEEE Trans. Cybern.* **2021**, *52*, 8574–8586. [CrossRef] [PubMed]
35. Li, X.; Wang, W.; Wu, L.; Chen, S.; Hu, X.; Li, J.; Tang, J.; Yang, J. Generalized focal loss: Learning qualified and distributed bounding boxes for dense object detection. *Adv. Neural Inf. Process. Syst.* **2020**, *33*, 21002–21012.
36. Zhang, Y.F.; Ren, W.; Zhang, Z.; Jia, Z.; Wang, L.; Tan, T. Focal and efficient IOU loss for accurate bounding box regression. *Neurocomputing* **2022**, *506*, 146–157. [CrossRef]
37. He, J.; Erfani, S.; Ma, X.; Bailey, J.; Chi, Y.; Hua, X.S. Alpha-IoU: A family of power intersection over union losses for bounding box regression. *Adv. Neural Inf. Process. Syst.* **2022**, *34*, 20230–20242.
38. Haque, M.E.; Paul, M.; Rahman, A.; Tohidi, F.; Islam, M.J. Rice leaf disease detection and classification using lightly trained Yolov7 active deep learning approach. In Proceedings of the Digital Image Computing: Techniques and Applications (DICTA), Port Macquarie, NSW, Australia, 28 November–1 December 2023.
39. Available online: https://docs.ultralytics.com/models/yolov8/ (accessed on 19 January 2024).
40. Available online: https://docs.ultralytics.com/models/yolov7/ (accessed on 19 January 2024).
41. Available online: https://www.stereolabs.com/blog/performance-of-yolo-v5-v7-and-v8 (accessed on 19 January 2024).
42. Zhang, H.; Wang, Y.; Dayoub, F.; Sunderhauf, N. VarifocalNet: An IoU-aware Dense Object Detector. *arXiv* **2020**, arXiv:2008.13367.
43. Loshchilov, I.; Frank, H. SGDR: Stochastic Gradient Descent with Warm Restarts. *arXiv* **2016**, arXiv:1608.03983.

AgriEngineering

Article

Enhanced Deep Learning Architecture for Rapid and Accurate Tomato Plant Disease Diagnosis

Shahab Ul Islam [1,*], Shahab Zaib [2], Giampaolo Ferraioli [1], Vito Pascazio [1], Gilda Schirinzi [1] and Ghassan Husnain [3]

1 Department of Engineering, University of Naples Parthenope, 80133 Naples, Italy; giampaolo.ferraioli@uniparthenope.it (G.F.); vito.pascazio@uniparthenope.it (V.P.); gilda.schirinzi@uniparthenope.it (G.S.)
2 Department of Computer Science, IQRA National University, Peshawar 25000, Pakistan; szaib@polyu.edu.hk
3 Department of Computer Science, CECOS University of IT and Emerging Sciences, Peshawar 25000, Pakistan; ghassan.husnain@cecos.edu.pk
* Correspondence: shahabul.islam001@studenti.uniparthenope.it

Abstract: Deep neural networks have demonstrated outstanding performances in agriculture production. Agriculture production is one of the most important sectors because it has a direct impact on the economy and social life of any society. Plant disease identification is a big challenge for agriculture production, for which we need a fast and accurate technique to identify plant disease. With the recent advancement in deep learning, we can develop a robust and accurate system. This research investigated the use of deep learning for accurate and fast tomato plant disease identification. In this research, we have used individual and merged datasets of tomato plants with 10 diseases (including healthy plants). The main aim of this work is to check the accuracy of the existing convolutional neural network models such as Visual Geometry Group, Residual Net, and DenseNet on tomato plant disease detection and then design a custom deep neural network model to give the best accuracy in case of the tomato plant. We have trained and tested our models with datasets containing over 18,000 and 25,000 images with 10 classes. We achieved over 99% accuracy with our custom model. This high accuracy was achieved with less training time and lower computational cost compared to other CNNs. This research demonstrates the potential of deep learning for efficient and accurate tomato plant disease detection, which can benefit farmers and contribute to improved agricultural production. The custom model's efficient performance makes it promising for practical implementation in real-world agricultural settings.

Keywords: plant disease; tomato; convolutional neural network; machine learning; deep learning

Citation: Islam, S.U.; Zaib, S.; Ferraioli, G.; Pascazio, V.; Schirinzi, G.; Husnain, G. Enhanced Deep Learning Architecture for Rapid and Accurate Tomato Plant Disease Diagnosis. *AgriEngineering* **2024**, *6*, 375–395. https://doi.org/10.3390/agriengineering6010023

Academic Editors: Francesco Marinello, Ray E. Sheriff and Chiew Foong Kwong

Received: 26 December 2023
Revised: 30 January 2024
Accepted: 9 February 2024
Published: 12 February 2024

1. Introduction

One of the most significant industries is agriculture production because it directly affects a society's economy and social structure [1]. Agriculture production faces a significant issue in identifying plant diseases; thus, we need a quick and precise method to do so. We can create a reliable and accurate system with the most recent advancements in deep learning convolutional neural networks (CNNs) [2]. In this study, we used two (individual, combined) datasets of tomato plants with 10 illnesses, including healthy plants as well. The major objective of this study is to compare several tomato plant diseases utilizing some of the top convolutional neural network models, including Visual Geometry Group, Residual Net, and DenseNet. In addition, we have also developed our neural network architecture. Despite the complexity of datasets exceeding 18,000 and 25,000 images with 10 distinct classes each, ResNet, DenseNet, and our custom model achieved remarkable accuracy above 99%.

The quality and amount of a nation's production determine how much agriculture it can produce [3]. Any nation's economic and social structure is directly impacted by the production of agricultural goods [4]. Agriculture production is essential to the advancement of society and the economy [5]. This is the reason that diagnosing and treating plant diseases has drawn so much attention due to the large amount of territory that farms now cover; finding, assessing, and treating plant diseases is a tremendously difficult problem for farmers today [6]. Traditional disease detection is a highly difficult and time-consuming technique. Additionally, there is a high likelihood of errors, which might result in productivity loss. Additionally, early diagnosis and treatment of plant diseases are essential because they have a significantly higher rate of success [7].

Because the plant's leaves are the most apparent and susceptible to disease, they are used to recognize plant diseases [8]. We require a system that can accurately identify the condition using images. For diagnosis purposes, plant diseases and their treatments have been the subject of extensive research, but each has its limitations. Researchers have employed a variety of methodologies, including deep learning, image processing, and machine learning. As a result, these studies were fruitful, and the majority of the procedures produced results that were superior to those of human specialists, especially deep learning. Although there is still debate over whether neural architecture performs better, in this study, we compared several leading neural network architectures on tomato plant diseases using two datasets with over 1800 and 2500 sample images, respectively. Additionally, we have included our neural network design, which produced test accuracy results of 99.22 and 99.24.

2. Related Works

Current research suggests some effective methods for diagnosing plant diseases. This study begins with a review of studies on the many types of plant diseases and methods to accurately identify them to better comprehend the existing literature on their identification.

Analytical research published by Punitha, Kartikeyan, Gyanesh, and Shrivastava in 2022. In their research, they have used Deep Learning architectures AlexNet, GoogLeNet, and DenseNet with different optimizers via the stochastic gradient descent (SGD) algorithm and root mean square propagation (RMSProp) for tomato plant disease identification and classification to compare their accuracies. In this research, the best performer is GoogleLeNet, with an accuracy of 99.56% [9].

Amarjeeth Singh et al. (2022) published research in which they used vegetable crops to detect diseases like Scab, Early Blight, Leaf scorch, and Bacterial spot. In this research, deep learning and CNN architecture were used, and they achieved an accuracy of 98.87%. The accuracy achieved in this research is good, but it can still be improved [10].

Vishakha Kathole Mousami Munot (2022) performed a comparative survey of different deep learning CNN architectures like the VGG, GoogleLeNet, and AlexNet to detect plant disease using their leaves. The dataset used in this research has six different diseases and a healthy class [11].

Khalil Khan, Rehan Ullah Khan, Waleed Albattah, and Ali Mustafa Qamar published a paper in 2022. In this research, semantic segmentation is used to highlight the foreground (leaf) and the background (non-leaf) and to look through each pixel for labeling. A CNN-based model is used on the Plant Village dataset. In this research, they have achieved a total of 97.6% accuracy [12].

Swati S. Wadadare and H. S. Fadewar (2022) have researched tomato plant diseases using the latest deep learning technology. In this research, Inception V3 is used for tomato plant disease detection. The transfer learning technique is used with a training accuracy of 92.19% and a test accuracy of 93.03% [13].

Chen, Hsing-Chung, Agung Mulyo Widodo, et al. (2022) published their research on plant disease detection in which they used AlexNet. They have used 18, 345 training data and 4585 test data. In their research, they have achieved an accuracy of 98% using a releasing rate of 0.0005 and 75 epochs. The accuracy can be justified, but by using new models, it can be improved further [14].

In 2022, Khan, Muhammad Attique, et al. published their research on cucumber plant disease detection. They have used multi-level deep entropy ELM feature selection. They have used VGG16, ResNet50, ResNet101, and DenseNet201 deep learning CNN models. In their research, they achieved the best accuracy of 98.4% [15].

In 2022, Xie, Yiting, Darren Plett, and Huajian Liu published research on crown rot disease in wheat plants. In their research, they used image color and machine learning techniques. They successfully distinguished between healthy and infected plants 14 days earlier. In their research, the F1 scores for most datasets were higher than 0.80 [16].

Research by Kaur, Prabhjot, et al. in 2021 shows plant disease detection using a hybrid convolutional network by applying feature reduction. They have used grape plants with Leaf blight, Black rot, stable, and Black measles diseases. In this study, the researchers analyzed the Plant Village dataset and employed logistic regression to reduce the dimensionality of the extracted features. They have used state-of-the-art classifiers and achieved an accuracy of 98.7% after 92 epochs [17].

Almadhor, Ahmad, et al. (2021) published their research in which they used machine learning techniques for guava plant disease detection. In this research, they have used different color and textual techniques for feature extraction with machine learning classifiers like Fine KNN, Complex Tree, Boosted Tree, Bagged Tree, and Cubic SVM. In their results, the Bagged Tree Classifier has the highest accuracy of 99% on all four guava plant diseases [18].

Applalanaidu, Majji V., and G. Kumaravelan (2021) performed review-based research on machine learning and deep learning for plant disease detection. In this research, they have a different comparison for both machine learning and deep learning. For machine learning, the SVM model has the best performance with an accuracy of 97.3%, followed by the Random Forest classifier with 97% accuracy. For deep learning, Inception V3 has the best accuracy of 99.76%, followed by VGG with 99.53% accuracy [19].

B. Srinivas, P. Satheesh, P. Rama Santosh Naidu, and U. Neelima (2021) used deep learning to detect guava plant diseases. In this paper, the researchers have used Flask in Python, in which they have created their own CNN model. The confusion matrix is used for obtaining the accuracy, and the accuracy archived was between 65 and 85% [20].

In 2023, Murat Tasyurek and Recep Sinan Arslan published a study on a new CNN model called Real Time-Droid (RT-Droid). RT-Droid is based on YOLO V5 and can detect malware very quickly and accurately. To create RT-Droid, the authors first extracted features from Android manifest files and converted them into RGB images similar to QR codes. Leveraging transfer learning, the researchers trained VGGNet, Faster R-CNN, YOLO V4, and YOLO V5 models on these images. Notably, the YOLO V5 model achieved exceptional object detection accuracy in real time, surpassing the efficiency of other CNN models used in the study. The authors also compared the results of RT-Droid with VGG Net, Faster R-CNN, and YOLO V4 and found that it yielded better results [21].

Existing studies have made valuable contributions to tomato plant disease detection with deep learning, typically focusing on individual datasets. Our work builds on past research by incorporating merged datasets and developing a high-performing custom model, potentially paving the way for more robust and efficient tomato plant disease detection systems. After analyzing the performance of established CNN architectures like VGG, ResNet, and DenseNet, we designed a custom model. This model achieved high accuracy, as discussed in the results section, while also offering benefits like reduced training time and computational cost.

3. Methods and Data Collection

In this section, an in-depth methodology for tomato plant disease recognition is discussed. The methodology comprises four steps, as shown in Figure 1 below.

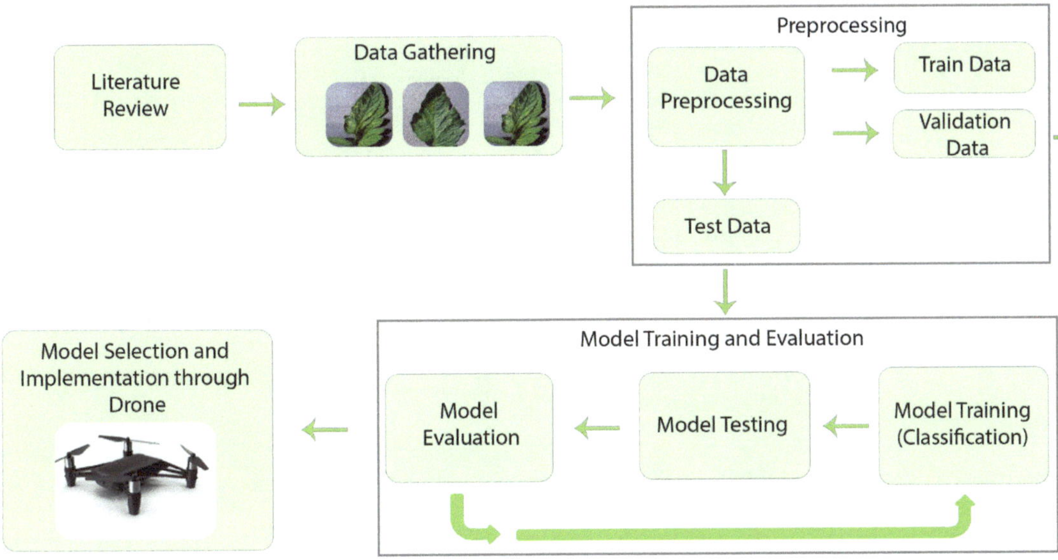

Figure 1. Methodology.

3.1. Data Collection

Data collection might seem like the simple act of gathering information, but the path can be riddled with obstacles. From biases to inconsistencies, let us dive into some of the common challenges that can trip up even the most seasoned data gatherers.

Inaccurate data: The collected data may not address the key variables or information needed to answer the intended research question.

Missing data: Gaps in the collected data, including empty fields and missing images for specific predictions, can negatively impact model performance.

Data imbalance: Unequal distribution of samples across categories can bias the model, hampering its performance for under-represented groups.

Data bias: Hidden within the data itself, even the most well-intentioned model can inherit and amplify biases based on factors like gender, political affiliation, age, or geographical origin. These biases, often subtle and difficult to detect, can lead to unfair and inaccurate predictions.

Several techniques can be applied to address those problems:

Freely available and pre-cleaned datasets. Pre-cleaned free data can be obtained from websites or vendors. They provide almost all sorts of data, and sources like Kaggle can offer pre-processed and clean data.

Leveraging web crawling and scraping techniques, automated tools such as bots and headless browsers can efficiently extract data from websites, providing valuable resources for training machine learning models.

Private data: In cases where publicly available data is scarce or insufficient, ML engineers can resort to generating their own synthetic data. This approach is particularly advantageous for models requiring smaller datasets to achieve optimal performance.

Custom data: Agencies can create or crowdsource the data for a fee.

In this research, we have collected our data from Kaggle for disease recognition. We have collected different types of image samples from Kaggle with the range from 10,000 to 40,000 images. The models are trained and tested using different data, and the performance is closely examined to make sure the collected data is correct.

3.2. Data Preprocessing

Data preprocessing is very important for machine learning and deep learning because the performance and accuracy of ML/DL models directly depend upon the quality of the data. Before the model is trained, we need to do preprocessing on the data and make sure the data is clean, relevant, and enough.

3.3. Data Cleaning

The data used in this research is different preprocessed image data taken from Kaggle; however, the data is pre-cleaned, but different data gave different results. This is why we trained and tested models using different data with different sizes, quality, and conditions, and images were captured. The different results were then compared to specify the most suitable dataset. The model was first trained with less data from 8000 samples with good results but not satisfactory results. The model trained and tested with 20,000–38,000 looks to be the best-suited numbers for deep learning. The dataset created was cleaned of any irregularities and irrelevant data.

3.4. Data Splitting

The data were split into three groups: Training, Validation, and Testing data. The training data is used to train the model. The model uses the training data to identify patterns and improve the accuracy. The validation data is used during the training process to validate the model. The validation data is entered in batches and works as test data during each step in the training process. The test data is used to test the model, and this is data that the model has never seen before. This data is used to find the accuracy of the model on new data. The training data is 70%. The validation data is 20%. The testing data is 10%.

3.5. Data Augmentation

The data augmentation is used to generate new data for the model; this technique works best for images in deep learning. Rotation, High contrast, Bright light, Low light, and image invert augmentation are used for the models that use augmented data.

Resize and Rescale

In this step, we resize the image to our desired size. We do resize to avoid the problem of images with different sizes because it can greatly affect the model's overall performance. By rescaling, we change the values of the image, e.g., we can convert the image into grayscale. We do this process for performance gain because it limits the image colors from 0 to 1.

Data Prefetch

In this step, we fetch the data into the short-term memory and fill the buffer. This is performed so that the data stays in the memory and the model can easily retrieve the data when required.

Label Encoding

Label encoding is used to convert text data into a format that the model can understand. We use label encoding to change the text data into arrays. These arrays are used to create machine learning models for text analysis and predictions.

Term Frequency Inverse Document Frequency

The term frequency inverse document frequency is used to filter out useless data. This technique is used to take all the important information in text data for better model performance.

3.6. Model Training

In this step, we train the model using training data. The model is a combination of statistical formulas that become optimized using numerical values. The training data is passed through to the model, and the model uses different functions such as Sigmoid, RMSE, TANH, and RELU for convergence. The model goes to global minima for the best accuracy. The model starts from low accuracy and high loss and slowly converges to the global minima to improve the accuracy and decrease the loss, as shown in Figure 2. Most of the time, the model becomes stuck at local minima because of some noise in the given data. To overcome the problem of local minima, we clean the data so it will converge to a more global minima. In this research, we have built different deep learning models and compared the accuracies to obtain the most suitable model.

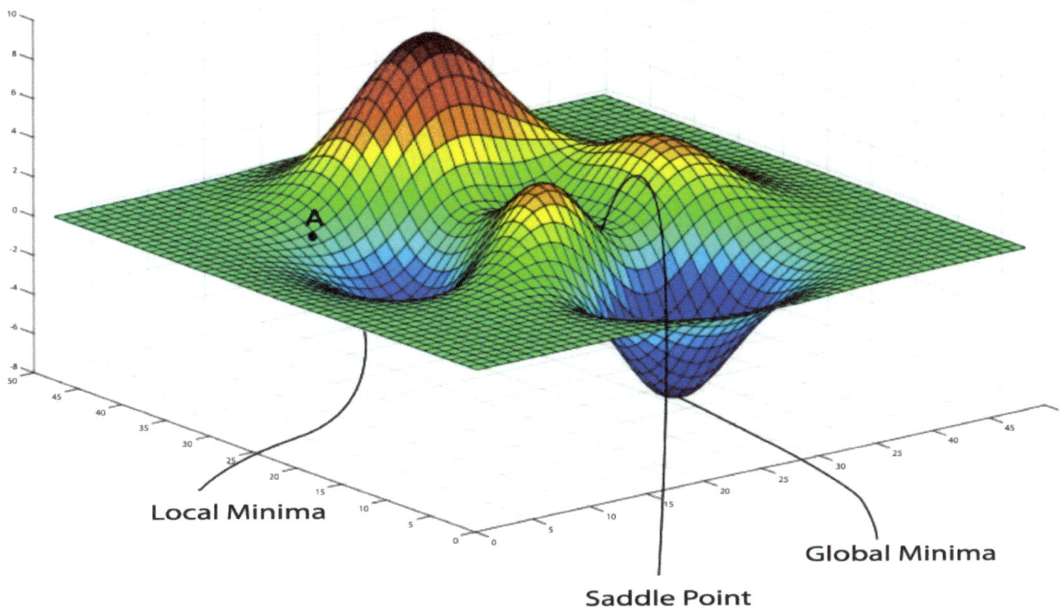

Figure 2. Model Convergence.

3.7. Model Testing

In this step, we test the model using new data that the model has not seen before. We used test data to evaluate the model's performance. In this step, we test all the models and performances to select 31 of the most suitable models. We give the model test data to make predictions and compare it with the existing results to evaluate the actual accuracy and performance.

3.8. Model Selection

In this step, we have to select the best model amongst the VGG, ResNet 50, DenseNet 121, DenseNet169, DenseNet201, and custom model. After we train the models and test them, we can demonstrate the performance of all the models.

We will have some models that will perform better than others. Here, we will evaluate the models' performances and choose the best-performing machine learning and deep learning model.

3.9. Implementation and Experimental Setup

Optimizing for both ease of development and performance, Python was chosen as the programming language for building the deep learning architectures. Leveraging the powerful combination of Keras, TensorFlow backend, and the CuDNN library, we achieved efficient training on an NVIDIA Quadro K2200 GPU (HP, Italy) with its 4 GB memory, 640 CUDA cores, and high bandwidth.

3.10. Deep Learning Architecture

Seeking to enhance plant disease classification accuracy, researchers explored improvements and modifications to existing deep learning architectures, demonstrating superior performance in identifying plant species ailments. Among them, we have considered VGG 16 [19], AlexNet [20], ResNet [21], and DenseNet 121 [21].

3.11. Visual Geometry Group (VGG)

The VGG16 architecture is a deep convolutional neural network (CNN) that was originally developed for image classification. It has been shown to be very effective for a variety of image classification tasks, including plant disease detection. The VGG16 architecture consists of 16 layers, as shown in Figure 3, each of which is a convolutional layer or a max pooling layer. The convolutional layers extract features from the input image, and the max pooling layers downsample the features to reduce the size of the model. The fully connected layers at the end of the network classify the features into different classes. The VGG16 architecture has been shown to be very effective for plant disease detection. In a study published in 2018, the VGG16 architecture achieved an accuracy of 93.5% on a dataset of tomato and potato plant images.

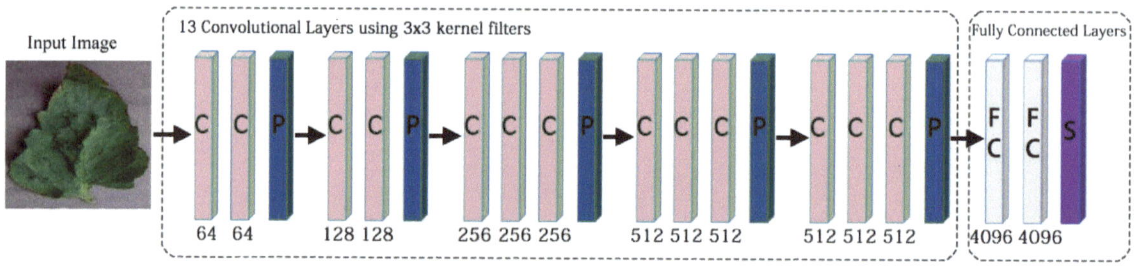

Figure 3. VGG Architecture.

3.12. Custom Model

This model is our own created model, as shown in Figure 4. This model has outperformed most of the models we have created with performance that is the same as VGG16.

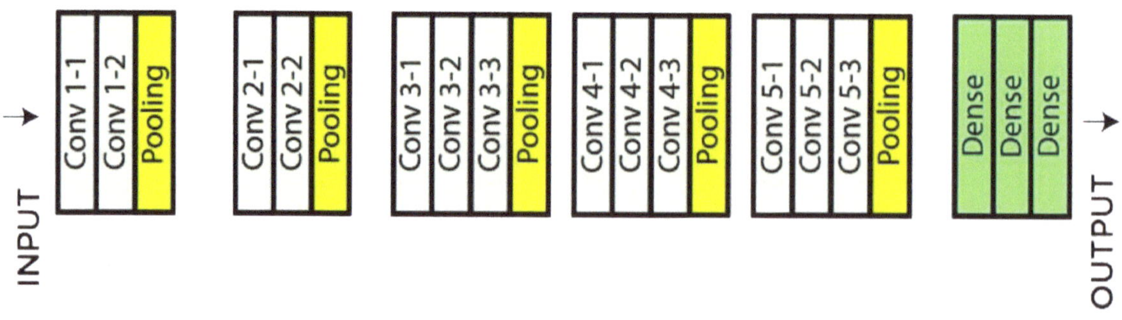

Figure 4. Custom Model Architecture.

- The first layer has 16 filters and a kernel size of 3 × 3.
- These layers have an average pooling of 2 × 2.
- The second and third layer has 32 filters and a kernel size of 3 ×3.
- These layers have an average pooling of 2 × 2.
- The fourth to sixth layer has 64 filters and a kernel size of 3 × 3.
- These layers have an average pooling of 2 × 2.
- The seventh to tenth layer has 512 filters and a kernel size of 3 × 3.
- These layers have an average pooling of 2 × 2.
- The first Dense layers have units of 256.
- The second Dense layers have units of 64.
- The third Dense layers have units of 32.

With total layers, ten layers, and batch normalization.

4. Results

In this research, we have built different deep learning models and compared the accuracies to obtain the most suitable model.

4.1. Visual Geometry Group 16

The VGG16 model has performed very well. It has a validation accuracy of 99 and a test accuracy of 98. The model's performance is very good, but if we compare it to other models that we have used in this research, then almost all of them have outperformed the VGG16. The VGG16 has performed very poorly on the merged data set, with an accuracy of just 92%. Table 1 shows the model is fighting to maintain constant validation accuracy. The training accuracy and loss are significantly better than the validation accuracy and loss. However, the validation accuracy on the merged data is just 92.06, which is a very bad performance.

Table 1. Visual Geometry Group 16 individual and Merged Data.

Visual Geometry Group 16 Individual Data				Visual Geometry Group 16 Merged Data			
Accuracy	Loss	Val Accuracy	Val Loss	Accuracy	Loss	Val Accuracy	Val Loss
98.96	0.0323	98.77	0.0448	99.41	0.0199	92.29	0.6432
98.76	0.0382	97.88	0.0709	99.80	0.0094	92.24	0.6590
98.90	0.0380	99.15	0.0295	99.95	0.0015	91.95	0.6751
99.24	0.0271	97.95	0.0784	99.82	0.0058	92.11	0.6922
99.32	0.0200	98.10	0.0618	99.91	0.0037	92.06	0.6915

Figure 5 shows the convergence of the model. The convergence rate is not that good because the model looks stuck at the elbow rules, which state that the model should be stopped from where it starts the elbow shape. In the test result, the model successfully predicted 90% of the test subjects correctly. Only one prediction out of nine is incorrect, as shown in Figure 6. The overall confidence that the model has shown is 98 to 100%.

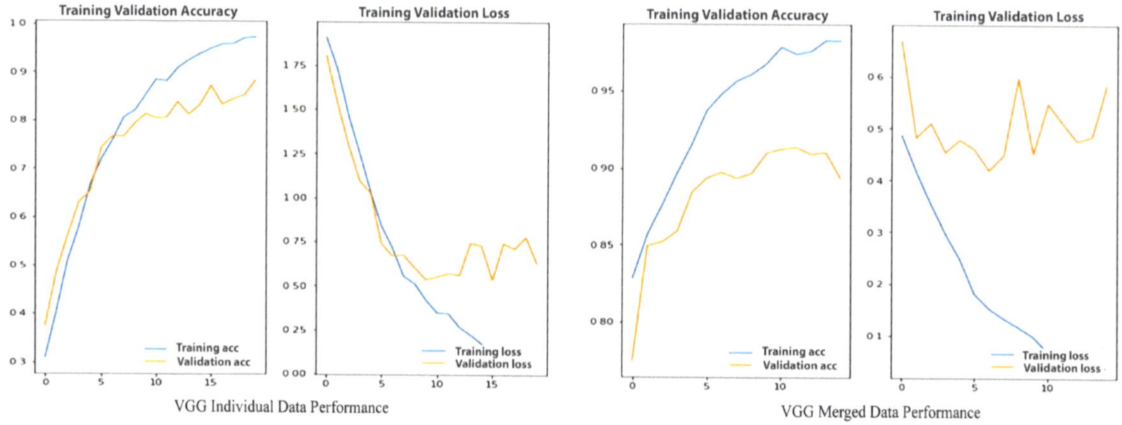

Figure 5. VGG Individual and Merged Data Performance.

Figure 6. VGG Individual and Merged Data Results.

4.2. Custom Model

This is our custom-built model, and it has performed well. This model has outperformed the VGG16 model, but it is still not the best we have used in this research. Other models have outperformed this model. The positive point of this model is that it has a very simple network architecture and gives very good results in very little time at less computation cost. The model has a constant 100 training accuracy, and the validation accuracy changes, but this model has a more stable performance than VGG16. The model validation accuracy is very stable; the only drop occurred at the end. The validation loss performance is one of the best, as shown in Table 2.

Table 2. Custom Model Individual and Merged Data.

Custom Model Individual Data				Custom Model Merged Data			
Accuracy	**Loss**	**Val Accuracy**	**Val Loss**	**Accuracy**	**Loss**	**Val Accuracy**	**Val Loss**
100	1.7998×10^{-4}	98.78	0.0483	95.86	0.1574	70.52	1.4740
100	1.2213×10^{-4}	98.78	0.0481	99.72	0.0110	79.61	1.0076
100	9.4443×10^{-5}	98.82	0.0486	99.96	0.0032	99.01	0.0328
100	7.4930×10^{-5}	98.89	0.0497	100	0.0012	99.11	0.0355
100	6.0184×10^{-5}	98.75	0.0487	100	9.1173×10^{-4}	99.22	0.0289

These are the performance graphs of the custom model. The graph has converged fast, but the stability is not that good compared to the other models we have used in this research. As you can see, the training accuracy and loss are very stable, but the validation accuracy and loss change rapidly. However, the model performed better on the merged dataset, as shown in Figure 7.

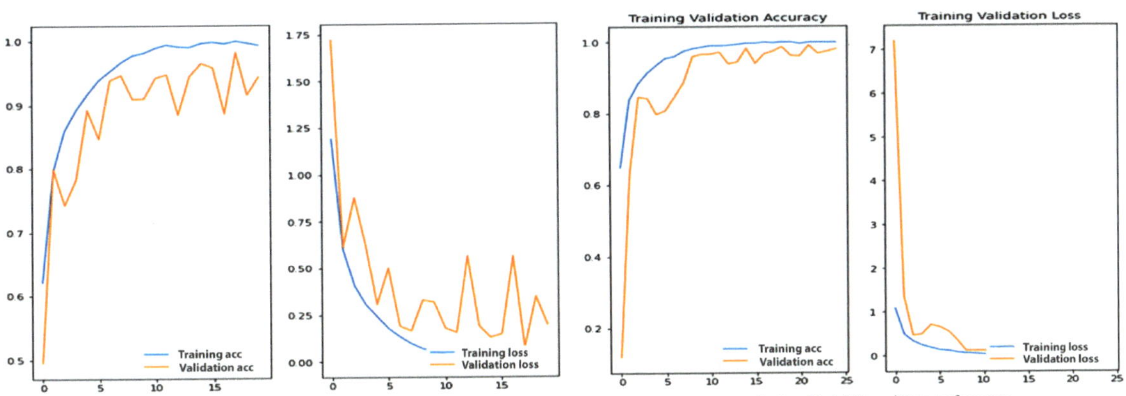

Custom Model Individual Data performance

Custom Model Merged Data performance

Figure 7. Custom Model Individual and Merged Data performance.

The results of nine images that we have taken from the test data are shown below. The model has successfully predicted all the diseases accurately. The confidence of the model is averaging almost 99% except for the on-test subject, where the model confidence is very low at 50%, but the prediction is correct on the individual data while keeping the accuracy and confidence to almost 100% on the merged data from Figure 8.

Figure 8. Custom Model Individual and Merged Data Results.

4.3. Residual Net

The best-performing model we have used in this research so far has an accuracy of more than 99.50%. ResNet is the most stable model we have used. It has the best validation accuracy of 99.44% and the best test accuracy of 99.78% on individual data. On the merged dataset, ResNet50 has a very good performance with 99.34% test accuracy and is just behind Dense20, as shown in Table 3.

Table 3. Residual Net Individual and Merged Data.

ResNet Individual Data				**ResNet Merged Data**			
Accuracy	**Loss**	**Val Accuracy**	**Val Loss**	**Accuracy**	**Loss**	**Val Accuracy**	**Val Loss**
99.99	5.2742×10^{-4}	99.44	0.0196	100	1.6210×10^{-4}	99.48	0.0282
100	2.0346×10^{-4}	99.44	0.0191	100	1.6100×10^{-4}	99.43	0.0286
100	1.5636×10^{-4}	99.44	0.0191	100	1.5992×10^{-4}	99.45	0.0285
100	1.3528×10^{-4}	99.44	0.0190	100	1.5885×10^{-4}	99.48	0.0285
100	1.2024×10^{-4}	99.44	0.0192	100	1.5780×10^{-4}	99.45	0.0284

The test and validation accuracy and loss are shown here. Using these graphs, we can demonstrate the performance of the model, as shown in Figure 9. The model fully converged at epoch 4, which is extremely fast. The ResNet model has the best accuracy, as we can see in Figure 10, where the model has successfully identified all the diseases correctly. Mostly, the confidence of the model in these test cases is 100%, and the average confidence is more than 99%.

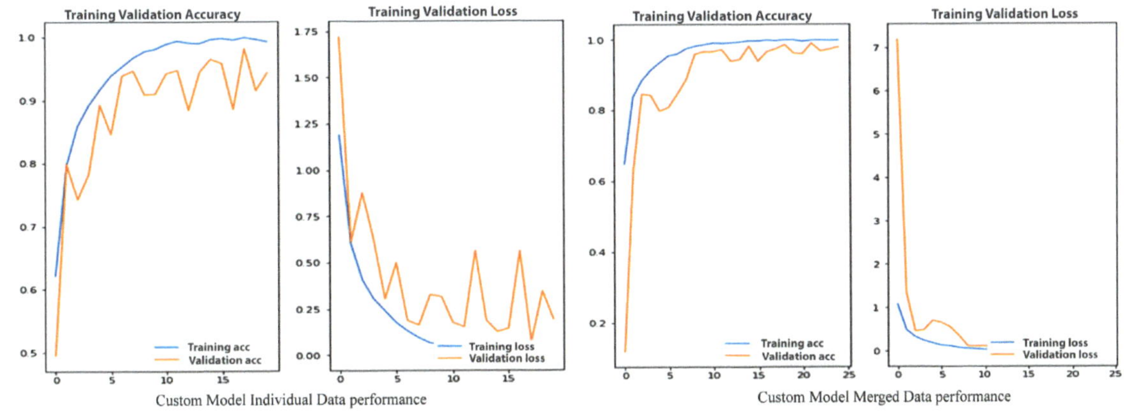

Figure 9. ResNet Individual and Merged Data Performance.

Figure 10. ResNet Individual and Merged Data Results.

4.4. DenseNet121

DenseNet121's performance is also very good. This model has outperformed all the models we have used in this research; it just falls behind ResNet. The DenseNet model has a very good accuracy of 99.55% on the individual dataset. The model has not performed as well on the merged dataset. The model has a very stable validation and training accuracy and loss performance at each epoch. The model is converging, and the accuracy is better. We have stopped the models at this point because the convergence rate has dropped very low; thus, we will obtain a very low-performance gain at a high cost, as shown in Table 4.

Table 4. DenseNet121 Individual and Merged Data.

DenseNet 121 Individual Data				DenseNet121 Merged Data			
Accuracy	Loss	Val Accuracy	Val Loss	Accuracy	Loss	Val Accuracy	Val Loss
98.26	0.0547	98.96	0.0319	98.75	0.0420	99.40	0.0207
98.69	0.0439	99.27	0.0277	99.05	0.0328	99.43	0.0219
99.00	0.0348	99.24	0.0225	99.22	0.0274	99.45	0.0207
99.10	0.0300	99.31	0.0211	99.46	0.0213	99.37	0.0218
99.25	0.0272	99.41	0.0203	99.46	0.0193	99.24	0.0209

Figure 11 shows the training validation accuracy and loss graph. The model has a very stable graph, but the problem is that the convergence drops very quickly on both datasets, as shown in the figures below.

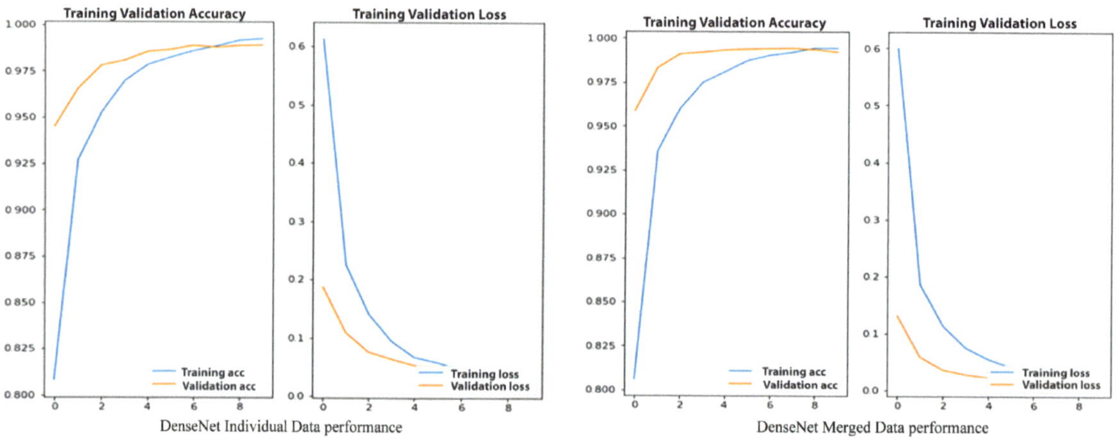

Figure 11. DenseNet121 Individual and Merged Data Performance.

The DenseNet 121 model has the second-best accuracy, as we can see in Figure 12, where the model has successfully identified all the diseases correctly. The confidence of the model in these test cases gives an average confidence of more than 99%.

4.5. DenseNet169

We have achieved a test accuracy of 99.78%, the same accuracy as ResNet50. The DenseNet169 performance was one of the best in this research. We believe that ResNet has performed a little better than DenseNet169 because the test accuracy is the same for both models, but the training and validation accuracy of ResNet50 is better, and also the parameters are fewer in ResNet. As we can see, models are very stale. The model has been almost fully trained because there are no more significant improvements, as shown in Table 5 below. The model has fully converged, as shown in Figure 13, and does not need any more epochs.

Figure 12. DenseNet121 Individual and Merged Data Results.

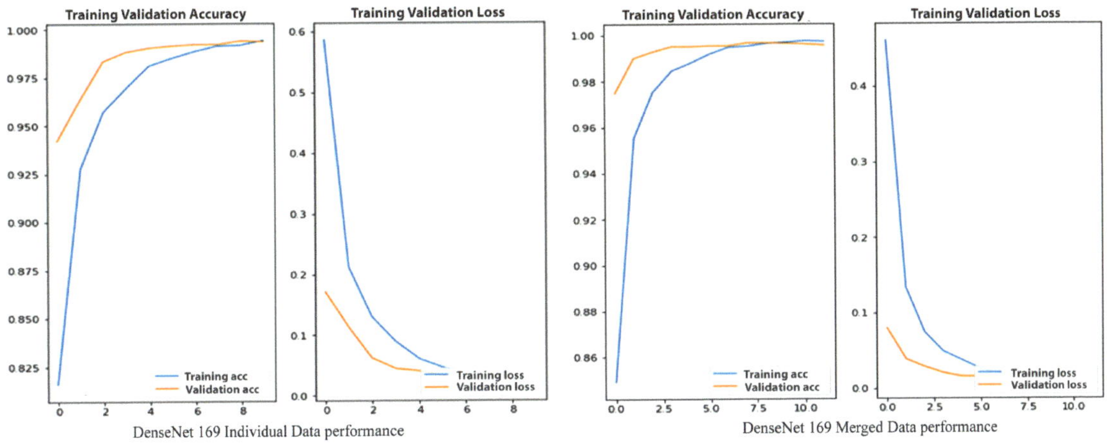

Figure 13. DenseNet169 Individual and Merged Data Performance.

Table 5. DenseNet169 Individual and Merged Data.

Dense Net 169 Individual Data				DenseNet169 Merged Data			
Accuracy	Loss	Val Accuracy	Val Loss	Accuracy	Loss	Val Accuracy	Val Loss
98.49	0.0470	99.13	0.0331	99.57	0.0184	98.95	0.0346
98.84	0.0371	99.20	0.0280	99.65	0.0161	98.87	0.0388
99.12	0.0296	99.20	0.0283	99.61	0.0141	98.87	0.0374
99.15	0.0266	99.37	0.0246	99.75	0.0112	98.98	0.0368
99.40	0.0205	99.34	0.0273	99.84	0.0092	99.02	0.0358

DenseNet169 has successfully identified all of the nine test subjects correctly. The confidence in the model is very high, averaging 99%. Only for one subject is the model confidence at 97%, which is below the average, but the confidence is almost 100% in most of the other subjects in Figure 14.

4.6. DenseNet201

DenseNet201 is a type of DenseNet model series. In this research, the models have performed well with 99.78% training accuracy, 99.33% validation accuracy, and 99.67% test accuracy. There are only two other models that have performed better than DensNet201 in this research. Table 6 shows that the training accuracy and loss are improving at each epoch, but the main issue is that the validation accuracy and loss are not improving. For this reason, we have ended the model training.

Figure 14. DenseNet169 Individual and Merged Data Results.

Looking at these accuracy and loss graphs, we can see that the model has stopped the convergence. By using these graphs, we can understand that the model cannot improve anymore, and further training can cause overfitting and gradient vanishing, as shown in

Figure 15. These are some of the results that we have taken from testing the model. The model has successfully identified all the diseases with almost 100% confidence. There are no false–true or true–false predictions, but on more testing data, this model does not have the best accuracy, as shown in Figure 16.

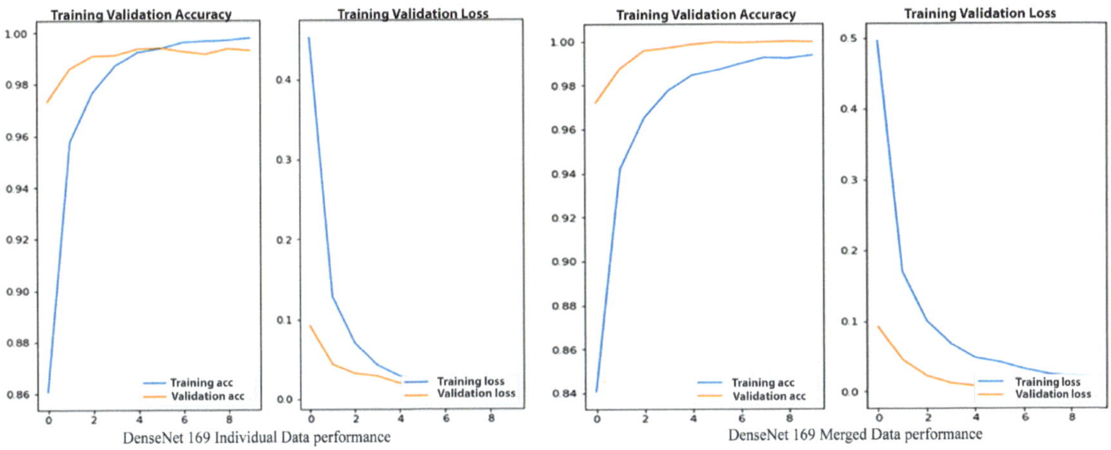

Figure 15. DenseNet201 Individual and Merged Data Performance.

Figure 16. DenseNet201 Individual and Merged Data Results.

Table 6. DenseNet201 Individual and Merged Data.

Dense Net 201 Individual Data				DenseNet201 Merged Data			
Accuracy	Loss	Val Accuracy	Val Loss	Accuracy	Loss	Val Accuracy	Val Loss
99.38	0.0240	99.41	0.0201	98.69	0.0418	99.97	0.0050
99.61	0.0171	99.27	0.0235	98.99	0.0322	99.95	0.0036
99.66	0.0148	99.17	0.0233	99.27	0.0250	99.97	0.0021
99.70	0.0131	99.37	0.0191	99.24	0.0229	100	0.0019
99.78	0.0098	99.31	0.0211	99.37	0.0204	99.97	0.0018

5. Comparative Analysis and Discussions

Unlike most prior research using transfer learning on diverse plant data, our study's targeted focus on tomatoes enabled the development of a custom model that delivers superior accuracy while simultaneously minimizing both model complexity and computational requirements. This research investigated the use of deep learning for accurate and fast tomato plant disease identification. Existing convolutional neural networks (CNNs) like VGG, ResNet, and DenseNet were evaluated for their performance, and a custom deep learning model was developed and tested. The custom model achieved over 99% accuracy in identifying 10 tomato plant diseases (including healthy plants). This high accuracy was achieved with less training time and lower computational cost compared to other CNNs. This research demonstrates the potential of deep learning for efficient and accurate tomato plant disease detection, which can benefit farmers and contribute to improved agricultural production. The custom model's efficient performance makes it promising for practical implementation in real-world agricultural settings. The research used large datasets of tomato plant images, which helped achieve high accuracy. Different deep learning and machine learning models have been developed in this study. This stage involves comparing all of the models. Here, we have to look at the models that were employed in this study, ranked by performance from worst to best. Tables 7 and 8 show us the model, no. of layers, the optimizer, the no. of parameters, the no. of epochs used for training the model, the training, testing, and validation accuracies and loss for the signal dataset that we have used.

Table 7. Comparison Table with Individual Data.

Model	Optimizer	Layers	Parameters	No. of Epochs	Training Accuracy	Training Loss	Validation Accuracy	Validation Loss	Test Accuracy
VGG16	Adamax	16	27,514,698	30	99.32%	0.0200	98.10%	0.0618	98.79%
Custom	Adamax	10	8,598,090	30	100%	6.0618×10^5	98.75%	0.0487	99.22%
ResNet 50	SGD	50	6,315,018	10	100%	1.202×10^4	99.44	0.0203	99.78%
DenseNet121	SGD	121	51,914,250	10	99.25%	0.0272	99.41%	0.0203	99.55%
DenseNet169	SGD	169	84,028,170	05	99.15%	0266	99.37%	0.0246	99.78%
DenseNet201	SGD	201	96,873,738	05	99.78%	0098	99.31%	0.0211	99.67%

Table 8. Comparison Table of Merged Data.

Model	Optimizer	Layers	Parameters	Training Accuracy	Training Loss	Validation Accuracy	Validation Loss	Test Accuracy
VGG16	Adamax	16	27,514,698	99.91%	0.0037	92.06%	0.6915	92.06%
Custom	Adamax	14	8,598,090	100%	9.1173×10^{-4}	99.22%	0.0289	99.24%
ResNet 50	SGD	50	6,315,018	100%	1.5780×10^{-4}	99.45%	0.0284	99.34%
Dense Net121	SGD	121	51,914,250	99.46%	0.0193	99.24%	0.0209	98.93%
Dense Net169	SGD	169	84,028,170	99.84%	0.0092	99.02%	0.0358	98.87%
Dense Net201	SGD	201	96,873,738	99.37%	0.0204	99.97%	0.0018	100%

5.1. Comparison Table of Individual Data

For each model created using a specific dataset, the accuracy and loss are shown in Table 7 for training, testing, and validation. ResNet50 and DenseNet169 have the highest test accuracies, according to the table, but ResNet50 is more effective. The VGG16 also has the lowest level of accuracy overall.

5.2. Comparison Chart of Individual Data

Here is a graph that shows how the models have performed. The DenseNet201 model is the best model, followed by ResNet50 in this study, as shown by the graphic. Figure 17 shows that ResNet50 has the highest overall training validation and test accuracy.

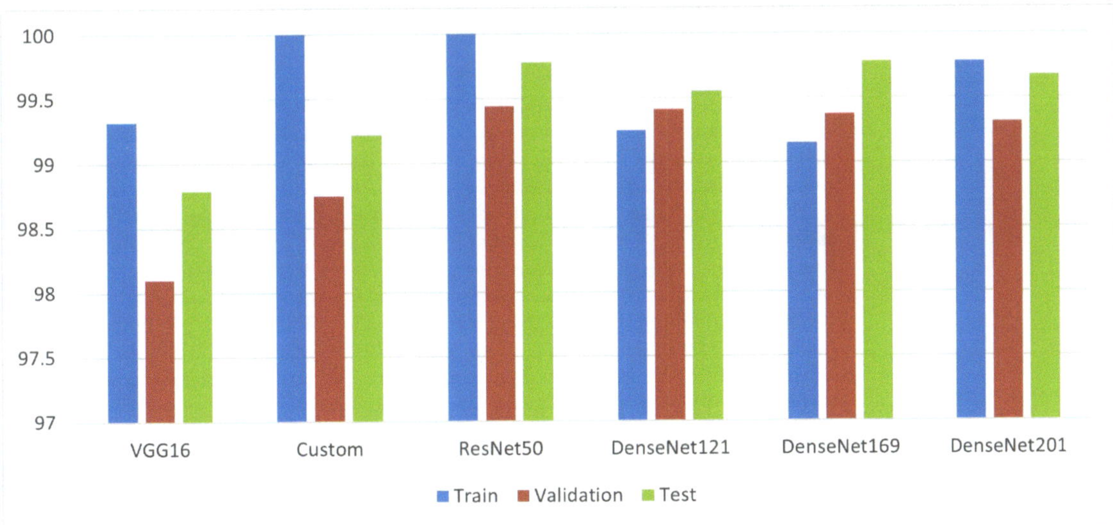

Figure 17. Comparison Chart of Individual Data.

5.3. Comparison Table of Merged Data

For all of the models created using the combined dataset, the training, testing, and validation accuracy and loss are shown in Table 8. According to the table, DenseNet201 and DenseNet169 have the best test accuracies, but DenseNet201 is superior. Additionally, the VGG16 once more has the lowest overall accuracy.

5.4. Comparison Chart of Merged Data

The graph below demonstrates that, except for the VGG16, most models have fared extremely well. With Dense201, the ResNet50 has performed best, and our model is closely following, as shown in Figure 18.

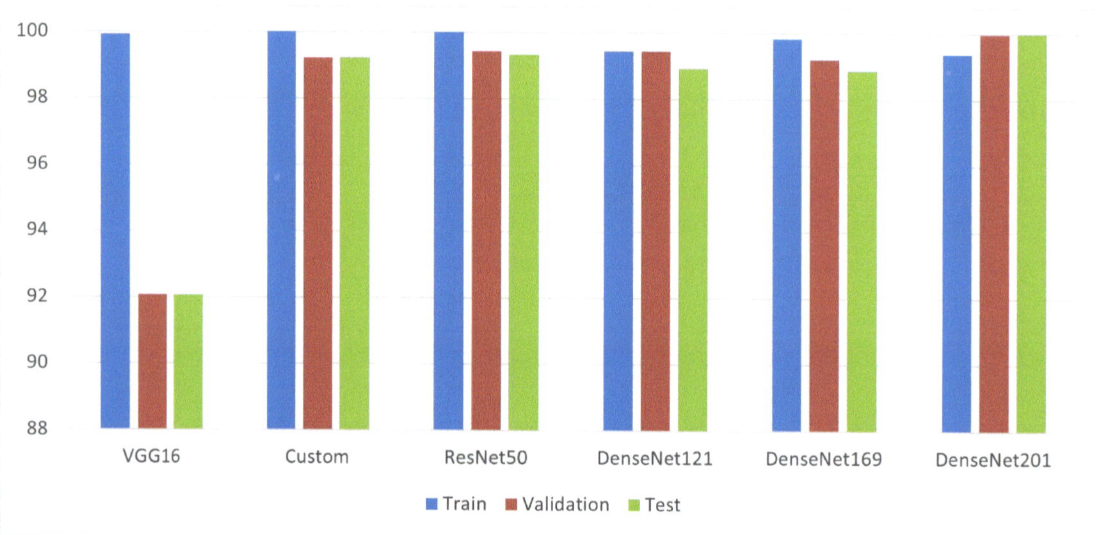

Figure 18. Comparison Chart of Merged Data.

6. Conclusions and Future Work

In this research, we have used different CNN architectures developed using different datasets. When the results were compared, we concluded that all of the DenseNet models had performed better, but in comparison to both datasets, DenseNet201 has extraordinary results, followed by DenseNet169 and ResNet50. However, our own deployed model also has accuracies of over 99% on both datasets, which is a strong reason that this model can also be used for plant disease detection. In this research, we have discussed how we can detect and recognize plant disease using its leaves and treatment for these plant diseases. We have used the ResNet deep learning model and the decision tree classifier model using a Python programming language. We have achieved very high accuracy in both models. The problems we faced related to our system were data gathering and model selection. There is a lot of data available, but most of it is outdated or false data. There are also a lot of machine learning and deep learning models, and we had to build and select the best model. Models like VGG, DenseNet, and ResNet are very computationally expensive and time-consuming because we have to train on a lot of parameters. This model will be implemented on a drone device along with the object detection model, which will detect the tomato leaf, and this model will detect the disease in the tomato leaf. After the disease is detected, the system will automatically prescribe the specified spray for the detected disease.

Author Contributions: S.U.I. conceived the presented idea. S.U.I. designed and developed the model and the computational framework and analyzed the data. S.Z. and G.H. proposed training and testing the model on the merged dataset and verifying the analytical methods. V.P. encouraged the investigation of CNN architectures and supervised the findings of this work. G.F. and G.S. review this work. All authors discussed the results and contributed to the final manuscript. All authors have read and agreed to the published version of the manuscript.

Funding: This research received no external funding.

Data Availability Statement: The data underlying the results presented in the study are available within the manuscript.

Conflicts of Interest: The authors declare that they have no conflicts of interest to report regarding the present study.

References

1. Punitha, K.; Shrivastava, G. Review on Emerging Trends in Detection of Plant Diseases using Image Processing with Machine Learning. *Int. J. Comput. Appl.* **2021**, *174*, 39–48.
2. Andreas, K.; Prenafeta-Boldú, F.X. A review of the use of convolutional neural networks in agriculture. *J. Agric. Sci.* **2018**, *156*, 312–322.
3. Sukhvir, K.; Pandey, S.; Goel, S. Plants disease identification and classification through leaf images: A survey. *Arch. Comput. Methods Eng.* **2019**, *26*, 507–530.
4. Kumar, S.S.; Raghavendra, B.K. Diseases detection of various plant leaf using image processing techniques: A review. In Proceedings of the 5th International Conference on Advanced Computing & Communication Systems (ICACCS), Coimbatore, India, 15–16 March 2019; pp. 313–316.
5. Arnal, B.J.G. Impact of dataset size and variety on the effectiveness of deep learning and transfer learning for plant disease classification. *Comput. Electron. Agric.* **2018**, *153*, 46–53.
6. Devanur, G.; Mallikarjuna, P.B.; Manjunath, S. Segmentation and classification of tobacco seedling diseases. In Proceedings of the Fourth Annual ACM Bangalore Conference, Bangalore, India, 25–26 March 2011; pp. 1–5.
7. Konstantinos, F. Deep learning models for plant disease detection and diagnosis. *Comput. Electron. Agric.* **2018**, *145*, 311–318.
8. Hammad, S.M.; Potgieter, J.; Arif, K.M. Plant disease classification: A comparative evaluation of convolutional neural networks and deep learning optimizers. *Plants* **2020**, *9*, 1319.
9. Punitha, K.; Shrivastava, G. Analytical study of deep learning architectures for classification of plant diseases. *J. Algebr. Stat.* **2022**, *13*, 907–914.
10. Amarjeeth, S.; Jaisree, H.; Aishwarya, D.; Jayasree, J.S. Plant disease detection and diagnosis using deep learning. In Proceedings of the International Conference for Advancement in Technology (ICONAT), Goa, India, 21–22 January 2022; pp. 1–6.
11. Vishakha, K.; Munot, M. Deep learning models for tomato plant disease detection. In *Advanced Machine Intelligence and Signal Processing*; Springer: Singapore, 2022; pp. 679–686.
12. Khalil, K.; Khan, R.U.; Albattah, W.; Qamar, A.M. End-to-end semantic leaf segmentation framework for plants disease classification. *Complexity* **2022**, *55*, 1076–2787.
13. Wadadare, S.S.; Fadewar, H.S. Deep learning convolution neural network for tomato leaves disease detection by inception. In *Applied Computational Technologies: Proceedings of ICCET*; Springer: Singapore, 2022; pp. 208–220.
14. Chen, H.-C.; Widodo, A.M.; Wisnujati, A.; Rahaman, M.; Lin, J.C.-W.; Chen, L.; Weng, C.-E. AlexNet convolutional neural network for disease detection and classification of tomato leaf. *Electronics* **2022**, *11*, 951. [CrossRef]
15. Khan, M.A.; Alqahtani, A.; Khan, A.; Alsubai, S.; Binbusayyis, A.; Ch, M.M.I.; Yong, H.-S.; Cha, J. Cucumber leaf diseases recognition using multi-level deep entropy-ELM feature selection. *Appl. Sci.* **2022**, *12*, 593. [CrossRef]
16. Yiting, X.; Plett, D.; Liu, H. Detecting crown rot disease in wheat in controlled environment conditions using digital color imaging and machine learning. *AgriEngineering* **2022**, *4*, 141–155.
17. Kaur, P.; Harnal, S.; Tiwari, R.; Upadhyay, S.; Bhatia, S.; Mashat, A.; Alabdali, A.M. Recognition of leaf disease using hybrid convolutional neural network by applying feature reduction. *Sensors* **2022**, *22*, 575. [CrossRef] [PubMed]
18. Almadhor, A.; Rauf, H.T.; Lali MI, U.; Damaševičius, R.; Alouffi, B.; Alharbi, A. AI-driven framework for recognition of guava plant diseases through machine learning from DSLR camera sensor based high resolution imagery. *Sensors* **2021**, *21*, 3830. [CrossRef] [PubMed]
19. Applalanaidu, M.V.; Kumaravelan, G. A review of machine learning approaches in plant leaf disease detection and classification. In Proceedings of the Third International Conference on Intelligent Communication Technologies and Virtual Mobile Networks (ICICV), Tirunelveli, India, 4–6 February 2021; pp. 716–724.

20. Satheesh, S.B.P.; Naidu, P.R.S.; Neelima, U. Prediction of guava plant diseases using deep learning. In Proceedings of the 3rd International Conference on Communications and Cyber Physical Engineering (ICCCE), Hyderabad, India, 9–10 April 2021; pp. 1495–1505.
21. Murat, T.; Arslan, R.S. RT-Droid: A novel approach for real-time android application analysis with transfer learning-based CNN models. *J. Real-Time Image Process.* **2023**, *20*, 55.

AgriEngineering

Article

A Transfer Learning-Based Deep Convolutional Neural Network for Detection of Fusarium Wilt in Banana Crops

Kevin Yan [1,*], Md Kamran Chowdhury Shisher [2] and Yin Sun [2,*,†]

1 Auburn High School, 1701 E Samford Ave, Auburn, AL 36830, USA
2 Department of Electrical and Computer Engineering, Auburn University, Broun Hall, 341 War Eagle Way, Auburn, AL 36849, USA; mzs0153@auburn.edu
* Correspondence: kyan7472@gmail.com (K.Y.); yzs0078@auburn.edu (Y.S.); Tel.: +1-334-524-1842 (K.Y.)
† Current address: Department of Electrical and Computer Engineering, Auburn University, 345 W Magnolia Ave, Auburn, AL 36849, USA

Abstract: During the 1950s, the Gros Michel species of bananas were nearly wiped out by the incurable Fusarium Wilt, also known as Panama Disease. Originating in Southeast Asia, Fusarium Wilt is a banana pandemic that has been threatening the multi-billion-dollar banana industry worldwide. The disease is caused by a fungus that spreads rapidly throughout the soil and into the roots of banana plants. Currently, the only way to stop the spread of this disease is for farmers to manually inspect and remove infected plants as quickly as possible, which is a time-consuming process. The main purpose of this study is to build a deep Convolutional Neural Network (CNN) using a transfer learning approach to rapidly identify Fusarium wilt infections on banana crop leaves. We chose to use the ResNet50 architecture as the base CNN model for our transfer learning approach owing to its remarkable performance in image classification, which was demonstrated through its victory in the ImageNet competition. After its initial training and fine-tuning on a data set consisting of 600 healthy and diseased images, the CNN model achieved near-perfect accuracy of 0.99 along with a loss of 0.46 and was fine-tuned to adapt the ResNet base model. ResNet50's distinctive residual block structure could be the reason behind these results. To evaluate this CNN model, 500 test images, consisting of 250 diseased and healthy banana leaf images, were classified by the model. The deep CNN model was able to achieve an accuracy of 0.98 and an F-1 score of 0.98 by correctly identifying the class of 492 of the 500 images. These results show that this DCNN model outperforms existing models such as Sangeetha et al., 2023's deep CNN model by at least 0.07 in accuracy and is a viable option for identifying Fusarium Wilt in banana crops.

Keywords: convolutional neural network; Fusarium wilt; transfer learning; ResNet-50; banana crop

Citation: Yan, K.; Shisher, M.K.C.; Sun, Y. A Transfer Learning-Based Deep Convolutional Neural Network for Detection of Fusarium Wilt in Banana Crops. *AgriEngineering* **2023**, *5*, 2381–2394. https://doi.org/10.3390/agriengineering5040146

Academic Editors: Chiew Foong Kwong and Ray E. Sheriff

Received: 23 September 2023
Revised: 3 December 2023
Accepted: 6 December 2023
Published: 11 December 2023

1. Introduction

Banana (*Musa* spp.) is one of the most widely produced cash crops in the tropical regions of the world and the fourth most important crop among developing nations. Over 130 countries export bananas, contributing to a total revenue of 50 billion dollars per year [1]. Fusarium wilt or Panama disease is caused by the *Fusarium oxysporum* f. sp. cubense tropical race 4 (TR4), and is a well-known threat to global banana production [2,3]. TR4 infected hundreds of thousands of hectares of banana plantations throughout countries like China, India, the Philippines, Australia, and Mozambique [4].

Four to five weeks after inoculation with Foc, banana crops begin to exhibit the main symptom of Fusarium wilt, yellowing of their leaves. TR4 can spread through flowing water, farm equipment, infected plant material, and soil contamination. Approximately four to five weeks after being inoculated with TR4, banana crops start displaying the primary symptom of Fusarium wilt: the yellowing of their leaves. These once-vibrant leaves gradually begin to droop and eventually collapse, forming a ring of lifeless foliage encircling the pseudo-stem of the crop [5]. Over time, an increasing number of leaves experience wilting and collapse, resulting in the entire canopy of the crop being composed solely of withering and deceased leaves.

Currently, the main method used to manage Fusarium wilt is to inspect banana plantations with manual labor in hopes of identifying these yellowing leaves. However, this process requires enormous amounts of time and money to perform [6]. Also, manual inspection quality is affected by the experience and expertise of the farmer performing the inspection, meaning accuracy of diagnosis cannot be ensured [7]. Additionally, there are currently no chemical or physical treatments available that can effectively control Fusarium Wilt. Once the signs of this disease are identified, the only viable treatment option is the rapid removal of the crop in order to prevent a large-scale infection from occurring [8].

Alternatively, Remote sensing has been used in recent studies to detect the presence of Fusarium wilt. For example, a DJI Phantom 4 quadcopter (DJI Innovations, Shenzhen, China) equipped with a MicaSense RedEdge M^{TM} five-band multi-spectral camera to perform remote sensing surveys of banana plantations in China [6]. The multi-spectral images were then used to calculate different vegetation indices, attempting to find the infection status of Fusarium wilt in the plantation. The same approach was used in [9] to detect Fusarium wilt on banana plantations.

The main drawback to these remote sensing detection systems, despite their ability to detect Panama disease, is similar to that of conducting manual inspections: the high cost. The high-quality multispectral and hyperspectral cameras required by these types of detection systems can range in cost from several thousands of dollars to tens of thousands of dollars. Given that Fusarium wilt infections mostly occur in developing countries on the continents of South America, Africa, and South Asia [10], it is clear that farmers in those regions cannot afford high-quality spectral cameras and the computational power used to analyze spectral images.

Presently, convolutional neural networks have been used to detect many different plant diseases through their symptoms within the visible light spectrum. In computer vision, deep convolutional neural networks (CNNs) can achieve excellent performance in image classification tasks [11]. CNNs are a variant of deep neural networks that are designed to mimic the cognitive process of human vision. CNNs receive an input, usually in the form of an image, which is then fed through layers of neurons that perform nonlinear operations, lastly, the output is in the form of a list of scores between 0 and 1, each of which represents the likelihood of the image belonging to an image class. The nonlinear operations at these neurons are optimized through a training procedure [12].

Transfer learning is a technique used in the design of deep learning models to reduce the need for large training datasets and high computational costs for training. Transfer learning essentially works by integrating the knowledge of a previously trained CNN model into a new CNN model designed for a specific task [13]. Transfer learning approaches have been used in plant classification, sentiment classification, software defect prediction and more [14]. There are several pre-trained models that can be selected as the base model for transfer learning, such as ResNet, AlexNet, Inception V-3, VGG16, and ImageNet. In [11], the authors evaluated the aforementioned four pre-trained models in a transfer learning-based plant disease detection task. It was found that ResNet-50 is the best model and achieves an accuracy of 0.9980.

In this paper, we present a transfer learning-based deep CNN model for the detection of Fusarium Wilt in banana crops. The key contributions of this paper are summarized as follows:

- We employed ResNet-50 as the base model, which is accessible in Keras with pre-trained weights on the TensorFlow backend. Originally trained to identify 1000 distinct object classes from ImageNet, the ResNet-50 architecture was adapted for our binary classification task. To do this, we substituted the original final fully connected layer of 1000 neurons with a single neuron fully connected layer to fit our binary output. Then, the ResNet-50 layers were fine-tuned to adapt the knowledge of the base model to the Fusarium wilt detection task. Our CNN model is illustrated in Figure 1.

- The training dataset was obtained from [15], consisting of 72 images of Fusarium wilt-infected banana leaves and 84 images of healthy banana leaves. To augment the dataset size and mitigate overfitting, random flips, rotations, and noise injection, were applied to each image. After training for 100 epochs, the model demonstrated near-perfect performance.
- Furthermore, for testing purposes, we created a new data set from Google searches, comprising 500 distinct images. This testing dataset will provide a robust platform for further research into Fusarium wilt detection.
- Our transfer learning-based CNN model achieved high accuracy and an F-1 score of 0.98, correctly classifying 492 out of 500 images (see Figures 2–5). We also demonstrate the impact of both transfer learning and data augmentation by evaluating our model's performance with and without these techniques. Figure 6 illustrates that our CNN model achieves an accuracy of 0.7685 without transfer learning and data augmentation, while it attains an accuracy of 0.9823 using the transfer learning and data augmentation methods.
- To evaluate the model's discriminative ability on the banana leaf image dataset, we employed a Receiver Operating Characteristic (ROC) curve (refer to Figure 7). The results, depicted by the ROC curve, affirm that our trained model can effectively distinguish between Fusarium wilt-infected and healthy banana plants.
- We conducted a performance comparison between our ResNet-50 model based CNN model and other CNN models: VGG16, AlexNet, and DenseNet121. As illustrated in Figure 8, ResNet-50 exhibited superior performance compared to these models. Therefore, our transfer learning-based ResNet-50 model emerges as a promising candidate for the detection of Fusarium wilt in banana crops.

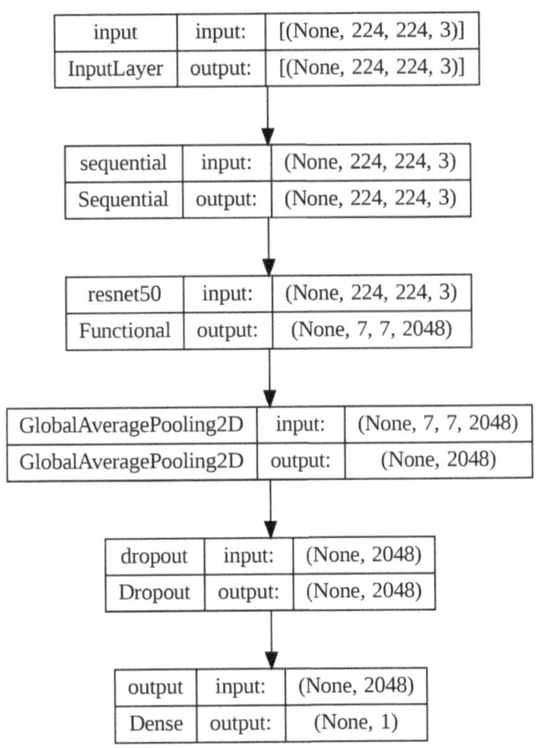

Figure 1. The diagram displays the dimensions of the inputs and outputs of each layer of our CNN model along with the names of each layer.

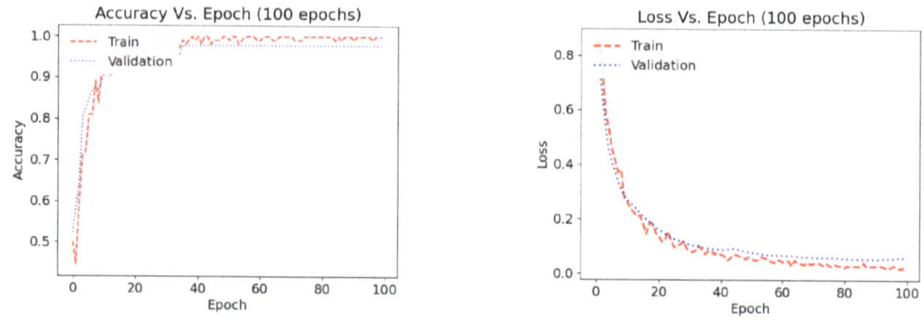

(**a**) (**b**)

Figure 2. (**a**) Accuracy (percentage of correctly predicted samples) vs. Epoch and (**b**) Loss (binary cross-entropy loss) vs. Epoch. Visualization of the learning process reveals improvement in both training and validation loss and accuracy for our CNN model with successive epochs.

```
Epoch 1/5
2/2 [==============================] - 55s 49s/step - loss: 0.0050 - binary_accuracy: 1.0000 - val_loss: 0.0232 - val_binary_accuracy: 0.9783
Epoch 2/5
2/2 [==============================] - 1s 285ms/step - loss: 0.0041 - binary_accuracy: 1.0000 - val_loss: 0.0226 - val_binary_accuracy: 0.9783
Epoch 3/5
2/2 [==============================] - 1s 280ms/step - loss: 0.0026 - binary_accuracy: 1.0000 - val_loss: 0.0205 - val_binary_accuracy: 1.0000
Epoch 4/5
2/2 [==============================] - 1s 285ms/step - loss: 0.0024 - binary_accuracy: 1.0000 - val_loss: 0.0182 - val_binary_accuracy: 1.0000
Epoch 5/5
2/2 [==============================] - 1s 298ms/step - loss: 0.0025 - binary_accuracy: 1.0000 - val_loss: 0.0169 - val_binary_accuracy: 1.0000
```

Figure 3. These outputs are the values of accuracy and loss of the training and validation sets during fine-tuning.

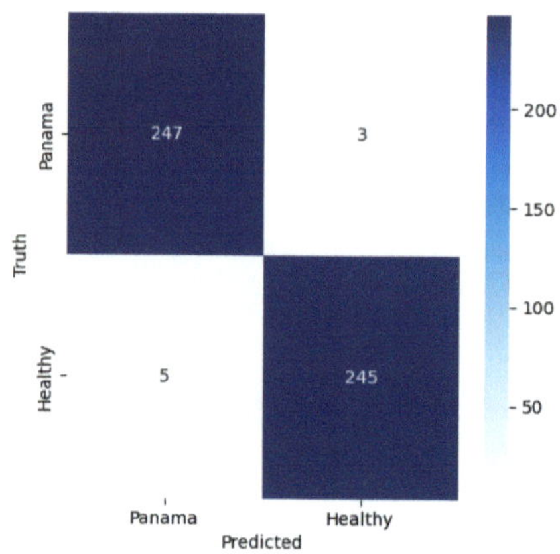

Figure 4. The confusion matrix represents the predictions made by the CNN model (500 total predictions).

	precision	recall	f1-score	support
Healthy	0.99	0.98	0.98	250
Panama	0.98	0.99	0.98	250
accuracy			0.98	500
macro avg	0.98	0.98	0.98	500
weighted avg	0.98	0.98	0.98	500

Figure 5. The classification report shown here depicts the precision, recall, F-1 score, accuracy, and the macro and weighted averages of these values.

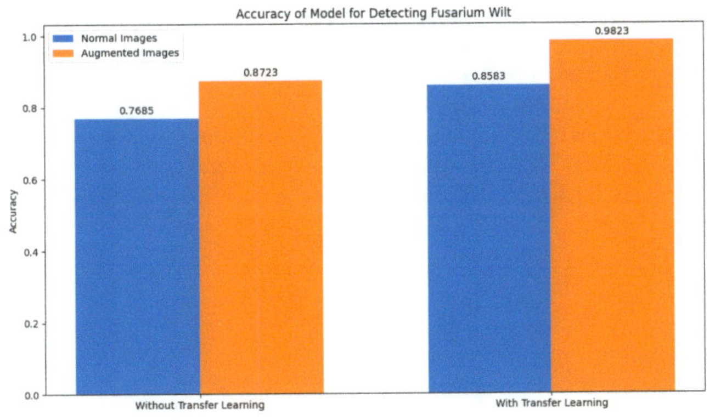

Figure 6. The graph shows the different accuracy values as a result of the different evaluation conditions of the model.

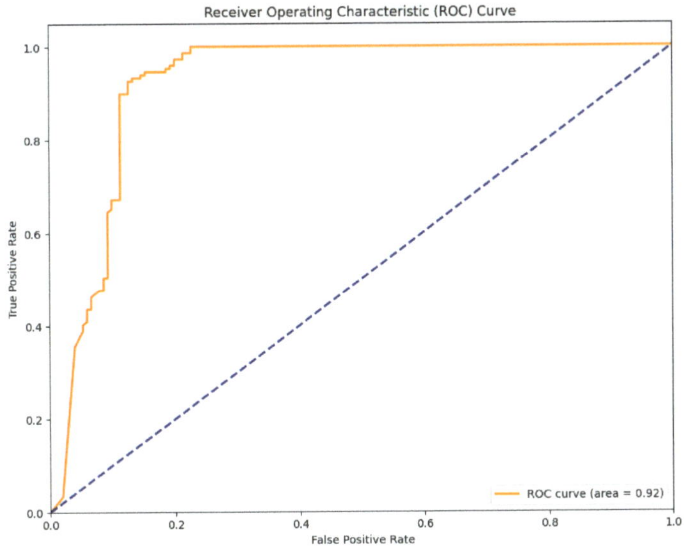

Figure 7. ROC curve for the CNN model trained on the Fusarium wilt dataset. The area under the curve (AUC) of 0.92 signifies that the model has good discriminatory power between the positive and negative classes.

1.1. Literature Review

CNNs have been employed for the detection of various plant diseases, such as soybean plant diseases [16], apple black rot, grape leaf blight, tomato leaf mold, cherry powdery mildew, potato with early blight, and bacterial spots on a peach [14]. In a study by [17], the authors developed a CNN model capable of detecting 58 classes of plant leaves from diverse crops, including aloe vera, apple, banana, cherry, citrus, corn, coffee, grape, paddy, peach, pepper, strawberry, tea, tomato, and wheat. In another work, Ref. [1] presented a deep learning-based neural network model for identifying Fusarium wilt. Their model achieved an accuracy of 0.9156, evaluated on 700 samples of diseased and healthy banana leaf images. Notably, the accuracy reported in this study surpassed that achieved in [1] by approximately 0.065, reaching 0.98. The study [6] proposed a remote sensing-based detection model for Fusarium wilt on banana plantations in China. Their model utilized changes in vegetation indices caused by Fusarium wilt for disease detection. Meanwhile,

Ref. [18] implemented the Yolo v4 neural network model in a Raspberry Pi for Fusarium wilt detection in banana leaves. The handheld system demonstrated an on-field accuracy of 0.90, offering portability and independence from internet connectivity. The paper [19] introduced a Support Vector Machine (SVM) classification framework for identifying four types of banana diseases in India, including sigatoka, cmv, bacterial wilt, and Fusarium wilt. The SVM classification achieved an average accuracy of approximately 0.85, with accuracy of 0.84, 0.86, 0.85, and 0.85 for the four diseases, respectively.

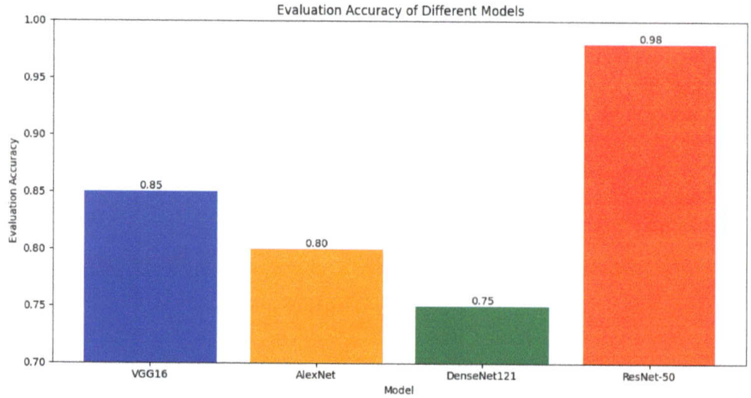

Figure 8. Accuracy comparison across different deep learning architectures. ResNet-50 demonstrates superior performance in Fusarium wilt detection.

In a recent development, Ref. [20] conducted an in-depth study leveraging the MobileNetV2 architecture, an efficient CNN model designed for mobile vision applications. In [20], the authors demonstrated the capability of lightweight neural networks when augmented with deep transfer learning techniques to achieve notable accuracy in classifying a diverse array of fruit images, including apples, oranges, and bananas. The practical implications of their study suggest the viability of implementing such models on edge devices within agricultural frameworks. Continuing the exploration of transfer learning potentials, Ref. [21] provided an extensive comparative analysis on the detection of diseases in sunflower crops. The study [21] highlighted the significant performance benefits of applying transfer learning to plant disease detection tasks. By using a CNN pre-trained on extensive datasets and subsequently fine-tuning it for sunflower disease identification, they underlined the robustness of transfer learning in achieving precise disease identification in agricultural applications. In a comprehensive review, Ref. [22] addressed the complications inherent to citrus fruit disease classification. The review [22] critically assessed the methodologies from initial image acquisition to advanced pre-processing and classification via CNN. Similarly, Ref. [23] investigated enhanced methods of data preparation. The study [23] put forth innovative annotation techniques that aid in the efficient training of CNN models for fruit classification tasks, with a specific focus on minimizing manual labeling efforts while enhancing the precision of dataset annotations. In [24], authors presented a novel CNN-based approach for seed classification. In their publication [24], they outlined the development of a CNN model adept at categorizing seeds with substantial accuracy, demonstrating the versatility of CNNs beyond disease identification and into the quality assessment and sorting of agricultural products.

1.2. Objectives

The main objectives of this study were to (a) develop and train a transfer learning-based convolutional neural network model for identifying Fusarium wilt in banana crops, (b) utilize a ResNet-50 pre-trained model as the foundational architecture, and (c) evaluate the precision of this model in Fusarium wilt identification.

2. Materials and Methods

2.1. Dataset

In [15], the authors compiled an extensive dataset featuring diverse images of banana diseases and healthy plant samples. For the purpose of training our CNN model, we utilized a subset of their data that comprised 72 images of Fusarium wilt-infected banana leaves and 84 images featuring healthy banana leaves. The dataset compiled by [15] is accessible at (https://data.mendeley.com/datasets/4wyymrcpyz/1 (accessed on 5 December 2023)) This dataset includes images that allow one to distinguish Fusarium wilt-infected banana leaves from healthy banana leaves. These images were captured at different times of the day and under varying environmental conditions, ensuring a broad range of scenarios for increased model robustness. The original images in the dataset had dimensions of 256 × 256 pixels. However, to align with the input layer dimensions of the CNN model, the images underwent resizing to a standardized size of 224 × 224 pixels. This resizing process ensured uniform dimensions for all input images, facilitating seamless processing within the CNN model.

2.2. Data Pre-Processing and Augmentation

To expand the initial dataset and address the risk of overfitting due to limited data, we employed data augmentation techniques. This involved applying random flipping, rotation, and noise addition to each image in the training dataset. The resulting augmented dataset provided the model with a more diverse and enriched sample for training, enhancing its generalization capabilities beyond the original dataset.

The dataset was divided into two separate subsets: a training dataset and a validation dataset. Specifically, 70% of the dataset was assigned to the training dataset, while the remaining 30% formed the validation dataset. By establishing a clear distinction between the training and validation datasets, the model's performance was effectively assessed and monitored during the training process. The separation facilitated a robust evaluation of the model's capacity to generalize predictions to previously unseen data, ensuring its effectiveness in detecting and classifying Fusarium wilt-infected banana leaves beyond the images it had been trained on.

2.3. Transfer Learning

Our study focused on creating a transfer learning-based CNN model specialized in detecting Fusarium wilt in banana crops. We aimed to develop a model that balances computational efficiency with ease of deployment. For this task, we selected Keras [25], a renowned high-level neural network API, in conjunction with TensorFlow [26], a leading machine learning framework. In our search for a suitable pre-trained model, we opted for ResNet-50 [27], a well-established CNN architecture in computer vision. We chose ResNet-50 for its proven effectiveness in image classification tasks and its innovative use of residual blocks to address the vanishing gradient problem in deep networks. The architecture ensures gradient information preservation across layers, enabling optimal weight learning throughout its depth.

We employed a ResNet-50 model pre-trained on the ImageNet dataset with 1000 object categories. Then, we modified its architecture by replacing the original 1000-neuron fully connected layer with a single-neuron layer tailored to our binary classification task.

While newer CNN architectures exist, many of them demand substantial computational resources, potentially hindering deployment in resource-constrained environments. ResNet-50 strikes a balanced trade-off, offering excellent performance without the hefty computational demands of more recent architectures. This aligns well with our goal of creating a model that is powerful and practical for widespread use.

2.4. Neural Network Architecture

A CNN model has a sequential structure, stacking each layer on the previous one. This configuration facilitates a systematic flow of information through the network. It also helps in extracting increasingly abstract features from the input image as it progresses through the layers. Figure 1 illustrates the architecture of our CNN model, depicting the step-by-step transformation of the input image.

2.5. Input Layer

Our model initiates with the input layer, which receives an image sized at 224 × 224 pixels with three RGB values, forming a 3-dimensional matrix. Each cell in this matrix contains a pixel value representing the intensity of the pixel for grayscale images or, for RGB images, three matrices or channels corresponding to the red, green, and blue components. Subsequently, the input image is fed into the ResNet-50 based layer.

2.6. ResNet-50 Layer

ResNet, short for Residual Network, represents a specific type of CNN introduced in [27]. ResNet-50, specifically, is a convolutional neural network with 50 layers (48 convolutional layers, one MaxPool layer, and one average pool layer). Residual neural networks are a subtype of artificial neural networks (ANNs) constructed by stacking residual blocks. The residual block is defined as:

$$y = f(x, W) + x, \tag{1}$$

where x, y, W, and $f(\cdot)$ denote the input, output, weights, and the residual function of the block, respectively. Refer to [27] for detailed information.

In Figure 1, the MaxPool layer and the average pool layer are separated from the ResNet-50 Layer, and details about these two layers are provided below.

2.7. Max Pooling Layer

After the 48 convolutional layers of ResNet-50, a max pooling layer is introduced to downsample the feature map. This process, which reduces the spatial dimensions, extracts essential features, enhancing computational efficiency. The downsizing aids in emphasizing the most salient information while discarding irrelevant details.

2.8. Global Average Pooling Layer

Once the information passes through the residual blocks and the max pooling layer, a global average pooling layer is applied. This layer reduces the spatial dimensions of the feature map by computing the average value of each feature. By summarizing information across the entire feature map, the global average pooling layer preserves the most relevant features while discarding spatial details. This process focuses on the most descriptive aspects of the image.

2.9. Dropout Layer

Following the global average pooling layer, a dropout layer is incorporated into the CNN model. This layer randomly drops out a fraction of the outputs generated by the previous layer during training, serving as a regularization technique. Its purpose is to enhance the model's ability to generalize by reducing the likelihood of overfitting.

2.10. Fully Connected Dense Layer

Finally, the CNN model concludes with a fully connected dense layer that conducts linear transformations on the input received from the preceding layers. Because our task (binary classification) is to distinguish between a healthy banana leaf and a Fusarium wilt-infected banana leaf, we include a fully connected output layer with one neuron. This neuron incorporates a sigmoid function, expressed as:

$$\sigma(\mathbf{z}) = \frac{1}{1 + e^{-(\mathbf{W}^T \mathbf{z} + b)}}, \tag{2}$$

where $\mathbf{z}$ denotes the input vector, the vector $\mathbf{W}$ contains the weights, and b is the bias.

The sigmoid Function (2) produces a value between 0 and 1. This characteristic allows the model to interpret the output as a probability, indicating the likelihood of a given input belonging to a specific class. For instance, if the output of the last layer exceeds 0.5, we predict the input as a healthy leaf; otherwise, the leaf is Fusarium wilt-infected.

3. Results

3.1. Training

Prior to training, the ResNet-50 architecture's layers were initialized with weights pre-trained on ImageNet. The weights of the last fully connected layer (i.e., the output layer) were initialized using a uniform random distribution, as detailed in [28]. During training, the pre-trained layers of the ResNet-50 architecture remained frozen, while the output layer was trained using the Adam optimization algorithm with a learning rate of 0.001. The standard binary cross-entropy loss function was employed in training, and two metrics were recorded: *accuracy*, representing the percentage of correctly predicted samples, and *loss*, representing the binary cross-entropy loss.

Let $y_i \in \{0, 1\}$ denote the actual label of the i-th input image, where $y_i = 0$ indicates a Fusarium wilt-infected leaf; otherwise, $y_i = 1$ indicates that the leaf is healthy. The output of the neural network is denoted by p_i, where $0 \leq p_i \leq 1$ represents the probability of the i-th input being healthy. Let N be the number of images in a dataset. The binary cross-entropy loss (BCE) of the neural network on the given dataset is determined by

$$\text{BCE} = -\frac{1}{N}\sum_{i=1}^{N} y_i \log p_i + (1 - y_i)\log(1 - p_i). \tag{3}$$

The accuracy of the neural network on a given dataset is determined by

$$\text{Accuracy} = \frac{1}{N}\sum_{i=1}^{N} y_i \mathbf{1}(p_i \leq 0.5) + (1 - y_i)\mathbf{1}(p_i > 0.5), \tag{4}$$

where $\mathbf{1}(\cdot)$ is an indicator function.

Utilizing our dataset, the CNN model underwent training on the Google Colab platform. The computation was expedited via the cloud-based Graphics Processing Unit (GPU). The model's training consisted of 100 epochs, and the observed metrics exhibited convergence tendencies.

As depicted in Figure 2, post-training, (a) the CNN model registered an accuracy of 100% and a BCE loss of 0.0247 on the training dataset; (b) the CNN model registered an accuracy of 97.83% and a BCE loss of 0.0351 on the validation dataset.

3.2. Fine Tuning

After the initial training phase, the last 10 pre-trained layers of the ResNet-50 architecture were unfrozen for model fine-tuning. Then, the weights of all the unfrozen layers of the CNN model were optimized using the Adam optimization algorithm. The learning rate of the model was also adjusted to 10^{-4}. This is conducted to ensure that the weights of the model are not drastically altered during this process. By fine-tuning, we are essentially adapting the knowledge of the base model to this identification task.

As shown in Figure 3, the outputs of the fine-tuning process show that the model was able to reach a peak accuracy of 100% on both the training dataset and the validation dataset, indicating the pre-trained model's weights having adapted to this classification task successfully.

3.3. Model Evaluation

In order to determine the ability of the CNN model to identify Fusarium wilt infections, it was tested on a dataset consisting of 500 images (https://www.dropbox.com/

scl/fo/uvcgie7p8izwwqf3norp7/h?rlkey=7066yyc9w5ezcju4m3gu594uc&dl=0 (accessed on 5 December 2023)). The images, previously unseen by the model, were taken from a Google search. Using the OpenCV-Python3 library, the data augmentation techniques of rotation, noise addition, and resizing were used to expand the 200 healthy and diseased images to 500 images.

The metrics used to determine the overall performance of the model were: accuracy, precision, recall, and F-1 score. These measures are calculated using Equations (5)–(8), respectively. Accuracy is a straightforward metric that measures the overall correctness of the model's predictions. Precision is a metric that focuses on the positive predictions ($y_i = 1$) made by the model. It is defined as the ratio of true positive predictions (correctly predicted positive samples) to the total number of positive predictions (both true positives and false positives). Recall, also known as sensitivity or true positive rate, measures the ability of the model to correctly identify positive samples. It is defined as the ratio of true positive predictions to the total number of actual positive samples (true positives and false negatives). The F-1 score is the harmonic mean of precision and recall. It provides a balanced measure that considers both precision and recall simultaneously. Additionally, a confusion matrix was visualized through the OpenCV library packages to see the exact number of true and false positive and negative identification classes. It should be noted that the accuracy of random guessing identification for Fusarium wilt will yield an accuracy of 0.50. It should also be noted that there is no method to truly quantify the accuracy of a farmer in identifying Fusarium wilt as this depends on many variables including experience, age, genetics and more.

$$\text{Accuracy} = \frac{\text{TP} + \text{TN}}{\text{TP} + \text{TN} + \text{FP} + \text{FN}}. \tag{5}$$

$$\text{Precision} = \frac{\text{TP}}{\text{TP} + \text{FP}}. \tag{6}$$

$$\text{Recall} = \frac{\text{TP}}{\text{TP} + \text{FN}}. \tag{7}$$

$$\text{F1 Score} = \frac{2 \times \text{Precision} \times \text{Recall}}{\text{Precision} + \text{Recall}}. \tag{8}$$

As depicted in Figure 4, the CNN model with augmented images in the training data and with transfer learning-based design was able to correctly identify 492 out of the 500 total test images to the correct classification class. The eight errors consisted of five false positive and three false negative identifications. The greater number of false positives in this case is considered a more desirable outcome because the farmer can verify that the banana crop is not truly infected before any potential removal of the entire plant. If the number of false negatives was higher, it would pose a larger issue of missing infections, thus, allowing a greater spread of Fusarium wilt.

The classification report shown in Figure 5 depicts the precision, recall, F-1 score, and accuracy of the testing dataset. The CNN model was able to perform with an accuracy of 0.99 on the healthy images and 0.98 on the Fusarium wilt-diseased images. These measures combined with the 0.98 F1 score show the model was able to generalize its prior training to new data. As a comparison for metrics, Ref. [29] trained a deep learning model to identify Fusarium wilt and was only able to achieve 0.9156 accuracy, 0.9161 precision, 0.8856 recall, and 0.8156 score. Their model was also able to substantially outperform existing models with an accuracy nearly 0.20 higher than the second-best model compared in their study. The model in this paper was able to outperform all the metrics of their model; however, this may be due to the original size of the evaluation set being substantially less than the set in their paper. In future real-world testing, this CNN model may yield slightly lower metrics; however, it is still expected that the model will outperform most existing models based on its stellar performance on this dataset.

3.4. Comparative Analysis: Impact of Data Augmentation and Transfer Learning

In this section, we analyze the impact of data augmentation and transfer learning on model performance. The evaluation was conducted with four different conditions: (a) model performance without data augmentation and transfer learning, (b) model performance with data augmentation and without transfer learning, (c) model performance without data augmentation and with transfer learning, and (d) model performance with data augmentation and with transfer learning. The results of the evaluations are illustrated in Figure 6.

3.5. Receiver Operating Characteristic

A Receiver Operating Characteristic (ROC) curve was also used to evaluate the performance of the model on the same dataset of banana leaf images. The ROC curve is a graphical representation of the true positive rate (TPR) against the false positive rate (FPR) at various threshold settings. It provides an aggregate view of the model's performance across all possible classification thresholds.

$$\text{TPR (Sensitivity)} = \frac{\text{TP}}{\text{TP} + \text{FN}} \quad (9)$$

$$\text{FPR (1 - Specificity)} = \frac{\text{FP}}{\text{FP} + \text{TN}} \quad (10)$$

where:

- TP: True Positives
- FN: False Negatives
- FP: False Positives
- TN: True Negatives

The high AUC value in Figure 7 indicates that the model can distinguish between affected and unaffected plants with high accuracy.

3.6. 10-Fold Cross Validation

To enhance the precision and robustness of the Convolutional Neural Network (CNN) model, a 10-fold cross-validation method was implemented. This technique's significance lies in assessing the CNN model's ability to generalize and make accurate predictions on unseen data.

Given a dataset D of size N, it is partitioned into 10 mutually exclusive subsets, $D_1, D_2, \ldots, D_{10}$. For each iteration i of the cross-validation: 1. D_i is reserved for validation, and the union of the remaining subsets $D - D_i$ is used for training. 2. The model is trained on $D - D_i$ and validated using D_i. 3. Performance metrics, namely training loss, training accuracy, validation loss, and validation accuracy, are computed.

The process is repeated for all subsets, ensuring each data point is used for validation precisely once.

$$\text{Average Validation Accuracy} = \frac{1}{10}\sum_{i=1}^{10} \text{Validation Accuracy}_i \quad (11)$$

$$\text{Average Training Accuracy} = \frac{1}{10}\sum_{i=1}^{10} \text{Training Accuracy}_i \quad (12)$$

ResNet-50's enhanced performance is attributed to its deep layers and the integration of residual blocks. These features help the model recognize complex patterns and avoid issues like vanishing gradients. This evaluation supports the choice of ResNet-50 for tasks requiring high classification accuracy.

An examination of the metrics in Table 1 reveals the model's consistent performance across distinct subsets. The gradual rise in training accuracy denotes the CNN's adaptative learning capability. In parallel, the upsurge in validation accuracy indicates proficient generalization on novel data vectors. This systematic 10-fold cross-validation solidifies the CNN model's efficacy in detecting Fusarium wilt, positioning it as a pivotal tool for disease identification and mitigation in agriculture.

Table 1. Metrics for the 10-fold cross-validation of the CNN model, segmented by subset.

Subset	Training Loss	Training Accuracy	Validation Loss	Validation Accuracy
1	0.7750	0.5188	0.9880	0.3125
2	0.6890	0.6250	0.8267	0.3750
3	0.7107	0.5813	0.6750	0.5312
4	0.5410	0.7375	0.5630	0.7188
5	0.3574	0.8627	0.4693	0.7812
6	0.2753	0.9500	0.4069	0.8438
7	0.2729	0.9375	0.3536	0.8438
8	0.2361	0.9688	0.3086	0.8750
9	0.2406	0.9125	0.2694	0.9688
10	0.1607	0.9804	0.2367	0.9688

3.7. Comparative Evaluation between Architectures

We assessed the performance of the ResNet-50 model in detecting Fusarium wilt and compared it with other CNN architectures: VGG16, AlexNet, and DenseNet121. All models were trained using the same dataset to ensure consistent evaluation conditions.

As presented in Figure 8, ResNet-50 achieved an accuracy of 0.98 on the testing dataset, outperforming the other architectures. Each model was set up using standard configurations to maintain fairness in the comparison.

4. Discussion

The accurate detection of Fusarium wilt in banana leaves using computer vision is a central challenge in agricultural tech research. Traditional methods often encounter difficulties in consistently differentiating between healthy and diseased banana leaves, particularly when symptoms are subtle. This study explored the application of transfer learning and deep convolutional neural networks to address this challenge.

The model introduced in this research, based on the CNN architecture, demonstrated commendable efficacy, surpassing several contemporary methods. In performance metrics, this model achieved an ROC AUC of 0.92, positioning it favorably against numerous existing algorithms. Such accuracy was achieved within a brief training duration. Training encompassed a mere 100 epochs and concluded within an hour, highlighting the efficient computational capabilities of the Google Colab GPU. An initial dataset of 156 images served as the starting point. Through data augmentation techniques, this dataset was expanded. The augmented dataset enabled the model to effectively differentiate between healthy and wilt-afflicted banana leaves.

When benchmarked against other solutions available, the model shows significant promise for integration into platforms like Unmanned Aerial Vehicles (UAVs) or mobile applications. Such integrations can significantly enhance agricultural practices by providing advanced tools for early Fusarium wilt detection and consequently reducing crop loss.

Future perspectives in computational research suggest the emergence of even more streamlined models tailored for real-world applications. For optimal performance, data collection from specific plantations remains a crucial aspect. Such targeted datasets ensure that models can cater to diverse environmental conditions specific to different agricultural regions. Techniques like federated learning can further optimize these models, allowing for on-field adjustments in alignment with unique crop conditions.

The potential of Generative Adversarial Networks (GANs) also warrants attention. GANs can aid in the generation of synthetic training datasets that mirror real-world conditions, thus enhancing the training process.

In summary, this research highlights the capabilities of CNNs and transfer learning in the realm of agricultural technology, offering a foundation for further advancements in the detection and management of crop diseases.

Author Contributions: Conceptualization, K.Y.; methodology, K.Y.; software, K.Y. and M.K.C.S.; validation, K.Y., M.K.C.S. and Y.S.; formal analysis, K.Y., M.K.C.S. and Y.S.; investigation, K.Y.; resources, K.Y., M.K.C.S. and Y.S.; data curation, K.Y.; writing—original draft preparation, K.Y.; writing—review and editing, K.Y., M.K.C.S. and Y.S.; visualization, K.Y. and M.K.C.S.; supervision, K.Y., M.K.C.S. and Y.S.; project administration, Y.S.; funding acquisition, Y.S. All authors have read and agreed to the published version of the manuscript.

Funding: This research was supported in part by the NSF under grant no. CNS-2239677 and by the USDA-NIFA, AFRI Competitive Program, Agriculture Economics and Rural Communities, under grant no. 2023-69006-40213.

Data Availability Statement: This study utilizes a modified data set from previous work (Medhi and Deb, 2022) for training neural networks. The data set used in evaluating the trained neural network is accessible via the following link: https://www.dropbox.com/scl/fo/uvcgie7p8izwwqf3norp7/h?rlkey=7066yyc9w5ezcju4m3gu594uc&dl=0 (accessed on 5 December 2023). We used the dataset compiled by [15] to train the neural network model. Any additional data not included in the main manuscript can be obtained from the corresponding author upon reasonable request.

Conflicts of Interest: The authors have no known competing financial or non-financial interests that are directly or indirectly related to the work submitted for publication.

References

1. Ploetz, R. Gone Bananas? Current and Future Impact of Fusarium Wilt on Production. In *Plant Diseases and Food Security in the 21st Century*; Springer: Cham, Switzerland, 2021.
2. Heslop-Harrison, J.S.; Schwarzacher, T. Domestication, Genomics and the Future for Banana. *Ann. Bot.* **2007**, *100*, 1073–1084. [CrossRef] [PubMed]
3. Shen, Z.; Xue, C.; Penton, C.R.; Thomashow, L.S.; Zhang, N.; Wang, B.; Ruan, Y.; Li, R.; Shen, Q. Suppression of banana Panama disease induced by soil microbiome reconstruction through an integrated agricultural strategy. *Soil Biol. Biochem.* **2019**, *128*, 164–174. [CrossRef]
4. Ordonez, N.; Seidl, M.F.; Waalwijk, C.; Drenth, A.; Kilian, A.; Thomma, B.P.; Ploetz, R.C.; Kema, G.H. Worse comes to worst: Bananas and Panama disease—When plant and pathogen clones meet. *PLoS Pathog.* **2015**, *11*, e1005197. [CrossRef] [PubMed]
5. Van den Berg, N.; Berger, D.K.; Hein, I.; Birch, P.R.J.; Wingfield, M.J.; Viljoen, A. Tolerance in banana to Fusarium wilt is associated with early up-regulation of cell wall-strengthening genes in the roots. *Mol. Plant Pathol.* **2007**, *8*, 333–341. [CrossRef] [PubMed]
6. Ye, H.; Huang, W.; Huang, S.; Cui, B.; Dong, Y.; Guo, A.; Ren, Y.; Jin, Y. Recognition of Banana Fusarium Wilt Based on UAV Remote Sensing. *Remote. Sens.* **2020**, *12*, 938. [CrossRef]
7. Mahlein, A.K. Plant disease detection by imaging sensors–parallels and specific demands for precision agriculture and plant phenotyping. *Plant Dis.* **2016**, *100*, 241–251. [CrossRef] [PubMed]
8. Lin, B.; Shen, H. *Fusarium oxysporum* f. sp. cubense. In *Biological Invasions and Its Management in China*; Springer: Berlin/Heidelberg, Germany, 2017; Volume 2, pp. 225–236.
9. Zhang, Z.; Xiang-Zhao, M.; Ran, J.; Gao, M.; Li, N.X.; Ma, Y.M.; Sun, Y.; Li, Y. *Fusarium oxysporum* infection-induced formation of agarwood (FOIFA): A rapid and efficient method for inducing the production of high quality agarwood. *PLoS ONE* **2022**, *17*, e0277136. [CrossRef] [PubMed]
10. Dita, M.; Barquero, M.; Heck, D.; Mizubuti, E.S.G.; Staver, C.P. Fusarium Wilt of Banana: Current Knowledge on Epidemiology and Research Needs Toward Sustainable Disease Management. *Front. Plant Sci.* **2018**, *9*, 1468. [CrossRef] [PubMed]
11. Mukti, I.Z.; Biswas, D. Transfer learning-based plant diseases detection using ResNet50. In Proceedings of the 2019 4th International Conference on Electrical Information and Communication Technology (EICT), Khulna, Bangladesh, 20–22 December 2019; pp. 1–6.
12. O'Shea, K.; Nash, R. An introduction to convolutional neural networks. *arXiv* **2015**, arXiv:1511.08458.
13. Shaha, M.; Pawar, M. Transfer Learning for Image Classification. In Proceedings of the 2018 Second International Conference on Electronics, Communication and Aerospace Technology (ICECA), Coimbatore, India, 29–31 March 2018; pp. 656–660. [CrossRef]
14. Geetharamani, G.; Pandian, A. Identification of plant leaf diseases using a nine-layer deep convolutional neural network. *Comput. Electr. Eng.* **2019**, *76*, 323–338.
15. Medhi, E.; Deb, N. PSFD-Musa: A dataset of banana plant, stem, fruit, leaf, and disease. *Data Brief* **2022**, *43*, 108427. [CrossRef] [PubMed]
16. Wallelign, S.; Polceanu, M.; Buche, C. Soybean plant disease identification using convolutional neural network. In Proceedings of the FLAIRS Conference, Oxford, UK, 19–21 May 2018; pp. 146–151.

17. Pandian, J.A.; Kumar, V.D.; Geman, O.; Hnatiuc, M.; Arif, M.; Kanchanadevi, K. Plant disease detection using deep convolutional neural network. *Appl. Sci.* **2022**, *12*, 6982. [CrossRef]
18. Ibarra, N.C.; Rivera, M.P.; Manlises, C.O. Detection of Panama Disease on Banana Leaves Using the YOLOv4 Algorithm. In Proceedings of the 2023 15th International Conference on Computer and Automation Engineering (ICCAE), Sydney, Australia, 3–5 March 2023; pp. 209–214.
19. Chaudhari, V.; Patil, M. Banana leaf disease detection using K-means clustering and Feature extraction techniques. In Proceedings of the 2020 International Conference on Advances in Computing, Communication & Materials (ICACCM), Dehradun, India, 21–22 August 2020; pp. 126–130.
20. Gulzar, Y. Fruit Image Classification Model Based on MobileNetV2 with Deep Transfer Learning Technique. *Sustainability* **2023**, *15*, 1906. [CrossRef]
21. Gulzar, Y.; Ünal, Z.; Aktaş, H.; Mir, M.S. Harnessing the Power of Transfer Learning in Sunflower Disease Detection: A Comparative Study. *Agriculture* **2023**, *13*, 1479. [CrossRef]
22. Dhiman, P.; Kaur, A.; Balasaraswathi, V.R.; Gulzar, Y.; Alwan, A.A.; Hamid, Y. Image Acquisition, Preprocessing and Classification of Citrus Fruit Diseases: A Systematic Literature Review. *Sustainability* **2023**, *15*, 9643. [CrossRef]
23. Mamat, N.; Othman, M.F.; Abdulghafor, R.; Alwan, A.A.; Gulzar, Y. Enhancing Image Annotation Technique of Fruit Classification Using a Deep Learning Approach. *Sustainability* **2023**, *15*, 901. [CrossRef]
24. Gulzar, Y.; Hamid, Y.; Soomro, A.B.; Alwan, A.A.; Journaux, L. A Convolution Neural Network-Based Seed Classification System. *Symmetry* **2020**, *12*, 2018. [CrossRef]
25. Chollet, F. Keras. Available online: https://github.com/fchollet/keras (accessed on 12 June 2023).
26. Abadi, M.; Agarwal, A.; Barham, P.; Brevdo, E.; Chen, Z.; Citro, C.; Corrado, G.; Davis, A.; Dean, J.; Devin, M.; et al. TensorFlow: Large-Scale Machine Learning on Heterogeneous Systems. Available online: https://www.tensorflow.org/ (accessed on 15 June 2023).
27. He, K.; Zhang, X.; Ren, S.; Sun, J. Deep residual learning for image recognition. In Proceedings of the IEEE Conference on Computer Vision and Pattern Recognition, Las Vegas, NV, USA, 27–30 June 2016; pp. 770–778.
28. Glorot, X.; Bengio, Y. Understanding the difficulty of training deep feedforward neural networks. In Proceedings of the 13th International Conference on Artificial Intelligence and Statistics (AISTATS), Sardinia, Italy, 13–15 May 2010; Volume 9, pp. 249–256.
29. Sangeetha, R.; Logeshwaran, J.; Rocher, J.; Lloret, J. An Improved Agro Deep Learning Model for Detection of Panama Wilts Disease in Banana Leaves. *AgriEngineering* **2023**, *5*, 660–679. [CrossRef]

AgriEngineering

Article

Multispectral Sensors and Machine Learning as Modern Tools for Nutrient Content Prediction in Soil

Rafael Felippe Ratke [1], Paulo Roberto Nunes Viana [1], Larissa Pereira Ribeiro Teodoro [1], Fábio Henrique Rojo Baio [1], Paulo Eduardo Teodoro [1], Dthenifer Cordeiro Santana [1], Carlos Eduardo da Silva Santos [2], Alan Mario Zuffo [3] and Jorge González Aguilera [4,*]

1 Department of Agronomic, Federal University of Mato Grosso do Sul, Rodovia MS-306, km 105, Zona Rural, Chapadão do Sul 79560-000, MS, Brazil; rafael.ratke@ufms.br (R.F.R.); paulo.viana@ufms.br (P.R.N.V.); larissa_ribeiro@ufms.br (L.P.R.T.); fabiobaio@ufms.br (F.H.R.B.); paulo.teodoro@ufms.br (P.E.T.); dthenifer.santana@unesp.br (D.C.S.)

2 Federal Institute of Education, Science and Technology of the Tocantins, Quadra Ae 310 Sul, Av. NS 10, S/N-Plano Diretor Sul, Palmas 77021-090, TO, Brazil; carlosedu@ifto.edu.br

3 Department of Agronomic, State University of Maranhão, Praça Gonçalves Dias, s/n, Centro, Balsas 65800-000, MA, Brazil; alanzuffo@professor.uema.br

4 Department of Crop Science, State University of Mato Grosso do Sul, Cassilândia 79540-000, MS, Brazil

* Correspondence: jorge.aguilera@uems.br

Abstract: The combination of multispectral data and machine learning provides effective and flexible monitoring of the soil nutrient content, which consequently positively impacts plant productivity and food security, and ultimately promotes sustainable agricultural development overall. The aim of this study was to investigate the associations between spectral variables and soil physicochemical attributes, as well as to predict these attributes using spectral variables as inputs in machine learning models. One thousand soil samples were selected from agricultural areas 0–20 cm deep and collected from Northeast Mato Grosso do Sul state of Brazil. A total of 20 g of the dried and homogenized soil sample was added to the Petri dish to perform spectral measurements. Reflectance spectra were obtained by CROP CIRCLE ACS-470 using three spectral bands: green (532–550 nm), red (670–700 nm), and red-edge (730–760 nm). The models were developed with the aid of the Weka environment to predict the soil chemical attributes via the obtained dataset. The models tested were linear regression, random forest (RF), reptree M5P, multilayer preference neural network, and decision tree algorithms, with the correlation coefficient (r) and mean absolute error (MAE) used as accuracy parameters. According to our findings, sulfur exhibited a correlation greater than 0.6 and a reduced mean absolute error, with better performance for the M5P and RF algorithms. On the other hand, the macronutrients S, Ca, Mg, and K presented modest r values (approximately 0.3), indicating a moderate correlation with actual observations, which are not recommended for use in soil analysis. This soil analysis technique requires more refined correlation models for accurate prediction.

Keywords: machine learning; soil analysis; reflectance spectra

Citation: Ratke, R.F.; Viana, P.R.N.; Teodoro, L.P.R.; Baio, F.H.R.; Teodoro, P.E.; Santana, D.C.; Santos, C.E.d.S.; Zuffo, A.M.; Aguilera, J.G. Multispectral Sensors and Machine Learning as Modern Tools for Nutrient Content Prediction in Soil. *AgriEngineering* **2024**, *6*, 4384–4394. https://doi.org/10.3390/agriengineering6040248

Academic Editors: Chiew Foong Kwong and Ray E. Sheriff

Received: 10 October 2024
Revised: 10 November 2024
Accepted: 18 November 2024
Published: 21 November 2024

1. Introduction

Ferralsols constitute 32.9% of Brazil's population [1]. Ferralsols are characterized by low levels of plant nutrients such as Ca, Mg, and P and high levels of Al [2]. Soybean is most commonly grown in the Ferralsols of Brazil [3].

Modern agriculture requires automation for analysis and management processes. Agricultural automation is well known for the ability to diagnose and observe pests and diseases [4,5]. Automation in soil analysis processes is complex and requires further study.

The assessment of soil properties, including a series of chemical processes, aims to determine the ability of the soil to provide specific nutrients necessary for the plant pathway cycle. The most common approach adopted for this purpose consists of extracting a chemical solution using extractants that simulate the absorption of nutrients by plants. However, the use of this traditional method of soil analysis raises environmental challenges, notably owing to the inadequate disposal of waste generated during the assessment of soil properties, including a range of chemical processes, which aims to determine the ability of the soil to provide specific nutrients essential for the plant growth cycle [6].

The current method not only presents risks for the professionals involved, but also has adverse environmental impacts due to the presence of chemical residues, in addition to incurring high costs related to the acquisition of chemical reagents and requiring a considerable time lag. In this context, sustainable and efficient analytical approaches that minimize environmental impacts and optimize the effectiveness of the soil assessment process are needed to guarantee more efficient agricultural practices [7].

Multispectral soil analysis involves the use of various spectroscopic techniques to evaluate soil properties [8]. Different methods have been used for soil analysis, such as microspectrophotometry, X-ray fluorescence spectroscopy, and laser-induced breakdown spectroscopy [9]. These techniques provide information on the composition, purity, and elemental content of soils, helping to discriminate between different types and sources of soils [10]. Data fusion methodologies have been applied to improve classification accuracy, especially when combining information from several spectroscopic analyzers. Multispectral analysis can capture the spectral dimensionality of soils, providing valuable information on the variability of soil elements, despite limitations in the resolution of narrowband absorption.

Soil spectral analysis calibration is crucial for accurately predicting soil properties. Soil parameters such as potassium, phosphorus, and organic matter are already being evaluated by visible-near-infrared (Vis-NIR) spectroscopy, but it requires calibration [11]. Calibration methods involve preprocessing transformations, variable selection techniques, and regression algorithms to increase prediction accuracy [10]. Methods of calibration are employed to calibrate soil spectral data, including preprocessing transformations, variable selection techniques, and regression algorithms [12]. Using spectral libraries and reducing sample processing levels have shown potential for lowering costs and time implications for predicting soil properties such as organic carbon, clay, and pH [13]. Overall, proper calibration methods are essential for leveraging soil spectral analysis to monitor soil properties effectively and contribute to precision agriculture.

The use of indirect analysis through multispectral sensors has enabled expeditious, economically viable, and ecologically sustainable monitoring of elementary soil levels. Moreover, the integration of machine learning (ML) algorithms has proven crucial for obtaining reliable estimates in this context [14]. The synergistic combination of these approaches provides effective and agile monitoring of the soil nutrient content, which is highly relevant for agricultural soil productivity, food security, and the promotion of sustainable agricultural development [15]. The convergence of these techniques not only optimizes the speed and efficiency of monitoring, but also contributes to mitigating the environmental challenges inherent to traditional soil analyses, such as the production of chemical residues.

The central problem of this study is the difficulty in quickly and accurately assessing and monitoring soil physicochemical attributes, which are crucial for the proper and sustainable management of agricultural resources. Traditional soil analysis methods, such as laboratory collection and analysis, are generally time-consuming, expensive, and limited in terms of spatial coverage, which makes large-scale and real-time monitoring difficult. In this context, the use of spectral variables as indirect indicators of soil properties emerges as a promising alternative. However, an important issue is the complexity of the relationship between spectral variables and soil physicochemical attributes, which can vary depending on the type of soil, moisture, and presence of organic matter, among other factors. Therefore, there is a technical challenge in building machine learning models capable of capturing these relationships in a robust and generalizable way so that they can be applied in different scenarios. The objective of the current investigation was to analyze the associations between the spectral and physicochemical variables of soil in addition to predicting the physicochemical attribute levels of soil via the use of spectral variables as inputs into machine learning models.

2. Materials and Methods

2.1. Sample Collection and Determination of Physicochemical Properties

Soil samples were collected at 0 to 20 cm depth from the municipalities of Cassilândia, Chapadão do Sul, Costa Rica, and Paraíso das Águas (18°46′26″ S 52°37′28″ W, average altitude of 810 m of sea level), with a coverage area of 16,130.84 km^2, located in the State of Mato Grosso do Sul (MS), Brazil. The regional climate is classified as humid tropical, with a rainy season in summer and a dry season in winter, with an average annual rainfall of 1.850 mm, an average annual temperature of 20.5 °C, and a variation of 7.5 °C.

The soil in the region is mostly classified as Rhodic Ferralsol [16]. A total of 33% of the 1000 samples analyzed were characterized as sandy, 25% as sandy loam, 24% as clay loam, and 20% as clay. The soil samples were collected with different augers, i.e., probe-type augers (20 mm diameter) and screw-type augers, at depths of 0–0.20 m. The soil samples were sieved through a 2 mm mesh and air-dried. The elements Ca, Mg, and K were analyzed in the Exata Brasil Laboratory located in Chapadão do Sul-MS.

KCl solution (1 mol L^{-1}) at a ratio of 1/10 (soil:solution) was used to extract Ca and Mg from the soil. The element potassium (K) was extracted from the Mehlich1 solution (0.05 mol L^{-1} HCl + 0.0125 mol L^{-1} H_2SO_4) at a ratio of 1/10 (soil:solution). The ammonium acetate solution in a proportion of 10 g of soil to 25 mL of the solution was used to extract S from the soil. Ca, Mg, K, and S contents in the soil extracts were measured via Argon Plasma Optical Atomic Emission Spectrometry (ICP-OES) (Perkin Elmer, Waltham, MA, USA).

Multispectral evaluations were carried out in a 20 g aliquot of each sieved, dried, and homogenized soil sample, which was subsequently added to a Petri dish for spectral measurements (Figure 1). The Petri dish was placed on a flat bench, and the sensor was installed 8 cm from the soil surface. The area of incidence of the spectral beam was 3 cm^2. Two external 50 W halogen lamps were positioned 35 cm from the Petri dish at a zenith angle of 30°, forming a 90° angle to each other following the method described by Franceschini et al. [14].

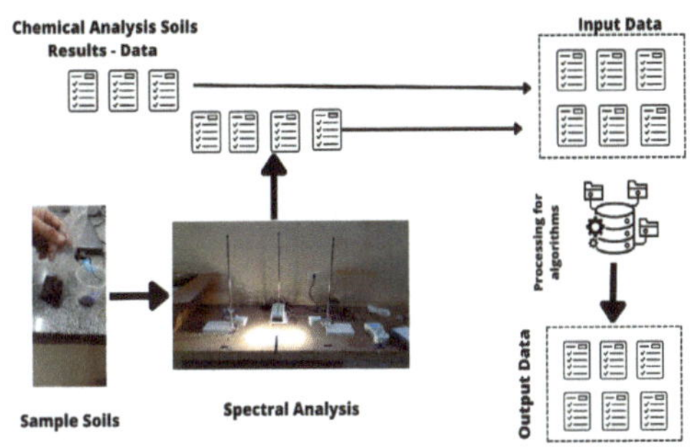

Figure 1. Illustration of spectral analysis and data processing by machine learning. Images are the property of the author.

The reflectance spectra were obtained with a CROP CIRCLE ACS-470 instrument (Holland Scientific, Inc., Lincoln, NE, USA). The six spectral bands used were green (532–550 nm), red (670–700 nm), and red edge (730–760 nm). The sensor was calibrated via FieldCal SC-1. The spectral bands were applied to the surface of the soil samples in 100 replicates for each band. Reflectances were recorded in spreadsheets, and reflectance averages were calculated for each spectral band.

2.2. Data Analysis via Computational Intelligence

The data were subjected to observation and compared via the WEKA (Waikato Environment for Knowledge Analysis) software version 3.9.3(c) 1999–2018, which was accessed by a computer with an AMD Phenom™ IIx4 B97 processor 3.20 GHz, installed memory RAM 4 GB, 32-bit operating system, Windows 7, using cross-validation with 10 folds (K = 10) and 10 repetitions (100 runs) in a spectral analysis of 690 samples with wavelength data obtained as input values and macronutrients as output values to be predicted for Ca, Mg, and K. The data prediction analysis used 370 samples for S.

The models tested were random forest (RF), multilayer perceptron (MLP), decision trees (M5P), REPTree (REPT), and random trees (RTs). All the parameters adopted were set to the default software configuration. The tested models were selected with applicability in other agronomic works according to Refs. [13,14]. MLP is a type of neural network that excels at solving supervised learning problems with multiple inputs. It consists of layers of neurons (or perceptrons) organized into an input layer, one or more hidden layers, and an output layer. Each neuron in a layer is connected to the neurons in the next layer by adjustable weights. In Weka software, the default MLP configuration includes a single hidden layer with a number of neurons defined by the average between the number of attributes and the number of classes, which generally provides a good balance between learning capacity and computational efficiency.

M5P provides more information on mathematical equations and addresses categorical and continuous variables and missing values. This model combines features of decision trees with linear regression, making it particularly useful for continuous and mixed-category data. It builds decision trees where the leaves contain linear equations that facilitate prediction. This allows M5P to handle both categorical and continuous variables and manage missing values, providing a more interpretable model by providing insight into the underlying mathematical relationships between variables.

The random forest (RF) algorithm uses multiple independent decision trees and combines their predictions. It is particularly effective for large-scale problems because of its robustness against overfitting, and facilitates data interpretation by allowing the assessment of variable importance. A random tree (RT) is used to build a decision tree with a random dataset through the division of nodes.

The REPTree algorithm builds decision trees via regression logic in multiple iterations. In each iteration, it evaluates several trees, selecting the best one on the basis of regression error criteria. This model allows a robust pruning approach, where the final tree is simplified to improve the generalizability of the model. The averages of the S, Mg, Ca, and K contents of the actual data and the predicted data of the samples randomly selected via machine learning were contrasted in scatter and line graphs via SigmaPlot 11.0 software.

The accuracy of the prediction models was evaluated by the correlation between the predicted and observed values (r) and the mean absolute error (MAE). The accuracy values of each of the tested models were subjected to analysis of variance to verify the existence of significant differences between the machine learning models. Subsequently, boxplots were generated for r and MAE for each model in the case of macronutrient prediction.

The means of the performance parameters were grouped via the Scott–Knott test at 5% probability. The boxplots and groupings of means were generated via the ggplot2 and ExpDes.pt packages of R software.

For the Pearson correlation coefficient (r), we applied the criterion adapted by Figueiredo Filho and Silva Júnior [15], which is classified into three categories—low, medium, and high—and is considered low when r is approximately 0.10–0.30, moderate when r is between 0.40 and 0.60, and high when r is 0.70–1.

3. Results

The averages of the S, Mg, K, and Ca contents in the analyzed soil show a disparity between the predicted data due to accumulated error (Figure 2). The S content in the soil predicted by machine learning analyses via MLP, M5P, RF, RT, and REPT exhibited significant dispersion compared with the chemically analyzed mean content (Figure 2A). Conversely, the Mg and K contents predicted by MLP, M5P, RF, and REPT showed low dispersion relative to the mean content, not aligning with the values obtained through chemical analysis (Figure 2B,C). In this context, the use of RT resulted in greater dispersion in the prediction of Mg, K, and Ca contents. However, the mean Ca content predicted by RT was the closest to the real value found in the chemical analyses (Figure 2D).

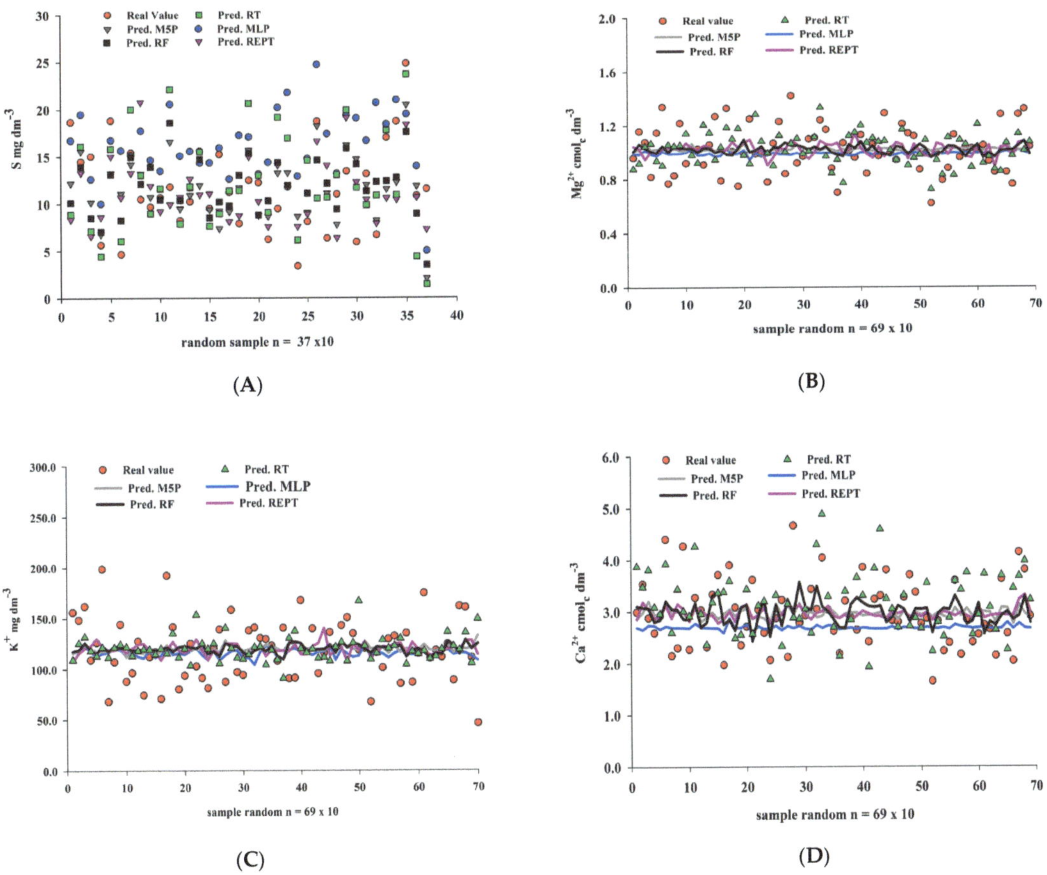

Figure 2. Boxplots of the Pearson correlation coefficient (r, to the left) and mean absolute error (MAE, on the right) for different sulfur-related machine learning models: random forest (RF), multilayer Perceptron (MLP), decision trees (M5P), REPTree (REPT), and random tree (RT). Mean levels of S (**A**), Mg^{+2} (**B**), K^{+} (**C**) and Ca^{+2} (**D**) of the soil chemically analyzed and predicted by different algorithms.

With respect to the prediction of the sulfur content, the M5P and RF algorithms outperformed the other algorithms (Figure 3), presenting high r values (higher than 0.6). This value guarantees the high accuracy of these algorithms in estimating the sulfur content on the basis of spectral reflectance. Another factor that contributes to the accuracy of both algorithms is the low MAE, indicating smaller errors in the prediction, ensuring greater precision of these algorithms when predicting S contents.

With respect to the performance of the algorithms in predicting the magnesium (Mg) content, the results revealed that the random forest (RF) algorithm was superior in terms of accuracy (Figure 4). This could be translated into greater consistency between the predictions and observed values, as revealed by the correlation coefficient (r), which surpassed those of the other algorithms. Additionally, the RF demonstrated a lower mean absolute error (MAE) value, denoting a significantly high precision in its estimates.

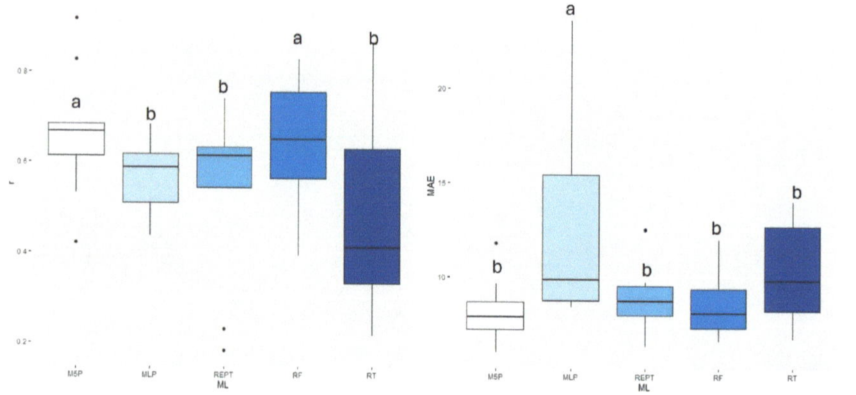

Figure 3. Boxplots of the Pearson correlation coefficient (r, to the left) and mean absolute error (MAE, on the right) for different sulfur-related machine learning models: random forest (RF), multilayer Perceptron (MLP), decision trees (M5P), REPTree (REPT), and random tree (RT). Different lowercase letters about the boxplots represent statistical differences at 5% probability by the Scott–Knott test.

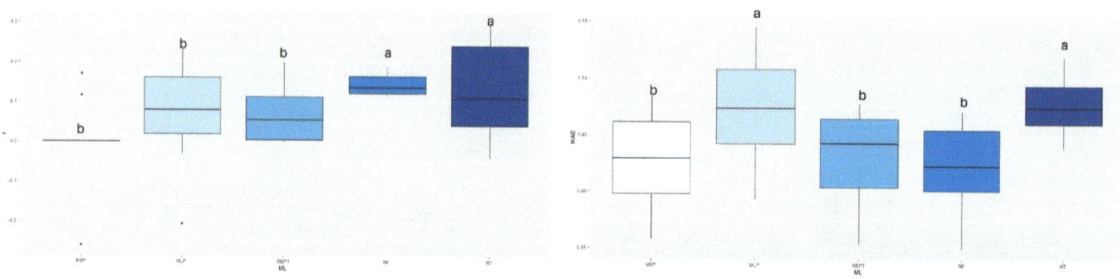

Figure 4. Boxplots of the Pearson correlation coefficient (r, to the left) and mean absolute error (MAE, on the right) for different magnesium-related machine learning models. Different lowercase letters about the boxplots represent statistical differences at 5% probability by the Scott–Knott test.

The RT algorithm exhibited similar behavior to that of the RF when evaluated in relation to the accuracy indicator (r). However, it is important to highlight that RT revealed a significantly high MAE value, which indicates that although it presents relative consistency in predictions, it is not an accurate algorithm for predicting magnesium content.

Statistically, the algorithms had the same behavior for the correlation coefficient (r) in potassium prediction. There was also no significant difference in terms of error. Therefore, using algorithms that maintain better performance for the other elements facilitates processing because, in the case of potassium, all the models have the same performance (Figure 5). The prediction value was approximately 0.3, indicating a moderate value that may be considerably adequate in terms of the variability and dynamicity of P in the soil.

On the other hand, in the analysis of the prediction of the calcium (Ca) content, the M5P algorithm demonstrated superior performance in relation to the other algorithms, as evidenced by correlation coefficient values (r) that approached 0.3 (Figure 6). Furthermore, notably, the M5P algorithm achieved notably lower mean absolute error (MAE) values, approximately 1.50, with the maximum MAE value for predicting this nutrient. These results emphasize the ability of the M5P algorithm to generate accurate estimates, with a relatively low level of error, which is extremely relevant for estimating the calcium content in soil samples via multispectral data.

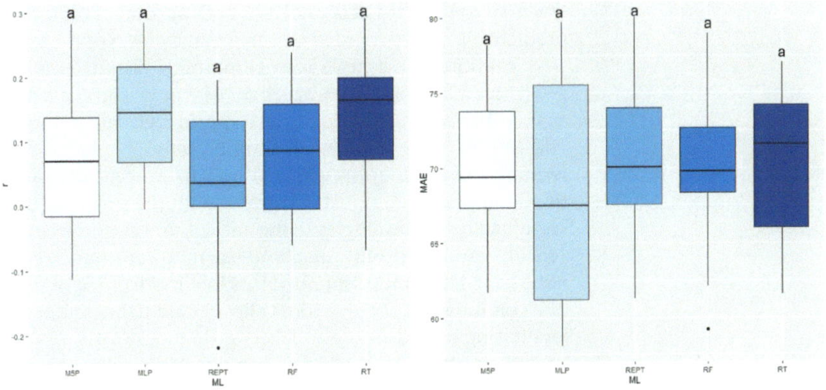

Figure 5. Boxplots of the Pearson correlation coefficient (r, to the left) and mean absolute error (MAE, on the right) for different potassium-related machine learning models. Different lowercase letters about the boxplots represent statistical differences at 5% probability by the Scott–Knott test.

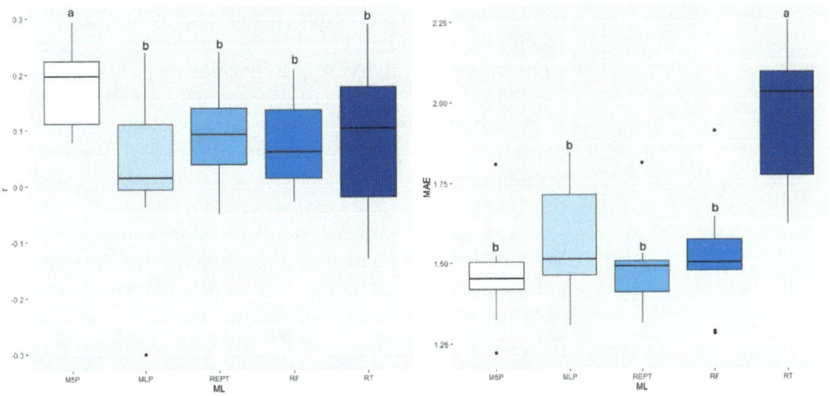

Figure 6. Boxplots of the Pearson correlation coefficient (r, to the left) and mean absolute error (MAE, on the right) for different calcium-related machine learning models. Different lowercase letters about boxplots represent statistical differences at 5% probability by the Scott–Knott test.

In short, the M5P and RF algorithms performed satisfactorily in terms of the correlation coefficient (r). Specifically, these algorithms achieve values remarkably close to the actual values, which is especially evident when the sulfur content is predicted, where the r value is close to 0.8. For the other elements analyzed, the accuracy remained at approximately 0.3, which is moderate and consistently shows that, owing to the dynamism and complexity of these soil elements, the use of multispectral reflectance to determine them is promising. Furthermore, the M5P and RF algorithms yielded lower MAE values, further reinforcing the reliability of the predictions generated by these algorithms. Reducing the error contributes to greater precision in the estimates, improving the accuracy of the predictions obtained via multispectral reflectance.

4. Discussion

The results indicate a significant disparity between the expected concentrations of sulfur (S), magnesium (Mg), potassium (K), and calcium (Ca) and the measurements derived from chemical analysis. The inconsistency of the predicted data is influenced by the quality of the data and noise in the accuracy of the predictive models [17]. Data variations amplify imprecision, particularly in more complex predictive models [18]. However, it is essential to validate data analyses with accurate predictive models. In this context, the RT predictive model provided better accuracy for S and Ca than did the actual analyzed data.

Traditional methods for estimating soil nutrients, such as laboratory analysis, are recognized for their accuracy, but they have significant limitations in terms of time and cost [10]. These methods require specialized labor, and the use of chemical reagents, in addition to being expensive, can pose environmental risks due to the generation of potentially contaminating waste [10]. This process becomes unfeasible for large-scale and timely application, which contrasts with the growing demand for fast, economical, and sustainable methodologies in the agricultural sector. In this context, the use of multispectral sensors combined with machine learning techniques has emerged as a promising alternative for estimating soil nutrients in a more efficient and environmentally friendly way. The combination of these tools allows data to be obtained noninvasively and in real time, facilitating continuous and large-scale monitoring of soil attributes. In this study, three specific multispectral bands were used to predict soil nutrient levels, employing different machine learning algorithms [6]. This approach offers a potentially faster and more accessible methodology that can contribute to more sustainable and precise agricultural practices, enabling more effective soil management in response to the needs of agricultural production [7].

Among the investigated nutrients, S had the highest predictive value, close to 0.80 for the correlation coefficient I, and notably low values of the mean absolute error (MAE), confirming its ability to offer highly accurate predictions (Figure 2). High prediction values were achieved by the M5P and RF algorithms, highlighting the robustness and reliability of the algorithms in the task of predicting S content. Both algorithms performed well because of their high r and low MAE values, and their use in other agricultural tasks, such as predicting soil organic carbon, stands out [19]. With satisfactory results in predicting soil nutrients, RF can be used to infer soil fertility [20]. The soil nutrients significantly influence the distribution of soil organic carbon [21].

The other predictions yielded median values, with a certain significance, highlighting the complexity of predicting potassium content through the evaluated algorithms, which is particularly relevant in agronomic and soil fertility contexts, where the precise estimation of Mg, K, and Ca contents is crucial for effective and sustainable agricultural practices. In magnesium prediction, the RF algorithm stands out as a superior choice, guaranteeing greater accuracy. These results highlight the relevance of careful selection of the appropriate algorithm for a given task. Dharumarajan et al. [22] reported that RF was the best model for most soil properties, from macro- to micronutrients, indicating that the RF model is better for solving multivariate adjustment problems since RF combines many trees to form an accurate prediction mechanism. In addition, the RF algorithm, when necessary, has fewer parameters to adjust.

In the potassium prediction results, there was homogeneity in the results regarding r, suggesting that the analyzed algorithms exhibited a moderate correlation with the real observations of potassium content, although they did not reach higher levels of correlation, indicating that the prediction of this element can be challenging. In a similar study, Forkuor et al. [23] reported that no machine learning algorithm works best for all global situations, and models must be tested to calibrate them to identify an accurate model for predicting soil properties, optimizing data processing.

The analysis of the calcium prediction (Ca) content revealed that the M5P algorithm performed better than the other algorithms did, as evidenced by correlation coefficient values that approached 0.3. This result suggests that M5P established a moderate and consistent r with actual observations of Ca content, indicating its ability to provide meaningful estimates.

The superior performance of M5P in predicting S and Ca, for which the algorithm stands out, highlights its applicability and usefulness in agronomic contexts, where accurate estimation of this nutrient is crucial for adequate soil management and the development of effective agricultural practices. This contributes to increased productivity and sustainability of agricultural activities. According to previous studies, this model performs well in several tasks in different areas, such as predicting cadmium in agricultural soils [24]. M5P also presents good results in the physical and chemical prediction of soil and water due to its greater accuracy and speed than the regression model [25]. This diversity of accuracies in different situations demonstrates the generalizability and robustness of the algorithm.

The complexity of S forms in soils contributes to variability in spectral measurements. Sulfur occurs in organic and inorganic forms in soil and is transformed by microorganisms in the soil [26]. In this sense, organosulfur compounds are stable over time, whereas others may decompose or convert to other forms, leading to variability in spectral measurements [27].

Ca, Mg, and K interact with each other and with the soil matrix, which changes their chemical bonding structures with other elements present and their spectral expressions [28]. Calcium and magnesium have the capacity to generate carbonates and additional compounds that affect the reflectance of soil samples [29]. Potassium influences the spectral properties of soil through interactions with clay [28].

Our findings demonstrated the effectiveness of the M5P and RF algorithms in predicting soil nutrients, particularly the S content, where the accuracy reached notable levels. The ability of these models to provide estimates close to real values has significant implications for agriculture and soil management, promoting decision-making on the basis of reliable data and contributing to more efficient and sustainable agricultural practices. The greatest contribution of these technologies is to reduce the work involved in analyzing samples and reducing the use of reagents in laboratory analyses, making this part of the procedure faster, requiring and dispensing with the use of expensive reagents from laboratories that report adequate disposal, which are not always served. With future research applying the algorithms found here, it will be possible to adapt and use such technologies with remote sensors or prototypes to be used in situ. Soil samples from other locations and a larger number of samples can be used to increase the accuracy of the algorithms. The use of hyperspectral sensors can also improve predictive value in addition to their application in agriculture, which enables real-time monitoring in agricultural scenarios.

5. Conclusions

In the present study, the use of the CROP CIRCLE ACS-470 multispectral sensor associated with machine learning was demonstrated to be a promising approach for predicting soil macronutrients, especially sulfur, with correlations between actual and estimated values above 0.6. However, regarding the macronutrients P, K, Ca, and Mg, the prediction accuracy reached values of approximately 0.3, indicating a moderate and coherent correlation with actual observations; however, the development of more refined models is needed to improve the results. These findings highlight the reliability and accuracy of predictions, thus strengthening the usefulness and effectiveness of the sensor in the context of soil analysis and agricultural decision-making.

The use of multispectral sensors and data prediction analysis via the M5P and RF algorithms derived from our results are directly applicable to areas with characteristics similar to those of Rhodic Ferralsol and Arenosol soils. For regions with soils of different compositions, we suggest conducting complementary studies to adapt the proposed practices to local conditions.

Author Contributions: Conceptualization, R.F.R., P.R.N.V., L.P.R.T., F.H.R.B. and P.E.T.; methodology, R.F.R., P.R.N.V., L.P.R.T., F.H.R.B., P.E.T., D.C.S. and C.E.d.S.S.; software, P.R.N.V., L.P.R.T., F.H.R.B., P.E.T. and D.C.S.; validation, R.F.R., P.R.N.V., L.P.R.T., F.H.R.B., P.E.T., D.C.S., C.E.d.S.S., A.M.Z. and J.G.A.; formal analysis, R.F.R., P.R.N.V., L.P.R.T., F.H.R.B., P.E.T., D.C.S. and C.E.d.S.S.; investigation, R.F.R., P.R.N.V., L.P.R.T., F.H.R.B., P.E.T., D.C.S. and C.E.d.S.S.; resources, R.F.R., A.M.Z. and J.G.A.; data curation, R.F.R., P.R.N.V., L.P.R.T., F.H.R.B., D.C.S., C.E.d.S.S., A.M.Z. and J.G.A.; writing—original draft preparation, R.F.R., P.R.N.V., L.P.R.T., F.H.R.B., P.E.T., D.C.S., C.E.d.S.S., A.M.Z. and J.G.A.; writing—review and editing, R.F.R., P.R.N.V., L.P.R.T., F.H.R.B., P.E.T., D.C.S., C.E.d.S.S., A.M.Z. and J.G.A.; visualization, R.F.R., P.R.N.V., L.P.R.T. and F.H.R.B. All authors have read and agreed to the published version of the manuscript.

Funding: This research did not receive external funding.

Institutional Review Board Statement: Approval for the study was not required in accordance with Brazilian legislation.

Data Availability Statement: Data are contained within the article.

Acknowledgments: The authors would like to thank the Federal University of Mato Grosso do Sul and the Coordenação de Aperfeiçoamento de Pessoal de Nível Superior—Brazil (CAPES). The company Exata Brazil provided the soil samples and results of the soil chemical analysis.

Conflicts of Interest: The authors declare that they have no conflicts of interest.

References

1. IBGE, Portal do IBGE. *Structural Provinces, Relief Compartments, Soil Types and Phytoecological Regions;* IBGE: Rio de Janeiro, Brazil, 2019.
2. Ratke, R.F.; Campos, A.R.; Inda, A.V.; Barbosa, R.S.; Jacques, Y.; Bezerra, A.; César, J.; Nóbrega, A.; Batista, J. Agricultural Potential and Soil Use Based on the Pedogenetic Properties of Soils from the Cerrado-Caatinga Transition. *Semin. Agrar.* **2020**, *41*, 1119–1134. [CrossRef]
3. Brannstrom, C.; Jepson, W.; Filippi, A.M.; Redo, D.; Xu, Z.; Ganesh, S. Land Change in the Brazilian Savanna (Cerrado), 1986-2002: Comparative Analysis and Implications for Land-Use Policy. *Land Use Policy* **2008**, *25*, 579–595. [CrossRef]
4. Dai, G.; Tian, Z.; Fan, J.; Sunil, C.K.; Dewi, C. DFN-PSAN: Multi-Level Deep Information Feature Fusion Extraction Network for Interpretable Plant Disease Classification. *Comput. Electron. Agric.* **2024**, *216*, 108481. [CrossRef]
5. Dai, G.; Fan, J.; Dewi, C. ITF-WPI: Image and Text Based Cross-Modal Feature Fusion Model for Wolfberry Pest Recognition. *Comput. Electron. Agric.* **2023**, *212*, 108129. [CrossRef]
6. Demattê, J.A.M.; Ramirez-Lopez, L.; Marques, K.P.P.; Rodella, A.A. Chemometric Soil Analysis on the Determination of Specific Bands for the Detection of Magnesium and Potassium by Spectroscopy. *Geoderma* **2017**, *288*, 8–22. [CrossRef]
7. Terra, F.S.; Demattê, J.A.M.; Viscarra Rossel, R.A. Spectral Libraries for Quantitative Analyses of Tropical Brazilian Soils: Comparing Vis-NIR and Mid-IR Reflectance Data. *Geoderma* **2015**, *255–256*, 81–93. [CrossRef]
8. Selkowitz, D.J. A Comparison of Multi-Spectral, Multi-Angular, and Multi-Temporal Remote Sensing Datasets for Fractional Shrub Canopy Mapping in Arctic Alaska. *Remote Sens. Environ.* **2010**, *114*, 1338–1352. [CrossRef]
9. Demattê, J.A.M. Characterization and Discrimination of Soils by Their Reflected Electromagnetic Energy. *Pesqui. Agropecu. Bras.* **2002**, *37*, 1445–1458. [CrossRef]
10. Breure, T.S.; Prout, J.M.; Haefele, S.M.; Milne, A.E.; Hannam, J.A.; Moreno-Rojas, S.; Corstanje, R. Comparing the Effect of Different Sample Conditions and Spectral Libraries on the Prediction Accuracy of Soil Properties from Near- and Mid-Infrared Spectra at the Field-Scale. *Soil Tillage Res.* **2022**, *215*, 105196. [CrossRef]
11. Guo, P.; Li, T.; Gao, H.; Chen, X.; Cui, Y.; Huang, Y. Evaluating Calibration and Spectral Variable Selection Methods for Predicting Three Soil Nutrients Using Vis-Nir Spectroscopy. *Remote Sens.* **2021**, *13*, 4000. [CrossRef]
12. Shepherd, K.D.; Ferguson, R.; Hoover, D.; van Egmond, F.; Sanderman, J.; Ge, Y. A Global Soil Spectral Calibration Library and Estimation Service. *Soil Secur.* **2022**, *7*, 100061. [CrossRef]
13. Mohammedzein, M.A.; Csorba, A.; Rotich, B.; Justin, P.N.; Melenya, C.; Andrei, Y.; Micheli, E. Development of Hungarian Spectral Library: Prediction of Soil Properties and Applications. *Eurasian J. Soil Sci.* **2023**, *12*, 244–256. [CrossRef]
14. Bellon-Maurel, V.; Fernandez-Ahumada, E.; Palagos, B.; Roger, J.-M.; McBratney, A. Critical Review of Chemometric Indicators Commonly Used for Assessing the Quality of the Prediction of Soil Attributes by NIR Spectroscopy. *TrAC Trends Anal. Chem.* **2010**, *29*, 1073–1081. [CrossRef]
15. Peng, Y.; Zhao, L.; Hu, Y.; Wang, G.; Wang, L.; Liu, Z. Prediction of Soil Nutrient Contents Using Visible and Near-Infrared Reflectance Spectroscopy. *ISPRS Int. J. Geo-Inf.* **2019**, *8*, 437. [CrossRef]
16. International Union of Soil Science (IUSS). *World Reference Base for Soil Resources (WRB);* World Soil; FAO: Rome, Italy, 2015.
17. Awais, M.; Naqvi, S.M.Z.A.; Zhang, H.; Li, L.; Zhang, W.; Awwad, F.A.; Ismail, E.A.A.; Khan, M.I.; Raghavan, V.; Hu, J. AI and Machine Learning for Soil Analysis: An Assessment of Sustainable Agricultural Practices. *Bioresour. Bioprocess.* **2023**, *10*, 90. [CrossRef]
18. Padarian, J.; Minasny, B.; McBratney, A.B. Machine Learning and Soil Sciences: A Review Aided by Machine Learning Tools. *SOIL* **2020**, *6*, 35–52. [CrossRef]
19. Chen, J.; Zhang, H.; Fan, M.; Chen, F.; Gao, C. Machine-Learning-Based Prediction and Key Factor Identification of the Organic Carbon in Riverine Floodplain Soils with Intensive Agricultural Practices. *J. Soils Sediments* **2021**, *21*, 2896–2907. [CrossRef]
20. Jeong, G.; Oeverdieck, H.; Park, S.J.; Huwe, B.; Ließ, M. Spatial Soil Nutrients Prediction Using Three Supervised Learning Methods for Assessment of Land Potentials in Complex Terrain. *CATENA* **2017**, *154*, 73–84. [CrossRef]
21. John, K.; Abraham Isong, I.; Michael Kebonye, N.; Okon Ayito, E.; Chapman Agyeman, P.; Marcus Afu, S. Using Machine Learning Algorithms to Estimate Soil Organic Carbon Variability with Environmental Variables and Soil Nutrient Indicators in an Alluvial Soil. *Land* **2020**, *9*, 487. [CrossRef]

22. Dharumarajan, S.; Lalitha, M.; Niranjana, K.; Hegde, R. Evaluation of Digital Soil Mapping Approach for Predicting Soil Fertility Parameters—a Case Study from Karnataka Plateau, India. *Arab. J. Geosci.* **2022**, *15*, 386. [CrossRef]
23. Forkuor, G.; Hounkpatin, O.K.L.; Welp, G.; Thiel, M. High Resolution Mapping of Soil Properties Using Remote Sensing Variables in South-Western Burkina Faso: A Comparison of Machine Learning and Multiple Linear Regression Models. *PLoS ONE* **2017**, *12*, e0170478. [CrossRef] [PubMed]
24. Agyeman, P.C.; Khosravi, V.; Michael Kebonye, N.; John, K.; Borůvka, L.; Vašát, R. Using Spectral Indices and Terrain Attribute Datasets and Their Combination in the Prediction of Cadmium Content in Agricultural Soil. *Comput. Electron. Agric.* **2022**, *198*, 107077. [CrossRef]
25. Morteza, R.; Ghasemnezhad, A.; Ghorbani, K.; Khodayar, H.; Abhari, A. Prediction of Saffron Flower and Stigma Yield Based on the Physical and Chemical Properties of Water and Soil Using Linear Multivariate Regression Models and M5 Decision Tree. *J. Saffron Res.* **2022**, *9*, 352–367.
26. Kar, G.; Schoenau, J.J.; Gillespie, A.W.; Dhillon, G.S.; Peak, D. Sulfur Species Formed in the Seed Row of Sulfur-Fertilized Soils as Revealed by K-Edge X-ray Absorption Near-Edge Structure Spectroscopy. *Soil Sci. Soc. Am. J.* **2019**, *83*, 1324–1332. [CrossRef]
27. Anunciado, M.B.; De Boskey, M.; Haines, L.; Lindskog, K.; Dombek, T.; Takahama, S.; Dillner, A.M. Stability Assessment of Organic Sulfur and Organosulfate Compounds in Filter Samples for Quantification by Fourier- Transform Infrared Spectroscopy. *Atmos. Meas. Tech.* **2023**, *16*, 3515–3529. [CrossRef]
28. Israr, M.A.; Abbas, Q.; Haq, S.-U.; Nadeem, A. Compositional Analysis of Soil Using Calibration-Free Laser-Induced Breakdown Spectroscopy. *Spectrosc. Lett.* **2022**, *55*, 350–361. [CrossRef]
29. Santos, H.S.; Nguyen, H.; Venâncio, F.; Ramteke, D.; Zevenhoven, R.; Kinnunen, P. Mechanisms of Mg Carbonates Precipitation and Implications for CO_2 Capture and Utilization/Storage. *Inorg. Chem. Front.* **2023**, *10*, 2507–2546. [CrossRef]

AgriEngineering

Article

Nutritional Monitoring of Rhodena Lettuce via Neural Networks and Point Cloud Analysis

Alfonso Ramírez-Pedraza [1,2], Sebastián Salazar-Colores [3], Juan Terven [1], Julio-Alejandro Romero-González [4], José-Joel González-Barbosa [1] and Diana-Margarita Córdova-Esparza [4,*]

1 CICATA Qro., Instituto Politécnico Nacional, Mexico City 76090, Mexico; pedro.ramirez@conahcyt.mx (A.R.-P.); jrtervens@ipn.mx (J.T.); jgonzalezba@ipn.mx (J.-J.G.-B.)
2 Dirección Adjunta de Desarrollo Científico, IxM, CONAHCyT, Alvaro Obregón, Mexico City 03940, Mexico
3 IA, Centro de Investigaciones en Óptica A.C., Loma del Bosque 115, León 37150, Mexico; sebastian.salazar@cio.mx
4 Facultad de Informática, Universidad Autónoma de Querétaro, Querétaro 76230, Mexico; julio.romero@uaq.mx
* Correspondence: diana.cordova@uaq.mx

Abstract: In traditional farming, fertilizers are often used without precision, resulting in unnecessary expenses and potential damage to the environment. This study introduces a new method for accurately identifying macronutrient deficiencies in Rhodena lettuce crops. We have developed a four-stage process. First, we gathered two sets of data for lettuce seedlings: one is composed of color images and the other of point clouds. In the second stage, we employed the interactive closest point (ICP) method to align the point clouds and extract 3D morphology features for detecting nitrogen deficiencies using machine learning techniques. Next, we trained and compared multiple detection models to identify potassium deficiencies. Finally, we compared the outcomes with traditional lab tests and expert analysis. Our results show that the decision tree classifier achieved 90.87% accuracy in detecting nitrogen deficiencies, while YOLOv9c attained an mAP of 0.79 for identifying potassium deficiencies. This innovative approach has the potential to transform how we monitor and manage crop nutrition in agriculture.

Keywords: model; greenhouses; Rhodena lettuce; diseased; macronutrient; point clouds

Citation: Ramírez-Pedraza, A.; Salazar-Colores, S.; Terven, J.; Romero-González, J.-A.; González-Barbosa, J.-J.; Córdova-Esparza, D.-M. Nutritional Monitoring of Rhodena Lettuce via Neural Networks and Point Cloud Analysis. *AgriEngineering* **2024**, *6*, 3474–3493. https://doi.org/10.3390/agriengineering6030198

Academic Editors: Chiew Foong Kwong and Ray E. Sheriff

Received: 18 August 2024
Revised: 11 September 2024
Accepted: 16 September 2024
Published: 23 September 2024

1. Introduction

Green leafy vegetables are recognized as vital sources of vitamins, minerals, and dietary fiber; therefore, they significantly contribute to human nutrition and health. They are particularly important in rural areas where they serve as a cheap and accessible source of essential nutrients [1,2]. Green leafy vegetables are found to be rich sources of both macro- and microelements. Calcium, phosphorus, and zinc levels in green leafy vegetables vary from 0.9 to 2.9, 0.4 to 1.2%, and 17.5 to 46.2 ppm, respectively [3].

Lettuce is a popular leafy green vegetable that is consumed worldwide. It is an excellent source of vitamin A, vitamin K, beta-carotene [4] (provitamin A), and vitamin C [5]. Shi et al. [6] emphasized the potential health benefits of consuming lettuce, including anti-inflammatory effects, improved cardiovascular health, and potential anticancer properties. The presence of antioxidants in lettuce is linked to enhanced immune function and overall health. However, lettuce also contains anti-nutrients such as nitrates, phytates, tannins, oxalates, and cyanogenic glycosides, which may have adverse effects. Additionally, alkaloids in lettuce can cause gastrointestinal tract and nervous system disorders. These anti-nutritional factors can influence nutrient absorption and health. Understanding these effects is crucial for optimizing nutrition.

Lettuces thrive in temperate and subtropical regions throughout the year, and the primary cultivation of lettuces occurs in Asia, North America, and Europe. According to the United Nations Food and Agriculture Organization (FAO), Asia produces 50% of the world's lettuce, followed by North America with 27% and Europe with 21%. Mexico's Agrifood and Fisheries Information Service (SIAP) reported a production of 523,000 tons of lettuce in 2023, and a production of 530,000 tons is expected for 2024. In particular, Mexico annually exports an average of 260,000 tons of lettuce to major partners such as the United States, Canada, Japan, Costa Rica, and Panama. This makes Mexico the third largest lettuce exporter globally, after the United States and Spain, as stated by the Spanish Federation of Associations of Producers Exporters of Fruits, Vegetables, Flowers, and Live Plants (FEPEX).

The Ministry of Agriculture, Livestock, Rural Development, Fisheries and Food (SADER) reports that lettuce is grown in 21 states of Mexico. Guanajuato is the top lettuce producer, accounting for 28% of the national production, followed by Zacatecas with 17.8% and Aguascalientes with 14.8%. Guanajuato's climate is ideal for vegetable production. Factors influencing the growth and development of greenhouse seedlings include crop irrigation, fertilizer use, substrate, and climate change.

Lettuce has two global cultivation trends: organic or hydroponic and inorganic. Hydroponics is the current trend being used to reduce fertilizer use [7–14], while inorganic cultivation is conventional and is used more widely around the world due to its low production and commercialization costs [15,16]. Organic or hydroponic cultivation emphasizes soil fertility and biological activity, minimizes the use of nonrenewable resources, and does not use fertilizers. In contrast, inorganic cultivation uses disproportionately nonrenewable resources and fertilizers, while focusing little on soil fertility and plant biological activity [17–22].

The process of detecting macroelements is costly for greenhouses in Mexico, primarily due to the expenses associated with laboratory analyses and the need for specialized equipment. Additionally, the time required to obtain and analyze the results from traditional laboratory tests can lead to delays in addressing macroelement deficiencies, thus potentially impacting crop health and overall yield. These challenges emphasize the necessity for a more efficient and cost-effective methodology for identifying and addressing macroelement deficiencies in greenhouse cultivation.

Artificial intelligence (AI) is transforming agricultural practices by safeguarding crop yields; reducing excessive use of water, pesticides, and herbicides; preserving soil fertility; automating tasks; and optimizing labor efficiency, thus improving productivity and improving quality [23]. This paper presents a technique for automating the identification of macroelements by processing point clouds and images with AI techniques, eliminating the need for laboratory analyses and providing valuable assistance to agronomists in forecasting crop yields.

The objective of this study is to identify macroelements to address practical challenges in greenhouses located in central Mexico. Our goal is to accomplish this using affordable equipment and without disrupting existing conditions. We also adapted the prototype and methodology to minimize production time and costs. Our solution is adaptable and can be customized to suit various greenhouse configurations.

Our methodology presents a cost-effective and minimally invasive alternative to conventional laboratory tests. By employing deep learning algorithms alongside classical methods, our approach aims to detect macroelement deficiencies in lettuce seedlings within greenhouse environments. This process integrates both expert simulation and laboratory testing to validate the presence of macroelement deficiencies. Point clouds are employed to capture the morphology of seedlings and identify nitrogen (N) deficiencies, while neural network imagery is utilized for the detection of potassium (K) deficiencies. This aims to minimize the consumption of fertilizers, water, and overall production costs. To this end, we propose the ScanSeedling v2.0 prototype for 3D scanning seedlings, as depicted in Figure 1.

Figure 1. ScanSeedling v2.0: prototype scan of entire crops. The sensor moves along the x-axis to 2 m, the y-axis to 6 m, and the z-axis to a height of 0.60 m.

2. Related Works

Automatic detection of macronutrient deficiencies in crops using artificial intelligence has been a significant area of research, with various studies exploring different methodologies and technologies. This section reviews related works and contrasts them with our approach, which involves using datasets of lettuce seedlings, point cloud registration, and neural networks for deficiency detection.

Several studies have used machine learning and computer vision techniques to detect nutrient deficiency in crops. For example, a study in maize plants used unsupervised machine learning algorithms to identify deficiencies, emphasizing the importance of image preprocessing and segmentation for accurate classification [24]. Similarly, MobileNet has been used to detect nutrient deficiencies with high precision, using depth-wise and point-wise convolutions for efficient processing [25].

In the realm of image processing, another study focused on using RGB color features and texture recognition to build datasets for training supervised machine learning models to identify nutritional deficiencies in crops [26]. This approach highlights the role of image features in enhancing model performance, which aligns with our use of 3D morphology features derived from point clouds.

Signal-based deep learning methods have also been explored, as seen in research involving lime trees. This study compared the performance of recurrent neural networks (RNNs) and multilayer perception (MLP) models, finding that MLP achieved higher accuracy in detecting nutrient deficiencies [27]; while our work primarily focuses on image and point cloud data, this study stresses the potential of alternative data sources and model architectures.

The importance of macronutrients such as nitrogen, phosphorus, and potassium (NPK) is well documented, and AI techniques are applied to identify deficiencies through image recognition and analysis [28]. Our research contributes to this field by providing a novel dataset and methodology tailored to lettuce crops, demonstrating high accuracy in detecting deficiencies.

A comprehensive survey of computer vision and machine learning approaches for the detection of nutrient deficiency underscores the use of remote sensing, UAVs, and IoT-based sensors [29]; while our study does not incorporate these technologies, it aligns with the broader trend of leveraging AI for noninvasive crop monitoring. In oil palm trees,

Landsat-8 imagery and machine learning were used to classify macronutrient levels, with varying success depending on the nutrient [30]. This study illustrates the challenges of remote sensing data, which contrasts with our direct imaging and point cloud approach, which offers more detailed morphological insights. Moreover, a review of plant nutrient deficiency detection emphasizes the need for continued observation and the role of AI in automating this process [28]. Our work advances this goal by integrating AI with novel data sources for real-time deficiency detection.

Traditional methods for detecting macronutrient deficiencies involve visual inspections, laboratory tests, and 2D image analysis, which are often limited by human error, time consumption [31,32], and the inability to capture the complete three-dimensional structure of plants [33,34]. These approaches also rely on invasive sampling techniques that can damage crops and are not practical for large-scale applications. In contrast, iterative closest point (ICP) [35–38] allows for precise point clouds registration, representing the 3D morphology of plants, such as lettuce seedlings [39]. By aligning these point clouds, ICP facilitates the extraction of detailed 3D morphological features, enabling accurate identification of nutrient deficiencies. When combined with image and point cloud data, this noninvasive approach allows continuous, real-time monitoring of crops without causing harm, making it scalable for large agricultural operations [40,41]. This capability supports early detection of deficiencies, enabling timely interventions to maintain crop health.

There are other approaches focused on food safety and public health, particularly in the context of agricultural practices. The study developed in [42] presents an effective model for detecting pesticide residues in edible parts of vegetables, specifically tomatoes, cabbages, carrots, and green peppers. A dataset consisting of 1094 images of both contaminated and uncontaminated vegetables was collected. The images were taken using an infrared thermal camera and underwent preprocessing steps such as noise removal and grayscale conversion. The researchers used a convolutional neural network (CNN) for feature extraction to detect specific pesticide residues (mancozeb, dioxacarb, and methidathion). Various transfer learning models, including Inception V3, VGG16, VGG19, and ResNet50, were tested, with Inception V3 achieving the highest classification accuracy of 96 77%. This research not only improves the ability to detect harmful chemical residues but also has implications for improving agricultural practices and consumer safety, thus contributing to better health outcomes and informed decision making in the agricultural sector.

On the other hand, models such as YOLO have been used in agriculture and livestock applications through object detection and instance segmentation techniques. The work developed in [43] presents a significant advancement in the field of agricultural technology by enhancing the efficiency and effectiveness of fruit detection systems. The study indicates that the model can accurately detect pitaya fruits under varying light conditions, which is crucial for agricultural applications where lighting can change significantly throughout the day. The YOLOv5s model was used to detect pitaya fruits in real time. Their model achieved a precision of 97.80%, a recall rate of 91. 20%, and a frame-per-second (FPS) rate of 139 with GPU. In the field of livestock, the approach from [44] developed and applied a YOLOv7 model for automated cattle detection and counting using drone imagery. This model integrates object detection and instance segmentation, allowing for high-speed and accurate identification of cattle in various environments. The authors demonstrate that Mask YOLOv7 significantly improves the accuracy of counting compared to traditional methods, achieving 93% accuracy in controlled environments and 95% in uncontrolled environments. This advancement showcases the potential of combining drone technology with deep learning algorithms for real-time monitoring of farm animals.

Compared to these studies, our research stands out by utilizing a unique combination of 3D point cloud data and neural networks, achieving notable accuracies with the decision tree classifier (DTC) and the YOLOv9c model. This approach not only provides a high level of precision but also offers a scalable solution for detecting macronutrient deficiencies in lettuce, potentially applicable to other crops as well.

3. Methods

The proposed system is designed to detect two specific macronutrient deficiencies in Rhodena lettuce: nitrogen (N) and potassium (K). These deficiencies manifest differently in crops, requiring different detection approaches.

Nitrogen deficiency in crops often results in stunted growth and reduced biomass [45]. To detect this, our system analyzes the morphology of lettuce seedlings using 3D point clouds and extracts key morphological characteristics, such as plant height and leaf area. We used these features as input to various machine learning methods to classify nitrogen deficiency.

However, potassium deficiency usually leads to abnormal leaf colorations, such as chlorosis or necrosis [46,47]. To identify these visual symptoms, we trained different YOLO models to detect and classify characteristic color changes associated with potassium deficiency.

The characteristics of two macronutrient deficiencies are illustrated in Figure 2. Figure 2a shows the lack of size due to a deficiency in nitrogen. Figure 2b,c show the yellow coloration of the veins and leaves attributed to potassium deficiency.

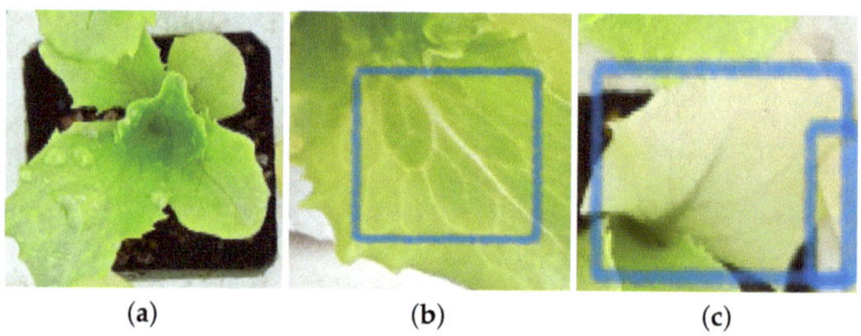

Figure 2. Diseases seedling Rhodena lettuce. Panel (**a**) shows nitrogen deficiencies, and (**b**,**c**) show potassium (K) deficiencies.

Figure 3 illustrates the methodology proposed in the following sections.

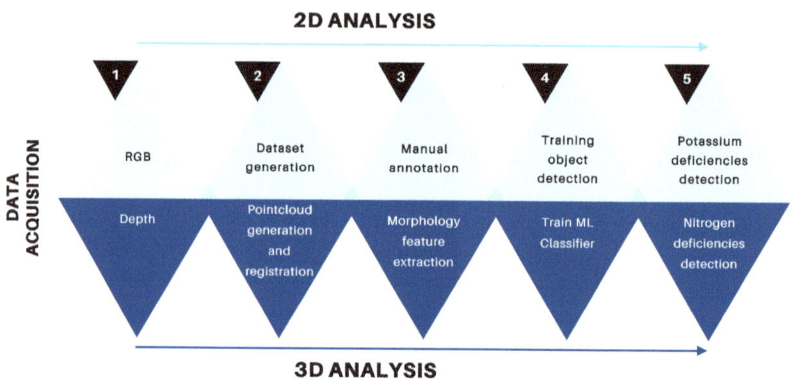

Figure 3. Overview of the proposed methodology for detecting macronutrient deficiencies in Rhodena lettuce seedlings. The top row illustrates the steps for 2D data analysis, which focuses on detecting potassium deficiencies using object detection models. The bottom row shows the steps for 3D data analysis, which are used to detect nitrogen deficiencies through morphological feature extraction and machine learning classifiers.

1. Data acquisition. In this step, we created two datasets of Rhodena lettuce seedling images labeled with macroelements and point clouds.

2. Point cloud registration. We used the iterative closest point (ICP) algorithm to register the different point clouds of the crops.
3. Training and evaluation. We trained and evaluated multiple machine learning models to detect the lack of nitrogen based on the seedlings morphology and we also trained multiple object detection architectures, including YOLOv8 [48], YOLOv9 [49], and YOLOv10 [50], to detect the lack of potassium based on the lettuces texture.
4. Results comparison: In this final step, we compare the results obtained from laboratory tests with those obtained from our system.

3.1. Data Acquisition

We collected two types of data: images and 3D point clouds using a Kinect v2.0 RGBD sensor mounted on a cart, as shown in Figure 1. The sensor is placed at a height of z = 60 cm relative to the seedling. It moves along the (x, y) axes at programmable distances based on the characteristics of the trays. The x axis has a distance of 4.20 m and the y axis has a distance of 1.70 m. During each movement of 40 cm along x and 30 cm along y, we captured a scan using the RGBD sensor to obtain different viewing angles of the crop.

We captured the data 15 days after germination. This timing allowed the agronomist to determine the crop yield and optimize fertilizer use. The entire crop includes 12 trays with 162 cavities, each containing Rhodena lettuce seedlings, for a total of 1944 cavities with lettuce seedlings.

We also acquired RGB images on the 15th and 28th days after germination at a resolution of 3000 × 3000 pixels, totaling 640 images from the crop. Then, we manually annotated macronutrient deficiency into two categories: healthy and disease.

3.2. Point Cloud Registration

Once the 3D scans of the crop were collected, we registered the 3D point clouds using the iterative closest point (ICP) as proposed by Zhang et al. [51]. In this technique, given two point sets, P and Q, the goal is to optimize a rigid transformation on P to align it with Q. ICP solves this problem using an iterative approach that alternates between two steps: correspondence and alignment.

However, the classical ICP is slow to converge to a local minimum due to its linear convergence rate. To address this issue, we used the Anderson acceleration [52]. This approach parameterizes a rigid transformation using another set of variables X, such that any value of X corresponds to a valid rigid transformation, and the ICP iteration can be rewritten as follows:

$$X(k+1) = G_{ICP}(X(k)) \tag{1}$$

Then, Anderson acceleration can be applied to the variable X by performing the following steps in each iteration:

1. From the current variable $X(k)$, recover the rotation matrix $R(k)$ and translation vector $t(k)$.
2. Perform the ICP update $(R_0, t_0) = G_{ICP}(R(k), t(k))$.
3. Compute the parameterization of (R_0, t_0) to obtain $G_{ICP}(X(k))$.
4. Compute the accelerated value X_{AA} with Anderson acceleration using $X(k-m), \ldots,$ $X(k)$ and $G_{ICP}(X(k-m)), \ldots, G_{ICP}(X(k))$.

   Where:

$$X(k+1)AA = G(X(k)) - \sum j = 1^m \theta_j^* (G(X(k-j+1)) - G(X(k-j))) \tag{2}$$

The registration process allowed us to apply the methodology proposed in [53] to extract morphological characteristics.

3.3. Detection of Nitrogen Deficiencies

Nitrogen deficiency in crops often leads to stunted growth and reduced biomass. Our system detects this by analyzing the morphology of lettuce seedlings using key morphological characteristics extracted from the registered point cloud.

To extract morphological characteristics, we measure the seedling size, the plant's germination stage, and the number of leaves and classify them into three stages: first, second, or third.

The height of the crop indicates a nitrogen deficiency, with an average crop height of 3.4 cm. A healthy fertilized seedling reaches an average height of 6 cm, which marks the first stage of growth. The nitrogen element N becomes apparent around day 15 after germination. In the first stage, the average height is 6 cm; in the second stage, the height ranges around 4 cm; in the third stage, the height is less than 3 cm; and the fourth stage corresponds to seedlings that did not germinate.

We annotated a total of 1944 cavities with lettuce seedlings and split the data into 70% for training and 30% for validation.

Then, we trained machine learning classifiers that included a support vector machine (SVM), neighborhood component analysis (NCA), decision tree classifier (DTC), and a linear model with stochastic gradient descent (SGD).

3.4. Detection of Potassium Deficiency

Potassium deficiency usually results in abnormal leaf colorations, such as chlorosis or necrosis. To recognize these visual symptoms, we compared various YOLO [54] models trained to detect and classify characteristic color changes related to potassium deficiency.

YOLO models feature a unified detector that utilizes regression, associated class probabilities, and detection bounding boxes. YOLO utilizes global features, and its main advantage is its speed, allowing real-time detection on standard hardware. In general, the YOLO algorithm can be summarized in three main components:

1. Single step grid predictions. It divides the image I into $M \times M$ grid cells, where each cell contains detection predictions.
2. Bounding box regression. It determines the bounding boxes, which correspond to the rectangles that contain the objects in the image. The attributes of these bounding boxes are determined using a single regression module as shown in Equation (3).

$$Y = [p, x, y, h, w, c] \tag{3}$$

   Here, Y is the vector representation of each bounding box, p is the probability score of each grid cell containing an object [0, 1], x, y are the coordinates of the center of the bounding box concerning the grid cell, h, w is the height and width of the bounding box to the cell, and c is the class for the n number of classes.
3. Nonmaximum suppression (NMS). An object may have several overlapped detections. NMS keeps the predictions with the highest detection confidence.

YOLO drawbacks include the potential failure to detect small objects and less likely prediction of false positives in the background. Early versions were prone to localization errors.

This work compares three of the most recent YOLO models: YOLOv8 [48], YOLOv9 [49], and YOLOv10 [50], described in the following sections.

3.4.1. YOLOv8

The architecture of YOLOv8 is designed to enhance real-time object detection capabilities with improved accuracy and efficiency. YOLOv8 is structured around three main components: the backbone, the neck, and the head. The backbone, which is responsible for feature extraction, utilizes CSPDarknet53, a variant of the Darknet architecture. This version incorporates cross-stage partial (CSP) connections, which improve information flow and gradient propagation during training. The neck of YOLOv8 employs a path aggregation network (PANet) to facilitate multiscale feature fusion, crucial for detecting objects of varying sizes within an image. The head consists of multiple detection heads that predict bounding boxes, class probabilities, and objectness scores, incorporating dynamic anchor assignment and a novel intersection over union (IoU) loss function to enhance detection accuracy and handle overlapping objects more effectively [55]. Figure 4 shows the architecture of YOLOv8.

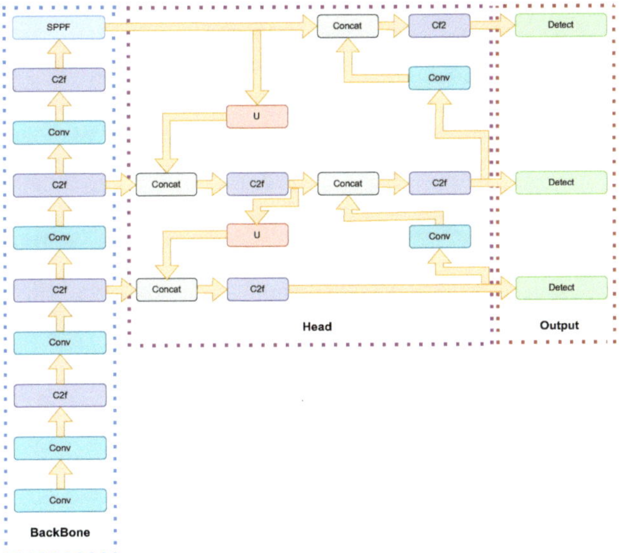

Figure 4. YOLOv8 network architecture: The diagram illustrates the structure of YOLOv8, divided into three main components: backbone, head, and output. The backbone is responsible for feature extraction and includes multiple convolutional (Conv) and C2f layers, along with a spatial pyramid pooling-fast (SPPF) block. The head performs feature aggregation and enhancement through multiple concatenation (Concat) and upsampling (U) operations, followed by additional C2f layers. Finally, the output section applies detection layers to produce the final object detection results.

YOLOv8 also introduces several advancements in training strategies and model variants to cater to different application needs. The model employs adaptive training techniques, including advanced data augmentation methods like MixUp and CutMix, to improve robustness and generalization. Additionally, YOLOv8 supports multiple backbones, such as EfficientNet and ResNet, allowing users to customize the architecture based on specific requirements. The availability of different model variants, such as YOLOv8-tiny and YOLOv8x, provides flexibility in balancing accuracy and computational efficiency [48].

3.4.2. YOLOv9

YOLOv9 represents a significant advancement in the YOLO series, introducing innovative techniques to enhance object detection performance. The architecture of YOLOv9 is built upon the foundation of YOLOv7, with substantial improvements through the integration of two key innovations: programmable gradient information (PGI) and the generalized efficient layer aggregation network (GELAN). PGI addresses the issue of information bottleneck, a common challenge in deep neural networks where crucial data can be lost as they progress through layers. By implementing PGI, YOLOv9 ensures efficient gradient flow and reliable gradient backpropagation, which enhances the model's learning capabilities and accuracy. This is achieved through an auxiliary reversible branch that provides additional pathways for gradient flow, thereby preserving essential information throughout the training process. Figure 5 shows the architecture of YOLOv9 [49].

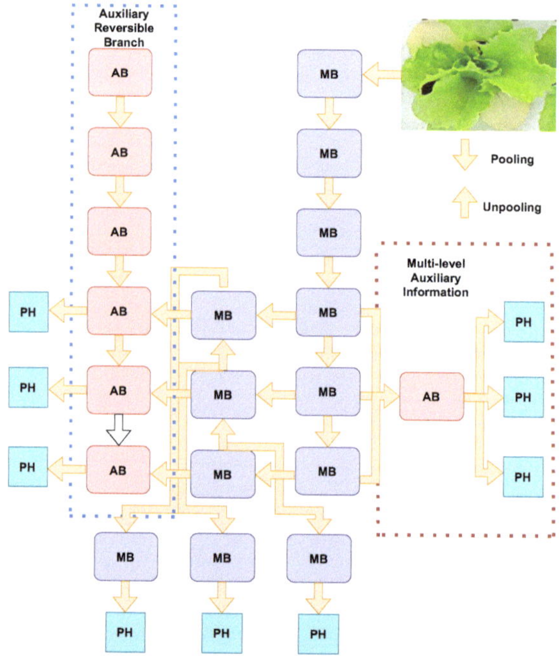

Figure 5. YOLOv9 programmable gradient information (PGI) architecture. The model is composed of an auxiliary reversible branch (left), a main processing branch (center), and a multilevel auxiliary information module (right). The auxiliary reversible branch utilizes a series of AB (auxiliary block) modules, while the main processing branch consists of MB (main block) modules. The multilevel auxiliary information module combines outputs from both branches and incorporates PH (pooling head) modules to integrate and process multiscale features. PGI enables dynamic information flow across different levels of the network, facilitating efficient detection and classification tasks.

GELAN further complements the PGI framework by focusing on efficient layer aggregation and parameter utilization. It combines principles from CSPNet and ELAN to optimize gradient path planning and feature integration, resulting in a lightweight yet robust architecture. This design allows YOLOv9 to maintain high inference speed and accuracy while using fewer parameters and computational resources compared to its predecessors. The architecture supports multiple model variants, such as YOLOv9-S, YOLOv9-M, YOLOv9-C, and YOLOv9-E, catering to different application needs from lightweight to performance-intensive scenarios [56].

3.4.3. YOLOv10

YOLOv10 introduces several architectural innovations that enhance both efficiency and accuracy in real-time object detection. One of the key advancements is the implementation of a consistent dual assignments strategy, which eliminates the need for nonmaximum suppression (NMS) during inference. This approach significantly reduces inference latency while maintaining competitive performance by combining one-to-one and one-to-many label assignments during training [50]. Furthermore, YOLOv10 adopts a holistic model design based on efficiency accuracy, optimizing various components to minimize computational overhead. This includes a lightweight classification head using depth-wise separable convolutions, spatial-channel decoupled downsampling to reduce information loss and computational cost, and a rank-guided block design to efficiently allocate computational resources based on stage redundancy. Figure 6 shows a summarized architecture of YOLOv10.

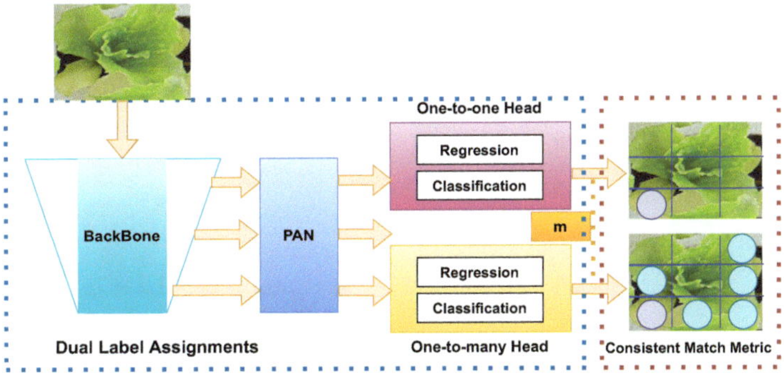

Figure 6. YOLOv10 NMS-free architecture with one-to-many matching. YOLOv10 eliminates the need for nonmaximum suppression (NMS) by incorporating a dual-label assignment framework. The backbone network extracts image features, which are processed by the path aggregation network (PAN). The model utilizes two heads: a one-to-one head for precise regression and classification, and a one-to-many head designed to handle a broader range of object sizes. The consistent match metric (CMM) ensures that the model consistently matches detected objects across different scales, enhancing the accuracy and robustness of detections.

To further enhance accuracy, YOLOv10 incorporates large-kernel convolutions and a partial self-attention (PSA) module. Large-kernel convolutions expand the receptive field, improving the model's ability to capture detailed features, particularly beneficial for small object detection. The PSA module integrates global representation learning with minimal computational cost by processing only a portion of the features through multihead self-attention, thus avoiding the overhead associated with full self-attention mechanisms. These architectural improvements allow YOLOv10 to achieve superior performance metrics, such as higher mean average precision (mAP) and reduced latency compared to previous YOLO versions and other state-of-the-art models, making it highly suitable for real-time applications.

3.5. Evaluation Metrics

The metrics used to evaluate potassium deficiencies are precision, recall, and mean average precision (mAP).

Mean average precision (mAP) is a widely used metric for evaluating the performance of object detection models. Combining precision and recall, it provides a comprehensive measure of a model's ability to detect objects accurately.

Precision measures the fraction of true positive detections out of all positive detections made by the model, and indicates how many of the detected objects are actually correct, as defined in Equation (4).

$$\text{Precision} = \frac{\text{TP}}{\text{TP} + \text{FP}} \tag{4}$$

Recall measures the fraction of true positive detections out of all actual objects in the image, and indicates how many of the actual objects were detected by the model, as defined by Equation (5).

$$\text{Recall} = \frac{\text{TP}}{\text{TP} + \text{FN}} \tag{5}$$

Intersection over union (IoU) is used to evaluate the accuracy of an object detector by comparing the overlap between the predicted bounding box and the ground truth bounding box. A higher IoU indicates better overlap and, thus, a more accurate detection.

Average precision (AP) is calculated for each class by integrating the area under the precision–recall curve. It provides a single score summarizing the precision–recall trade-off for that class. mAP is calculated by averaging the AP scores over all classes and possibly over different IoU thresholds. This ensures that the evaluation does not rely on a single IoU threshold, which can vary depending on the task and dataset. For example, in the COCO challenge, mAP is calculated by averaging AP over multiple IoU thresholds (from 0.5 to 0.95 in increments of 0.05) across all object classes.

A high mAP score indicates that the model is capable of detecting objects with both high precision and recall, which is crucial for safety and reliability in these applications.

4. Results

The nutritional needs of seedlings in a greenhouse are primarily dependent on the concentration of existing nutrients on the substrate and irrigation. However, the lack of macroelement detection in seedlings, especially lettuce at an early age, allows for appropriate treatment and avoids total loss of production. This work demonstrates that it is possible to match or mimic laboratory tests and proposes the use of AI algorithms to apply only the necessary fertilizers.

There are two types of cycles in cultivation: long and short. The long cycle lasts between 100 and 120 days in winter and 90 to 100 days in autumn. The short cycle lasts between 50 and 70 days in summer and 80 to 90 days in spring. This work was performed in the short spring cycle, which has a relatively warm climate with temperatures close to 30 °C. In addition, the hours of sunlight are longer, directly impacting the development of the seedlings. The greenhouse is located north of Celaya, Guanajuato, Mexico, at latitude 20.613112 and longitude −100.921866, at an altitude of 1756 m above mean sea level.

4.1. Laboratory Tests

The Rhodena lettuce crop was divided into two groups: healthy and diseased. The diseased group, which covered 80% of the crop, was only watered, while the healthy group (the remaining 20%) was watered and fertilized normally. To verify the values of macroelements, a laboratory conducted destructive tests on both groups of seedlings.

Table 1 shows the laboratory tests of the macroelements nitrogen, phosphorus, calcium, magnesium, and sulfur in percentages, as well as iron, zinc, manganese, copper, and boron in units of parts per million (ppm). The table compares healthy seedlings with diseased seedlings, indicating the necessary nutritional requirements for the seedlings.

Figure 7 shows macroelements, where 1 is very low, 2 low, 3 sufficient, 4 high, and 5 very high. The level of 3 is considered sufficient. In this work, we focus on detecting the lack of nitrogen and potassium, which are very low and the most prevalent in the crop.

Table 1. Test laboratory analysis of macroelement concentrations in Rhodena lettuce seedlings. The table presents the concentrations of multiple elements in parts per million (ppm). The nutritional sufficiency ranges for each element are also provided for reference.

Test Laboratory Lettuce Rhodena				
Element	**Result**		**Sufficiency**	**Unit**
	Healthy	**Diseased**		
Nitrogen	**4.91**	**2.05**	4.50–5.50	%
Phosphorus	0.47	0.32	0.35–0.40	%
Potassium	**6.17**	**2.75**	6.00–7.00	%
Calcium	1.89	0.78	1.00–2.00	%
Magnesium	0.81	1.48	0.30–0.40	%
Sulfur	0.51	0.35	0.20–0.60	%
Iron	1429	3956	50.0–100	ppm
Zinc	76.20	27.49	25.0–100	ppm
Manganese	93.49	51.39	20.0–200	ppm
Copper	6.24	11.25	5.00–10.0	ppm
Boron	52.25	43.28	25.0–80.0	ppm

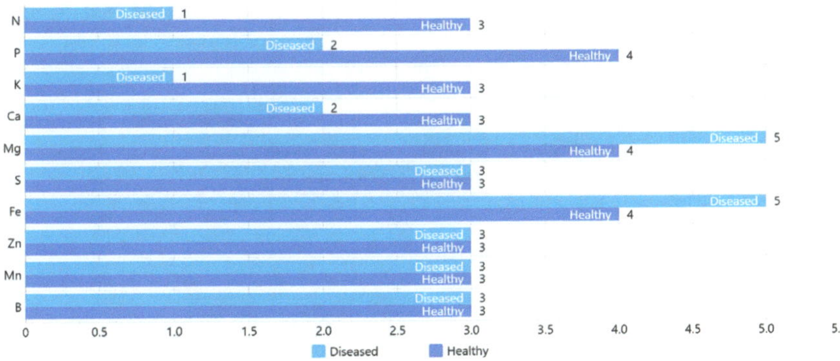

Figure 7. Comparative analysis of macroelement levels in healthy and diseased Rhodena lettuce seedlings. The bar chart categorizes macroelement concentrations into five levels: 1 (very low), 2 (low), 3 (sufficient), 4 (high), and 5 (very high), with level 3 indicating sufficiency. The analysis highlights that nitrogen (N) and potassium (K) are at very low levels, particularly in diseased seedlings.

4.2. Detection of Nitrogen Deficiencies

Using multiple scans taken on the 15th day after germination, we obtain a single point cloud combining multiple scans using ICP, as described in Section 3.2.

Figure 8 shows the alignment of 8 point clouds, each containing 12 trays with 162 cavities. The color variations along the z axis indicate the seedlings' height, and the gaps represent nongerminated seedlings.

From the registered point cloud, we extract the characteristics of each seedling, which serve as inputs for the classifier. These characteristics encompass leaf count, height, and growth stage.

To obtain the height, we use the algorithm presented in [53]. In this procedure, we normalize the points to the agrolite or substate and obtain the highest point point as (x, y). At the time of the scan, the Rhodena lettuce plant had developed to the second stage of growth. The height is used to determine nitrogen deficiency. The crop has an average height of 3.2 cm, while healthy seedlings have an average height of 4 cm. The seedlings in the first stage are those that have received fertilizer.

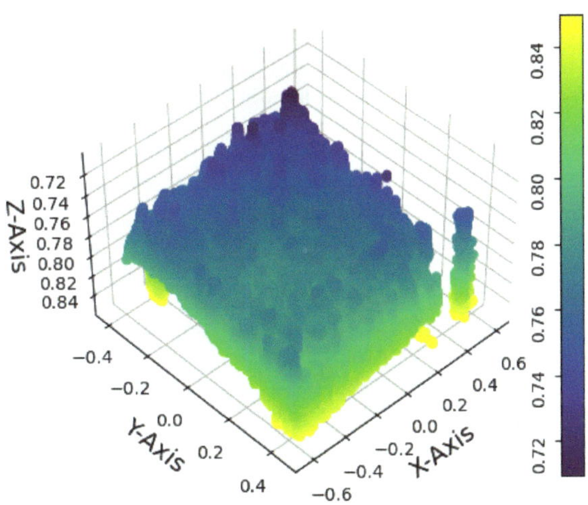

Figure 8. Point clouds registration. The image shows the whole crop area. The colors along the z axis represent the height of the seedlings, while the gaps reveal the seedlings that did not germinate.

The diagram in Figure 9a illustrates the height distribution (in cm) during three different growth stages of Rhodena lettuce. The tallest median height is observed at growth stage 0, while the lowest is seen at growth stage 2. The interquartile ranges vary among stages, indicating differences in height variability. Additionally, Figure 9b displays the distribution of the number of leaves at each growth stage. Growth stage 0 exhibits the highest median number of leaves, while growth stages 1 and 2 have similar median values but lower overall leaf counts.

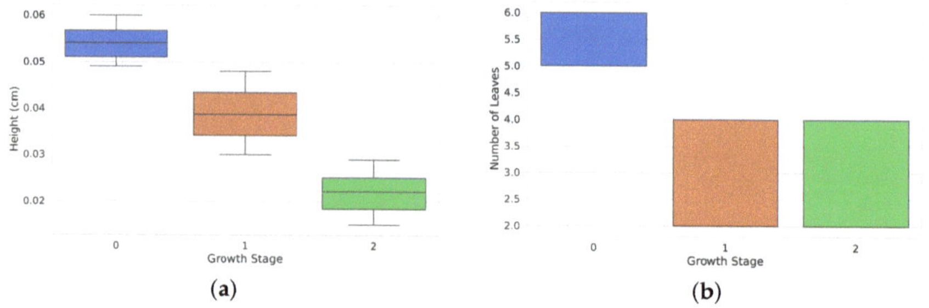

Figure 9. Heights and leaves distribution across development stages. Panel (**a**) shows the height distribution (in cm) of lettuces at different growth stages (0, 1, and 2). Height decreases progressively from Stage 0 to Stage 2. (**b**) Bar plot showing the number of leaves at each growth stage, with Stage 0 having the most leaves, and Stages 1 and 2 showing similar and lower leaf counts.

Figure 10a displays the distribution of plant heights in centimeters. The data show a relatively symmetrical spread with a peak of around 0.04 cm, indicating that most plants fall within this height range. The distribution shows some variability, with heights ranging from approximately 0.02 cm to 0.06 cm. Figure 10b illustrates the distribution of the number of leaves per plant. The distribution is bimodal, with two prominent peaks at 2 and 4 leaves, suggesting that most plants have 2 or 4 leaves, with fewer occurrences of other leaf counts. Figure 10c shows the frequency of plants in different stages of growth (0, 1, and 2). Growth stage 1 has the highest frequency, followed by Stage 2, while Stage 0 has the lowest frequency, indicating the prevalence of plants in intermediate growth stages.

The scatter plots in Figure 11 illustrate the three growth stages of the Rhodena lettuce seedling. In a healthy crop without diseases, most of the seedlings would be at this stage. However, our experiment shows that a high percentage of the monitored crop plants are in the second growth stage. Specifically, Figure 11a displays a lettuce plant with four true leaves and two cotyledons, standing at a height of 4 cm. Figure 11b depicts four leaves, two cotyledons, and two first true leaves with a height of 3 cm. Lastly, Figure 11c shows two cotyledons with a height of 2 cm. The first, second, and third stages represent 5.81%, 53.98%, and 25.45% of the crop, respectively, with the remaining 5.63% representing failures.

We trained multiple machine learning models to classify lettuce seedlings into four categories: first stage, second stage, third stage, and failures. The results of the machine learning classifiers to classify nitrogen deficiencies are shown in Table 2. Each cell in the table represents the percentage of seedlings at different stages. The sum of the percentages in the first, second, and third stages indicates the overall production. Adding all stages and failures gives the accuracy of the classifiers. The decision tree classifier (DTC) attained the highest accuracy at 90.87% and a crop production rate of 85.24% on the 15th day. The percentage of seedlings in the third and second stages indicates limited development of the lettuce plant.

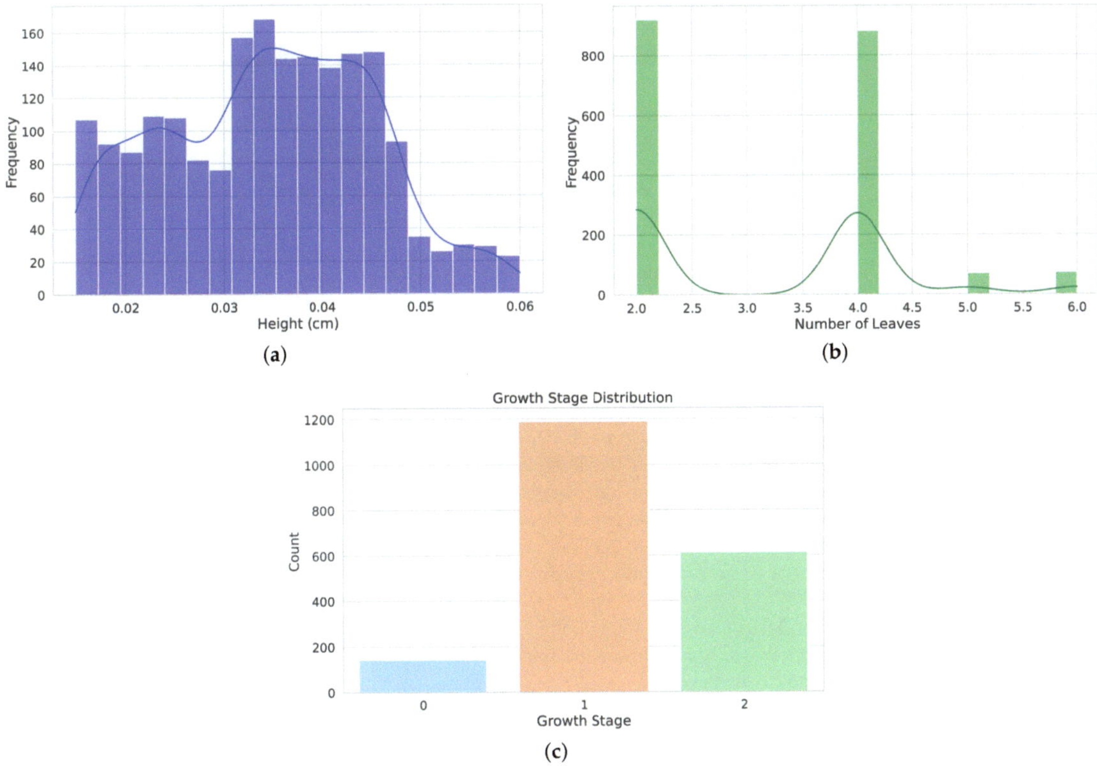

Figure 10. Distribution of morphological measurements and grown stages. (**a**) shows an histogram of plant height distribution in centimeters, showing a peak around 0.04 cm with a broad spread from 0.02 cm to 0.06 cm. (**b**) Histogram of the distribution of the number of leaves, highlighting two prominent peaks at 2 and 4 leaves. (**c**) Bar plot showing the distribution of growth stages.

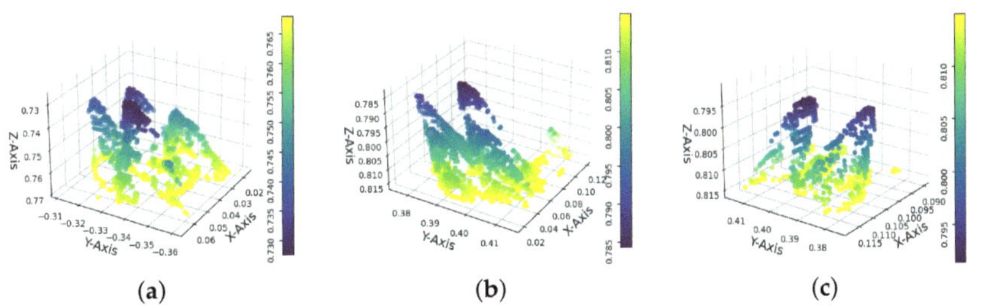

Figure 11. Developmental phases of lettuce. Panel (**a**) exhibits four true leaves, two cotyledons, and a height of 4 cm. Panel (**b**) presents two true leaves and two cotyledons with a height of 3 cm. Panel (**c**) displays two cotyledons with a height of 2 cm. According to the DTC technique, optimal results were achieved by delineating the plant morphology for each growth stage with an accuracy of 90.87% and a production forecast of 85.24%.

Table 2. Crop production accuracy by growth stage of each classifier production accuracy. The table presents the prediction results of various classifiers, detailing the percentage of lettuce seedlings classified into different growth stages (third, second, first, and failures). The overall production is calculated by summing the percentages across the first, second, and third stages. The accuracy of each classifier is determined by summing the percentages of all stages and failures.

	Classifier Crop Production %					
Stage	**NCA**	**SGD**	**DTC**	**Linear**	**RBF**	**Poly**
Third	30.1	27.42	25.45	23.49	23.9	22.01
Second	52.13	48.34	53.98	53.65	52.12	53.87
First	1.87	2.18	5.81	2.02	5.95	5.36
Fails	4.94	6.54	5.63	6.03	4.80	6.08
Accuracy	89.04	84.48	**90.87**	85.19	86.77	87.32
15th day Prod.	84.01	77.94	**85.24**	79.16	81.97	81.24

4.3. Detection of Potassium Deficiency

Lack of potassium in seedlings can cause yellow marks to appear in the veins of the leaves or on the leaves themselves, usually around day 20 after germination.

We conducted training using manually annotated images to train YOLOv8, YOLOv9, and YOLOv10 models for detecting healthy and diseased categories. Table 3 presents the precision, recall, and mAP50 for each model. Specifically, YOLOv9c achieved the highest mAP of 0.792 among the three detectors.

Table 3. Comparison of model accuracy across YOLO versions: The table presents the precision, recall, and mAP50 values for different versions of YOLO models—YOLOv8, YOLOv9, and YOLOv10—trained on a dataset containing manually annotated images with healthy and diseased categories. The highest-performing variant within each YOLO model family is highlighted in bold.

Detector	**Model**	**Precision**	**Recall**	**mAP50**
	n	0.750	0.701	0.736
	s	**0.633**	**0.782**	**0.747**
YOLOv8	m	0.788	0.658	0.729
		0.786	0.658	0.713
	x	0.627	0.765	0.737
	t	0.670	0.792	0.772
	s	0.702	0.778	0.784
YOLOv9	m	0.653	0.798	0.783
	c	**0.701**	**0.79**	**0.792**
	e	0.66	0.792	0.784
	n	0.739	0.703	0.716
	s	0.731	0.730	0.740
YOLOv10	m	0.800	0.639	0.740
	b	0.775	0.657	0.715
	l	**0.679**	**0.752**	**0.750**
	x	0.764	0.687	0.726

Figure 12 presents the precision/recall curves for the three most proficient neural network models examined in this study. The YOLOv8s model is depicted in blue with an mAP of 74.7%, YOLOv10l in green with 75%, and YOLOv9c in orange with 79.2%. Notably, the YOLOv9c model exhibited the superior performance.

Figure 13 presents the detection results obtained from the YOLOv8s, YOLOv9c, and YOLOv10l models. The initial row illustrates the manually annotated ground truth. Subsequently, the second, third, and fourth rows depict the predictions generated by the YOLOv8s, YOLOv9c, and YOLOv10l models, respectively.

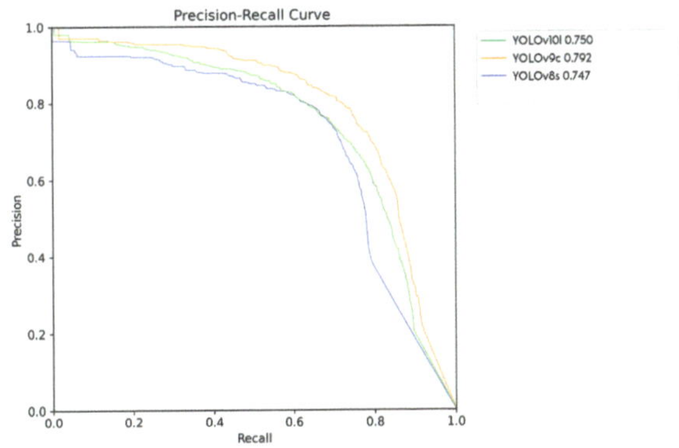

Figure 12. Precision/recall curves corresponding to the evaluation of the three most effective models.

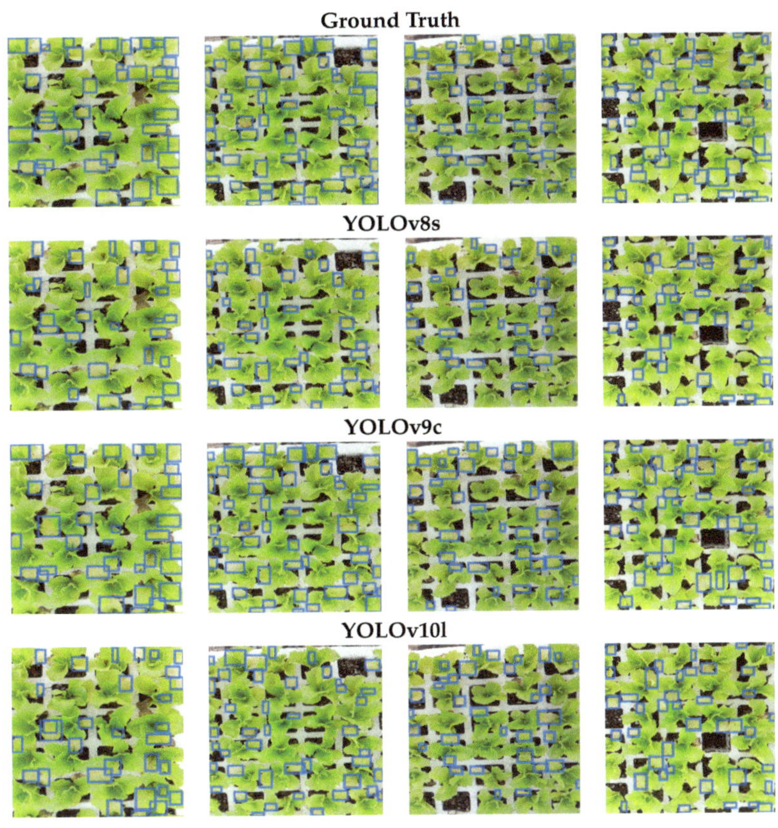

Figure 13. Potassium deficiency detection. The first row displays manually labeled ground truth detections. The second, third, and fourth rows present the predictions for YOLOv8s, YOLOv9c, and YOLOv10l, respectively.

5. Discussion and Conclusions

This study introduces a novel method for detecting micronutrient deficiencies in Rhodena greenhouse-grown lettuces via 2D and 3D analysis. We used a depth camera to capture 3D point clouds to estimate nitrogen deficiencies through morphological variables, while a deep learning detector analyzed 2D images to identify potassium deficiencies.

Conventional methods for detecting nutrient deficiencies predominantly depend on laboratory analyses and visual inspections, which are inherently time-intensive, costly, and susceptible to human error. Our methodology proposes a noninvasive, real-time alternative that markedly diminishes the necessity for laboratory testing. In identifying nitrogen deficiencies, we evaluated several machine learning techniques, with the decision tree classifier (DTC) attaining the highest accuracy at 90.87%. For the detection of potassium deficiencies, YOLOv9c demonstrated an accuracy of 79.20%. These findings emphasize the promise of AI-driven methodologies in agricultural applications.

The capability to precisely identify nutrient deficiencies at an early stage is indispensable for optimizing fertilizer use and ensuring the robust development of crops. The results indicate that the proposed methodology can be effectively utilized for monitoring lettuce crops, potentially diminishing the necessity for excessive fertilizer usage, thereby yielding economic and environmental advantages. Furthermore, this approach is adaptable to other crops, providing a scalable solution for nutrient monitoring across diverse agricultural contexts.

Although the results are promising, this study has limitations. The performance of the model is highly dependent upon the quality of the input data, specifically the resolution of the images and the precision of the point cloud registration. Likewise, the efficacy of the method in varying environmental conditions or with different varieties of lettuce has yet to be comprehensively examined. Future research should aim to evaluate the methodology in diverse settings and with a variety of crops to determine its broader applicability.

Furthermore, the existing system predominantly addresses deficiencies in nitrogen and potassium. Extending the model to identify other macronutrients, such as phosphorus and magnesium, would significantly increase its efficacy. Additionally, the incorporation of IoT devices for continuous monitoring and automated decision-making systems could be investigated to develop a more holistic crop management tool.

Author Contributions: A.R.-P. was involved in methodology, software development, prototype implementation, and writing—original draft. S.S.-C. was involved in data preprocessing and hand labeling. J.T. was involved in prototype implementation and data acquisition. D.-M.C.-E. and J.-A.R.-G. were involved in writing—original draft. J.-J.G.-B. was involved in prototype design and software validation. All authors have read and agreed to the published version of the manuscript.

Funding: This work was funded by Instituto Politecnico Nacional research grant SIP 20240760, and by the National Council of Humanities, Sciences, and Technologies (CONAHCYT) of Mexico for supporting project number 669 and CIR/026/2024.

Data Availability Statement: Dataset will be made available on request to the authors.

Acknowledgments: Thanks are given to the Greenhouse Solecito, especially to engineers and collaborators.

Conflicts of Interest: The authors declare no conflicts of interest.

References

1. Natesh, H.; Abbey, L.; Asiedu, S. An overview of nutritional and antinutritional factors in green leafy vegetables. *Hortic. Int. J.* **2017**, *1*, 58–65. [CrossRef]
2. Kumar, D.; Kumar, S.; Shekhar, C. Nutritional components in green leafy vegetables: A review. *J. Pharmacogn. Phytochem.* **2020**, *9*, 2498–2502. Available online: https://www.phytojournal.com/archives/2020.v9.i5.12718/nutritional-components-in-green-leafy-vegetables-a-review (accessed on 1 August 2024).
3. Gupta, K.; Barat, G.; Wagle, D.; Chawla, H. Nutrient contents and antinutritional factors in conventional and non-conventional leafy vegetables. *Food Chem.* **1989**, *31*, 105–116. [CrossRef]
4. Das, R.; Bhattacharjee, C. Chapter 9—Lettuce. In *Nutritional Composition and Antioxidant Properties of Fruits and Vegetables*; Jaiswal, A.K., Ed.; Academic Press: Cambridge, MA, USA, 2020; pp. 143–157. [CrossRef]
5. Medina-Lozano, I.; Bertolín, J.R.; Díaz, A. Nutritional value of commercial and traditional lettuce (*Lactuca sativa* L.) and wild relatives: Vitamin C and anthocyanin content. *Food Chem.* **2021**, *359*, 129864. [CrossRef]

6. Shi, M.; Gu, J.; Wu, H.; Rauf, A.; Emran, T.B.; Khan, Z.; Mitra, S.; Aljohani, A.S.M.; Alhumaydhi, F.A.; Al-Awthan, Y.S.; et al. Phytochemicals, Nutrition, Metabolism, Bioavailability, and Health Benefits in Lettuce—A Comprehensive Review. *Antioxidants* **2022**, *11*, 1158. [CrossRef]
7. Faran, M.; Nadeem, M.; Manful, C.F.; Galagedara, L.; Thomas, R.H.; Cheema, M. Agronomic Performance and Phytochemical Profile of Lettuce Grown in Anaerobic Dairy Digestate. *Agronomy* **2023**, *13*, 182. [CrossRef]
8. Frasetya, B.; Harisman, K.; Ramdaniah, N.A.H. The effect of hydroponics systems on the growth of lettuce. *IOP Conf. Ser. Mater. Sci. Eng.* **2021**, *1098*, 042115. [CrossRef]
9. Majid, M.; Khan, J.N.; Ahmad Shah, Q.M.; Masoodi, K.Z.; Afroza, B.; Parvaze, S. Evaluation of hydroponic systems for the cultivation of Lettuce (*Lactuca sativa* L., var. *longifolia*) and comparison with protected soil-based cultivation. *Agric. Water Manag.* **2021**, *245*, 106572. [CrossRef]
10. Sularz, O.; Smoleń, S.; Koronowicz, A.; Kowalska, I.; Leszczyńska, T. Chemical Composition of Lettuce (*Lactuca sativa* L.) Biofortified with Iodine by KIO3, 5-Iodo-, and 3.5-Diiodosalicylic Acid in a Hydroponic Cultivation. *Agronomy* **2020**, *10*, 1022. [CrossRef]
11. Mujiono, M.; Suyono, S.; Purwanto, P. Growth and Yield of Lettuce (*Lactuca sativa* L.) Under Organic Cultivation. *Planta Trop.* **2017**, *5*, 127–131. [CrossRef]
12. Cova, A.M.W.; Freitas, F.T.O.d.; Viana, P.C.; Rafael, M.R.S.; Azevedo, A.D.d.; Soares, T.M. Content of inorganic solutes in lettuce grown with brackish water in different hydroponic systems. *Rev. Bras. Eng. Agríc. Ambient.* **2017**, *21*, 150–155. [CrossRef]
13. Jose, V.; Gino, A.; Noel, O. Humus líquido y microorganismos para favorecer la producción de lechuga (*Lactuca sativa* var. *crespa*) en cultivo de hidroponía. *J. Selva Andin. Biosph.* **2016**, *4*, 71–83. Available online: http://www.scielo.org.bo/scielo.php?script=sci_arttext&pid=S2308-38592016000200004&nrm=iso (accessed on 1 August 2024). [CrossRef]
14. Monge, C.; Chaves, C.; Arias, M.L. Comparación de la calidad bacteriológica de la lechuga (*Lactuca sativa*) producida en Costa Rica mediante cultivo tradicional, orgánico o hidropónico. *Arch. Latinoam. Nutr. Super.* **2011**, *61*, 69–73. Available online: http://ve.scielo.org/scielo.php?script=sci_arttext&pid=S0004-06222011000100009&nrm=iso (accessed on 1 August 2024).
15. Saah, K.J.A.; Kaba, J.S.; Abunyewa, A.A. Inorganic nitrogen fertilizer, biochar particle size and rate of application on lettuce (*Lactuca sativa* L.) nitrogen use and yield. *All Life* **2022**, *15*, 624–635. [CrossRef]
16. Dávila Rangel, I.E.; Trejo Téllez, L.I.; Ortega Ortiz, H.; Juárez Maldonado, A.; González Morales, S.; Companioni González, B.; Cabrera De la Fuente, M.; Benavides Mendoza, A. Comparison of Iodide, Iodate, and Iodine-Chitosan Complexes for the Biofortification of Lettuce. *Appl. Sci.* **2020**, *10*, 2378. [CrossRef]
17. Liu, Q.; Huang, L.; Chen, Z.; Wen, Z.; Ma, L.; Xu, S.; Wu, Y.; Liu, Y.; Feng, Y. Biochar and its combination with inorganic or organic amendment on growth, uptake and accumulation of cadmium on lettuce. *J. Clean. Prod.* **2022**, *370*, 133610. [CrossRef]
18. Ahmed, Z.F.R.; Alnuaimi, A.K.H.; Askri, A.; Tzortzakis, N. Evaluation of Lettuce (*Lactuca sativa* L.) Production under Hydroponic System: Nutrient Solution Derived from Fish Waste vs. Inorganic Nutrient Solution. *Horticulturae* **2021**, *7*, 292. [CrossRef]
19. Anisuzzaman, M.; Rafii, M.Y.; Jaafar, N.M.; Izan Ramlee, S.; Ikbal, M.F.; Haque, M.A. Effect of Organic and Inorganic Fertilizer on the Growth and Yield Components of Traditional and Improved Rice (*Oryza sativa* L.) Genotypes in Malaysia. *Agronomy* **2021**, *11*, 1830. [CrossRef]
20. El-Mogy, M.M.; Abdelaziz, S.M.; Mahmoud, A.W.M.; Elsayed, T.R.; Abdel-Kader, N.H.; Mohamed, M.I.A. Comparative Effects of Different Organic and Inorganic Fertilisers on Soil Fertility, Plant Growth, Soil Microbial Community, and Storage Ability of Lettuce. *Agric. (Pol'nohospodárstvo)* **2020**, *66*, 87–107. [CrossRef]
21. Zhang, E.; Lin, L.; Liu, J.; Li, Y.; Jiang, W.; Tang, Y. The effects of organic fertilizer and inorganic fertilizer on yield and quality of lettuce. *Adv. Eng. Res.* **2022**, *129*, 624–635. [CrossRef]
22. Pavlou, G.C.; Ehaliotis, C.D.; Kavvadias, V.A. Effect of organic and inorganic fertilizers applied during successive crop seasons on growth and nitrate accumulation in lettuce. *Sci. Hortic.* **2007**, *111*, 319–325. [CrossRef]
23. Kutyauripo, I.; Rushambwa, M.; Chiwazi, L. Artificial intelligence applications in the agrifood sectors. *J. Agric. Food Res.* **2023**, *11*, 100502. [CrossRef]
24. Rameshwari, D. Classification of Macronutrient Deficiencies in Maize Plant using Machine Learning. *Int. J. Res. Appl. Sci. Eng. Technol.* **2021**, *9*, 4321–4326. [CrossRef]
25. Raju, S.H.; Adinarayna, S.; Prasanna, N.M.; Jumlesha, S.; Sesadri, U.; Hema, C. Nutrient Deficiency Detection using a MobileNet: An AI-based Solution. In Proceedings of the 2023 7th International Conference on I-SMAC (IoT in Social, Mobile, Analytics and Cloud) (I-SMAC), Kirtipur, Nepal, 11–13 October 2023; pp. 609–615. [CrossRef]
26. Krishna, V.; Raju, Y.D.S.; Raghavendran, C.V.; Naresh, P.; Rajesh, A. Identification of Nutritional Deficiencies in Crops Using Machine Learning and Image Processing Techniques. In Proceedings of the 2022 3rd International Conference on Intelligent Engineering and Management (ICIEM), London, UK, 27–29 April 2022; pp. 925–929. [CrossRef]
27. Suleiman, R.F.R.; Riduwan, M.K.; Kamal, A.N.M.; Wahab, N.A. Soil Nutrient Deficiency Detection of Lime Trees using Signal-based Deep Learning. In Proceedings of the 2022 International Visualization, Informatics and Technology Conference (IVIT), Kuala Lumpur, Malaysia, 1–2 November 2022; pp. 261–265. [CrossRef]
28. J, A.; K, M.; K, V. Plant Nutrient Deficiency Detection and Classification—A Review. In Proceedings of the 2023 5th International Conference on Inventive Research in Computing Applications (ICIRCA), Coimbatore, India, 3–5 August 2023; pp. 796–802. [CrossRef]

29. Sudhakar, M.; Priya, R.M.S. Computer Vision Based Machine Learning and Deep Learning Approaches for Identification of Nutrient Deficiency in Crops: A Survey. *Nat. Environ. Pollut. Technol.* **2023**, *22*, 1387–1399. [CrossRef]
30. Kok, Z.H.; Shariff, A.R.M.; Khairunniza-Bejo, S.; Kim, H.T.; Ahamed, T.; Cheah, S.S.; Wahid, S.A.A. Plot-Based Classification of Macronutrient Levels in Oil Palm Trees with Landsat-8 Images and Machine Learning. *Remote Sens.* **2021**, *13*, 2029. [CrossRef]
31. Sabri, N.; Kassim, N.S.; Ibrahim, S.; Roslan, R.; Mangshor, N.N.A.; Ibrahim, Z. Nutrient deficiency detection in Maize (*Zea mays* L.) leaves using image processing. *IAES Int. J. Artif. Intell. (IJ-AI)* **2020**, *9*, 304. [CrossRef]
32. Rahadiyan, D.; Hartati, S.; Nugroho, A. An Overview of Identification and Estimation Nutrient on Plant Leaves Image Using Machine Learning. *J. Theor. Appl. Inf. Technol.* **2022**, *100*, 1836–1852. Available online: https://www.jatit.org/volumes/Vol100No6/20Vol100No6.pdf (accessed on 1 August 2024).
33. Sowmiya, M.; Krishnaveni, S. Deep Learning Techniques to Detect Crop Disease and Nutrient Deficiency—A Survey. In Proceedings of the 2021 International Conference on System, Computation, Automation and Networking (ICSCAN), Puducherry, India, 30–31 July 2021; pp. 1–5. [CrossRef]
34. Kamelia, L.; Rahman, T.K.A.; Saragih, H.; Uyun, S. Survey on Hybrid Techniques in The Classification of Nutrient Deficiency Levels in Citrus Leaves. In Proceedings of the 2021 7th International Conference on Wireless and Telematics (ICWT), Bandung, Indonesia, 19–20 August 2021; pp. 1–4. [CrossRef]
35. Arun, K.S.; Huang, T.S.; Blostein, S.D. Least-squares fitting of two 3-D point sets. *IEEE Trans. Pattern Anal. Mach. Intell.* **1987**, *9*, 698–700. [CrossRef]
36. Besl, P.J.; McKay, N.D. Method for registration of 3-D shapes. In Proceedings of the Robotics'91: Sensor Fusion IV: Control Paradigms and Data Structures, Boston, MA, USA, 14–15 November 1991; SPIE: Bellingham, WA, USA, 1992; Volume 1611, pp. 586–606. [CrossRef]
37. Chen, Y.; Medioni, G. Object modelling by registration of multiple range images. *Image Vis. Comput.* **1992**, *10*, 145–155. [CrossRef]
38. Zhang, Z. Iterative point matching for registration of free-form curves and surfaces. *Int. J. Comput. Vis.* **1994**, *13*, 119–152. [CrossRef]
39. Zhang, K.; Chen, H.; Wu, H.; Zhao, X.; Zhou, C. Point cloud registration method for maize plants based on conical surface fitting—ICP. *Sci. Rep.* **2022**, *12*, 6852. [CrossRef] [PubMed]
40. Amritraj, S.; Hans, N.; Pretty, C.; Cyril, D. An Automated and Fine- Tuned Image Detection and Classification System for Plant Leaf Diseases. In Proceedings of the 2023 International Conference on Recent Advances in Electrical, Electronics, Ubiquitous Communication, and Computational Intelligence (RAEEUCCI), Chennai, India, 19–21 April 2023; pp. 1–5. [CrossRef]
41. Pandey, P.; Patra, R. A Real-time Web-based Application for Automated Plant Disease Classification using Deep Learning. In Proceedings of the 2023 IEEE International Symposium on Smart Electronic Systems (iSES), Ahmedabad, India, 18–20 December 2023; pp. 230–235. [CrossRef]
42. Nabaasa, E.; Natumanya, D.; Grace, B.; Kiwanuka, C.N.; Muhunga, K.B.J. A Model for Detecting the Presence of Pesticide Residues in Edible Parts of Tomatoes, Cabbages, Carrots, and Green Pepper Vegetables. In Proceedings of the Artificial Intelligence and Applications, Toronto, ON, Canada, 20–21 July 2024; Volume 2, pp. 225–232. [CrossRef]
43. Li, H.; Gu, Z.; He, D.; Wang, X.; Huang, J.; Mo, Y.; Li, P.; Huang, Z.; Wu, F. A lightweight improved YOLOv5s model and its deployment for detecting pitaya fruits in daytime and nighttime light-supplement environments. *Comput. Electron. Agric.* **2024**, *220*, 108914. [CrossRef]
44. Bello, R.W.; Oladipo, M.A. Mask YOLOv7-based drone vision system for automated cattle detection and counting. *Artif. Intell. Appl.* **2024**, *2*, 129–139. [CrossRef]
45. Broadley, M.; Escobar-Gutierrez, A.; Burns, A.; Burns, I. What are the effects of nitrogen deficiency on growth components of lettuce? *New Phytol.* **2000**, *147*, 519–526. [CrossRef]
46. Yang, R.; Wu, Z.; Fang, W.; Zhang, H.; Wang, W.; Fu, L.; Majeed, Y.; Li, R.; Cui, Y. Detection of abnormal hydroponic lettuce leaves based on image processing and machine learning. *Inf. Process. Agric.* **2023**, *10*, 1–10. [CrossRef]
47. Gao, H.; Mao, H.; Ullah, I. Analysis of metabolomic changes in lettuce leaves under low nitrogen and phosphorus deficiencies stresses. *Agriculture* **2020**, *10*, 406. [CrossRef]
48. Terven, J.; Córdova-Esparza, D.M.; Romero-González, J.A. A comprehensive review of yolo architectures in computer vision: From yolov1 to yolov8 and yolo-nas. *Mach. Learn. Knowl. Extr.* **2023**, *5*, 1680–1716. [CrossRef]
49. Wang, C.Y.; Yeh, I.H.; Liao, H.Y.M. YOLOv9: Learning What You Want to Learn Using Programmable Gradient Information. *arXiv* **2024**, arXiv:2402.13616.
50. Wang, A.; Chen, H.; Liu, L.; Chen, K.; Lin, Z.; Han, J.; Ding, G. YOLOv10: Real-Time End-to-End Object Detection. *arXiv* **2024**, arXiv:2405.14458.
51. Zhang, J.; Yao, Y.; Deng, B. Fast and Robust Iterative Closest Point. *IEEE Trans. Pattern Anal. Mach. Intell.* **2021**, *44*, 3450–3466. [CrossRef]
52. Anderson, D.G. Iterative procedures for nonlinear integral equations. *J. ACM* **1965**, *12*, 547–560. [CrossRef]
53. González-Barbosa, J.J.; Ramírez-Pedraza, A.; Ornelas-Rodríguez, F.J.; Cordova-Esparza, D.M.; González-Barbosa, E.A. Dynamic Measurement of Portos Tomato Seedling Growth Using the Kinect 2.0 Sensor. *Agriculture* **2022**, *12*, 449. [CrossRef]
54. Redmon, J.; Divvala, S.; Girshick, R.; Farhadi, A. You Only Look Once: Unified, Real-Time Object Detection. *arXiv* **2016**, arXiv:1506.02640.

55. Torres, J. YOLOv8 Architecture: A Deep Dive into its Architecture. 2024. Available online: https://yolov8.org/yolov8-architecture/ (accessed on 14 August 2024).
56. Mukherjee, S. YOLOv9: Exploring Object Detection with YOLO Model. 2024. Available online: https://blog.paperspace.com/yolov9-2/ (accessed on 14 August 2024).

AgriEngineering

Article

Performance of Neural Networks in the Prediction of Nitrogen Nutrition in Strawberry Plants

Jamile Raquel Regazzo [1,*], Thiago Lima da Silva [1], Marcos Silva Tavares [1], Edson José de Souza Sardinha [2], Caroline Goulart Figueiredo [1], Júlia Luna Couto [2], Tamara Maria Gomes [2], Adriano Rogério Bruno Tech [2] and Murilo Mesquita Baesso [2]

1 Luiz de Queiroz Higher School of Agriculture, University of São Paulo—USP, Piracicaba 13635-900, SP, Brazil
2 Faculty of Animal Science and Food Engineering, University of São Paulo—USP, Pirassununga 13418-900, SP, Brazil; julialuna@usp.br (J.L.C.); adriano.tech@usp.br (A.R.B.T.); baesso@usp.br (M.M.B.)
* Correspondence: jamile.regazzo@usp.br; Tel.: +55-14-99718-6130

Abstract: Among the technological tools used in precision agriculture, the convolutional neural network (CNN) has shown promise in determining the nutritional status of plants, reducing the time required to obtain results and optimizing the variable application rates of fertilizers. Not knowing the appropriate amount of nitrogen to apply can cause environmental damage and increase production costs; thus, technological tools are required that identify the plant's real nutritional demands, and that are subject to evaluation and improvement, considering the variability of agricultural environments. The objective of this study was to evaluate and compare the performance of two convolutional neural networks in classifying leaf nitrogen in strawberry plants by using RGB images. The experiment was carried out in randomized blocks with three treatments (T1: 50%, T2: 100%, and T3: 150% of recommended nitrogen fertilization), two plots and five replications. The leaves were collected in the phenological phase of floral induction and digitized on a flatbed scanner; this was followed by processing and analysis of the models. ResNet-50 proved to be superior compared to the personalized CNN, achieving accuracy rates of 78% and 48% and AUC of 76%, respectively, increasing classification accuracy by 38.5%. The importance of this technique in different cultures and environments is highlighted to consolidate this approach.

Keywords: *San Andreas*; fertilization; machine learning; ResNet-50

Citation: Regazzo, J.R.; Silva, T.L.d.; Tavares, M.S.; Sardinha, E.J.d.S.; Figueiredo, C.G.; Couto, J.L.; Gomes, T.M.; Tech, A.R.B.; Baesso, M.M. Performance of Neural Networks in the Prediction of Nitrogen Nutrition in Strawberry Plants. *AgriEngineering* **2024**, *6*, 1760–1770. https://doi.org/10.3390/agriengineering6020102

Academic Editors: Ray E. Sheriff and Chiew Foong Kwong

Received: 19 April 2024
Revised: 27 May 2024
Accepted: 28 May 2024
Published: 18 June 2024

1. Introduction

Strawberry production (*Fragaria × ananassa Duch., Rosaceae*), worldwide, is led by China and the United States, but Brazil stands out as the 8th largest producer, with production reaching 197 thousand tons [1]. From 2011 to 2021, global acreage grew by 20%, and production increased by 44%. In Brazil, the strawberry area grew to 6.280 ha in the 2022/23 crop year, up from 2.500 ha in 2012 [2]. This rise highlights Brazil's contribution to the global strawberry production scenario. Investments in research and new technologies have been made to meet this growing demand, with the aim of developing more productive varieties.

Nutritional management plays a fundamental role in crop development, making it a widely studied topic due to its direct influence on increasing productivity and fruit quality. Nitrogen is an essential element for plants, being the main constituent of chlorophyll, which is directly involved in photosynthetic activity. Therefore, the greenness of leaves is determined by the concentration of chlorophyll molecules [3]. Variations in green hues due to foliar nitrogen (N) content can result in class separation in an RGB image bank when it comes to computer vision. Thus, this pigment becomes essential in the context of precision agriculture because it serves as a basis for quantifying foliar N content [4–6]. A practical and quick way to determine the existence of nutrient deficiency in plants is through visual diagnosis. However, its accuracy is limited by the experience of the technician and it also requires a significant amount work in generating a prescription map to be used in localized management. As a result, new research is constantly emerging to estimate the ideal nutrient level in crops [7,8].

Several scientific studies and research efforts have highlighted the proven relationship between the concentration of green in the leaves, analyzed through computer vision, and the nitrogen content in plants [9–12]. In this context, nitrogen exerts a significant influence on plant growth, productivity and soil fertility [13]. Agricultural production, which is essential for global food security, relies largely on synthetic fertilizers that provide this key element [14]. Thus, studies on nitrogen strive to increase its efficiency of use.

With regard to the assessment of the nutritional status of plants, conventional approaches to assessing the nutritional status of plants are generally destructive, labor-intensive and require specific resources [15]. Considering the importance of strawberry cultivation and the need for more accessible, quick and efficient methods to determine nutritional status, there is an interest in the development of more effective methods [16].

In the field of precision agriculture, the application of predictive techniques, such as the use of nutrient sensors in plants, stands out [17], and remote sensing with drones in urban forest management has been developed [18]. In addition, there are approaches based on image processing and artificial intelligence, which stand out for their ability to identify nutritional deficiencies in crops in a non-invasive way. However, it is essential to improve N prediction techniques in order to ensure the necessary accuracy in agricultural decision-making, considering the complexity of the relationships between the availability of this element, plant growth, and fruit quality.

Among the AI (artificial intelligence) tools used for prediction and classification, deep learning has been applied in different areas, presenting satisfactory results in solving various problems. According to Ferentinos et al., [19], this feature refers to the use of artificial neural network architectures with multiple processing layers, distinguishing them from the more superficial approaches of traditional neural networks. Among the former class, the convolutional neural network (CNN) stands out among the deep learning algorithms and is widely used to extract features from images [20,21]. During image processing, CNNs allow the extraction of several characteristics, especially those related to color-based indices, which are relevant to nutritional applications. In the face of the challenge of accurately identifying nitrogen levels and the growing importance of precision agriculture for sustainable development, investing in innovative technologies is essential. Deep learning tools capable of handling RGB images can assist in nitrogen fertilization management in strawberry crops by classifying foliar nutritional status. This approach involves applying CNNs to process images and classify nitrogen levels in plants.

Therefore, the present study aimed to evaluate and compare the performance of two convolutional neural networks in the classification of leaf nitrogen levels in strawberry crops, based on RGB images. The analysis focused on performance metrics of the generated models, aiming to contribute to the advancement of knowledge and the practical implementation of more effective tools in the nutritional management of plantations.

2. Materials and Methods

An experiment was carried out to create an image bank, with strawberry cultivation as its theme. The experiment was carried out from August to December 2022 at the Faculty of Animal Science and Food Engineering (FZEA/USP), with strawberries (*Fragaria × ananassa Duch., Rosaceae*) of the *San Andreas* cultivar. The experiment took place in a 100 m^2 greenhouse, with climate control provided by an evaporative air system, located at the "Fernando Costa" campus, in Pirassununga. The geographic coordinates of the location are approximately 21°57′25.2″ S, 47°27′13.7″ W, with an average altitude of 610 m above sea level. The region has a CWA climate, with well-defined seasons, one, rainy, from October to March, and the other, dry, from April to September, with an average annual rainfall of 1298 mm and an average temperature of 20.8 °C. The strawberry seedlings were transplanted in October 2022 to 30 pots, comprising three treatments and five replications, as each replication was divided into two plots. The treatments consisted of applying different doses of nitrogen fertilizers as follows: 50% of the recommended fertilization (FR), by Passos et al., 2013 (T1); 100% FR (T2), control; and 150% FR (T3) [22]. This methodology has been used by different researchers [23,24].

The image database was generated and organized to serve as a resource of extractable characteristics for classification, and these characteristics were used as input to the network and their vectors used in the training of computational classifiers based on deep learning methods, having the classes of the different N doses as the output or target attribute of the model.

To obtain parameters for comparison of the model, leaf chemical analysis was performed at the Laboratory of Agrarian/Soil Sciences of the Faculty of Animal Science and Food Engineering of the University of São Paulo, according to the methodology described in the manual of chemical analysis of soils, plants and fertilizers published by EMBRAPA [25].

2.1. Obtaining and Processing the Images

The indicative leaves of strawberries were collected on the third expanded leaf, and were selected for uniformity of age and stage of development in all treatments and replications, to ensure consistency in the comparative analyses. After collection, at the Laboratory of Agricultural Machinery and Precision Agriculture (LAMAP), of the Department of Biosystems Engineering of the University of São Paulo, the leaves were prepared to obtain the images. This step involved removing any excess dust or dirt by using a paper towel. With a high-resolution flatbed scanner (HP Scanjet 3800), the images of the sheets were scanned in RGB (red–green–blue) color, in JPG (Joint Photographic Experts Group) format and with a resolution of 1200 DPI (dots per inch).

Using a script developed in the Matlab© R2021b software, the scanned images of the strawberry leaves were cropped, automatically extracting blocks of 224 × 224 pixels and thus generating an image bank with specific areas of interest [26] for further processing. This procedure aimed to obtain images that would provide an ideal visualization of the patterns generated by the doses on the leaves, as exemplified in Figure 1. The image blocks were scaled to a standardized size, as required in the input of the ResNet architecture models [27].

Figure 1. Example of the image bank formed by strawberry leaf cleavage blocks.

After the completion of the automatic cropping stage, the manual selection of image blocks was performed to compose the database. During this process, each image block was checked to ensure the quality and integrity of the data to be used. The selection followed specific criteria, requiring that the image of each block be completely filled by a region of the leaf, i.e., without characteristics of visible leaf damage or injury. This approach ensured the presence of only the areas relevant for analysis in each block. In addition, blocks that presented noise, such as white background or dirt on the leaf, were excluded, since such imperfections could interfere with the subsequent classification process. After manual selection, 50 blocks of images were chosen for each vessel, totaling 500 blocks per treatment. With three treatments analyzed, the final dataset consisted of 1500 blocks of images.

With the definition of the dataset, a script was developed in the Matlab© R2021b software to generate the classifier models of the fertilized nitrogen dosage, using two different CNN architectures, one customized and the other based on ResNet-50, respectively, as described in Tables 1 and 2.

Table 1. Personalized CNN architecture for leaf nitrogen level prediction.

Layers	Settings
2D Convolutional	256 filters with size 5 × 5, activation by ReLU
Batch Normalization	*Default*
2D MaxPooling	Size *pooling* = 2 × 2, *stride* = 2
2D Convolutional	128 filters with size 3 × 7, activation by ReLU
Batch Normalization	*Default*
2D MaxPooling	Size *pooling* = 2 × 2, *stride* = 2
2D Convolutional	128 filters with size 7 × 3, activation by ReLU
Batch Normalization	*Default*
2D Convolutional	32 filters with size 3 × 3, activation by ReLU
Batch Normalization	*Default*
Fully Connected	3 neurons, activation by *softmax*

Table 2. CNN ResNet-50 architecture for leaf nitrogen level prediction.

Layer Name	Output Size	50 Layers
conv1	112 × 112	7 × 7, 64, *stride* 2
conv2_x	56 × 56	3 × 3 *max pool*, *stride* 2 $\begin{bmatrix} 1 \times 1, & 64 \\ 3 \times 3, & 64 \\ 1 \times 1, & 256 \end{bmatrix} \times 3$
conv3_x	28 × 28	$\begin{bmatrix} 1 \times 1, & 128 \\ 3 \times 3, & 128 \\ 1 \times 1, & 512 \end{bmatrix} \times 4$
conv4_x	14 × 14	$\begin{bmatrix} 1 \times 1, & 256 \\ 3 \times 3, & 256 \\ 1 \times 1, & 1024 \end{bmatrix} \times 6$
conv5_x	7 × 7	$\begin{bmatrix} 1 \times 1, & 512 \\ 3 \times 3, & 512 \\ 1 \times 1, & 2048 \end{bmatrix} \times 3$
	1 × 1	*Average pool*, 3-d fc, *softmax*

The personalized CNN architecture is designed as a sequential model, with four layers of two-dimensional convolution (2D convolutional) using different numbers and sizes of filters. These layers form the basis of a CNN model, being responsible for extracting characteristics from the input data [28]. In the case of images, convolution layers allow one to learn visual patterns (such as colors, shapes, and textures), from the simplest to the most complex, successively [29]. Using the Rectified Linear Unit activation function in the convolution layers allows you to better deal with non-linearity issues at a lower computing cost [28]. The adoption of rectangular filters (3 × 7 and 7 × 3) in the second and third convolution layers aimed to capture possible vertical and horizontal patterns. Following each convolution layer is a normalization layer (Batch Normalization), which speeds up training, decreases overfitting and optimizes the learning process [30]. After the first and second normalization layers is a subsampling layer (2D MaxPooling), which is used to reduce the dimensionality of the activation maps, saving computing resources while preserving the most relevant data [28]. Finally, there is the output (Fully Connected), or classification layer. The softmax activation function used in this layer transforms its neurons into probabilities assigned to each class, effectively turning the model into a classifier.

The other CNN architecture adopted for this study was based on ResNet-50, as described in Table 2. A convolutional neural network proposed by He et al. [31], it consists of 50 layers, of which 49 are convolutional. Addressing the complexity of deep network training, ResNet-50 introduces the "identity shortcut connection", allowing one to skip one or more layers to form residual blocks. In addition, to optimize computational efficiency, the ResNet-50 utilizes a "bottleneck design" structure, reducing the number of parameters and operations on each residual block [31].

The ResNet-50 architecture allows the loading of a pre-trained version with more than one million images from the ImageNet database. This makes it possible to classify images into 1000 categories. With an image input of 224 × 224 pixels, this pre-trained neural network provides comprehensive representations for a variety of natural images, allowing you to reuse learned visual patterns in applications involving other categories of images. This technique of reusing learned knowledge, called learning transfer, has become a common practice, as it accelerates the training process, saves computational resources, and helps in the generalization and convergence capacity of new models, being especially useful in situations where the database of the problem addressed is insufficient for learning [32]. Table 2 shows the elements of the 50-layer ResNet architecture.

2.2. Analysis of the Models

To evaluate the reliability of the model, the K-fold cross-validation technique was used, in which the dataset is divided into distinct groups [33]. In this study, we opted for a five-fold cross-validation (K = 5), which entails dividing the dataset into five different groups. Each of these groups served as a test set in five independent iterations. This means that in each iteration, a different group was used for validation, while the other four (or K − 1) were employed for model training. At the end of the five iterations, the average precision was calculated, which then served as the final measure of the model's performance. During the cross-validation process, 80% of the images were used for training, which is equivalent to 1200 blocks of images. The remaining group, consisting of 20% of the images (300 images) was designated as a validation and test set to evaluate the performance of the model. This training and testing procedure was repeated five times, ensuring that each group would be used once as a test set. This strategy ensures that all images in the dataset are used for both training and testing, providing a more robust and reliable assessment of the model's performance.

For the model, we took into account the configurations, as shown in Table 3, in training the neural networks in which the hyperparameters are inserted, since they are responsible for defining how the model will be trained, and, consequently, how it will be able to perform the task for which it was designed [34].

Table 3. Training options used in the convolutional model.

Number of times	10
Mini-lot size	8
Option for data scrambling	every-epoch
Positive scaling of the initial learning rate	0.0001

For both the personalized convolutional neural network and ResNet-50, the number of epochs used was 10. In a comparative study, Too et al. [35] considered the fine-tuning of deep learning models for plant disease identification, and compared the behaviors for 10 and 30 epochs, observing that ResNet-50 and ResNet 101 work properly with fewer iterations, and, in addition, after fine-tuning, the models using 10 epochs obtained an accuracy above 90%.

Table 4 summarizes the formulas of the performance metrics and their expressions used in the present study.

Table 4. Description of performance evaluation metrics.

Metric	Formula	Description
Accuracy	$A_{cc} = \frac{TP+TN}{TP+TN+FP+FN}$	The overall efficiency of a model.
Sensitivity or Recall	$Se = \frac{TP}{TP+FN}$	The efficiency of a model as to positive samples.
Specificity	$Sp = \frac{TN}{TN+FP}$	The efficiency of a model as to negative samples.
Precision	$Pre = \frac{TP}{TP+FP}$	The proportion of actual positives out of all positives predicted by the model.
F1-score	$F1_{score} = \frac{2 * Pre * Se}{Pre+Se}$	The harmonic mean between precision and sensitivity.
G-mean	$G_{mean} = \sqrt{recall * specificity}$	Maximum accuracy in each of the classes.

TP: true positive; *TN*: true negative; *FN*: false negative; *FP*: false positive; *Pre*: accuracy; *Se*: sensitivity.

The area under the curve (AUC) reflects the performance of a classifier, illustrating how the true positive rate (recall/sensitivity) varies as the false positive rate (inverted specificity) changes [36]. This analysis provides insights into the model's ability to distinguish between classes in different decision-making contexts. Models with an AUC close to 1.0 are considered highly effective in classification tasks [37], thus highlighting the fundamental relevance of this metric in evaluating model performance.

3. Results and Discussion

3.1. Foliar Chemical Analysis

Figure 2 shows the boxplot test for the values of leaf nitrogen concentration in strawberry leaves in the flower induction stage. N concentrations ranged from 19.4 g kg^{-1} to 33.3 g kg^{-1}. The T1 (50% of the N dose) and T2 (100% of the N dose) treatments exhibit quite similar median and mean values, while the T3 treatment (150% of the N dose) shows a higher trend.

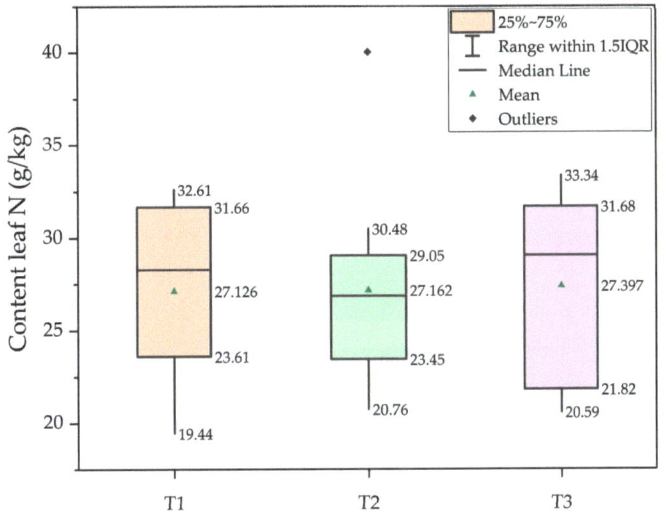

Figure 2. Foliar concentration of N in strawberry plants at the flower induction stage.

3.2. Analysis of the Models

The confusion matrix evaluates the correspondence between the output predicted by the model and the desired output, and then calculates the classification accuracy for each class in a specific class, thus revealing the strengths and weaknesses of the model [38]. The rating rate using personalized CNN was 48.1%, while for ResNet-50 it was 78.1%. Analysis of the matrices for the personalized CNN and ResNet-50 models, as described in Figure 3, reveals distinct patterns in the classification capacity of the treatments (T1, T2, and T3). For personalized CNN, the performance was not satisfactory in view of the fact that the model got more wrong than it got right. Strategies could be explored to improve the outcome of the model, such as optimizing the hyperparameters, including filter size, number of layers and learning rate [39]. As for ResNet-50, it performed better, as it extracts patterns from more complex spectral features [40]; the rate of correct answers was higher in this model, reaching 78% accuracy. For both networks, T1 was the one with the highest rate of correct answers, with 60.2% and 84.4%, respectively, for the personalized CNN and ResNet-50; this was possibly due to the symptom of nutritional deficiency, expressed as a uniform chlorosis, which allowed the model to classify more assertively.

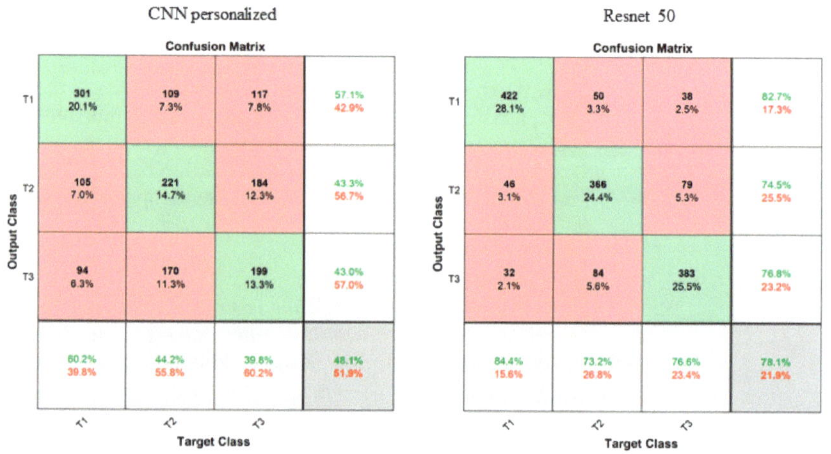

Figure 3. Confusion matrix between neural network-based models for classification of nitrogen doses in strawberry floral induction. In green (main diagonal) are the true positives of each class. In salmon, horizontally there are false negatives, and vertically the false positives. In light and dark gray are accuracies and errors by classes and total respectively.

Table 5 shows the performance of the classifiers in discriminating the different N doses (T1 = 50%, T2 = 100%, and T3 = 150%) in the phenological stage of floral induction based on neural networks. These metrics provide different perspectives on the model's performance and help in the overall understanding of its effectiveness.

Table 5. Personalized CNN convolutional neural networks and ResNet-50 performance metrics.

	Personalized CNN (%)			ResNet-50 (%)		
Metrics	**T1**	**T2**	**T3**	**T1**	**T2**	**T3**
Accuracy	48	48	48	78	78	78
Sensitivity	60	44	40	84	73	77
Specificity	42	50	52	75	81	79
Precision	34	31	29	63	65	64
Recall	60	44	40	84	73	77
F1-score	44	36	34	72	69	70
G-mean	50	47	46	80	77	78

The generated models used 1200 excerpts from the database for training. The customized CNN model achieved a mean accuracy of 48%, an unsatisfactory performance in distinguishing between the 50%, 100% and 150% treatments, as confirmed by the lower recall values and precision values of these classes. Possibly, the personalized CNN had an unsatisfactory performance due to the complexity of the standards expressed by the treatments, making categorization difficult, given this structure and the configuration of the CNN, in addition to the lack of generalization in relation to negative samples.

The best performance in the classification presented was with the use of ResNet-50 for the dose of 50% N, with an accuracy of 84%. For the T2 and T3 treatments, the accuracy, recall and F1-score metrics showed better results than those of the personalized CNN, but they are similar to each other. This suggests that the leaves in these treatments exhibited less variation in color intensity and distribution, making the classification of models more costly.

A comparison of the AUC analyses is shown in Figure 4. First, for the customized CNN, the result was not satisfactory for floral induction, remaining within a range of 0.57 to 0.64, except for T1. As the phenological stage of the plant advances, an increase in leaf chlorosis can be observed due to differences in nitrogen concentration between the treatments. The coloration of the plant's leaf can be influenced in a variety of ways, depending on the availability of nitrogen. In nitrogen deficiency, leaves tend to become paler, yellowish, or light greenish, due to the reduction in chlorophyll synthesis, which is responsible for the green coloration of leaves, and excess nitrogen can lead to a very dark green color of leaves and excessive vegetative growth [41].

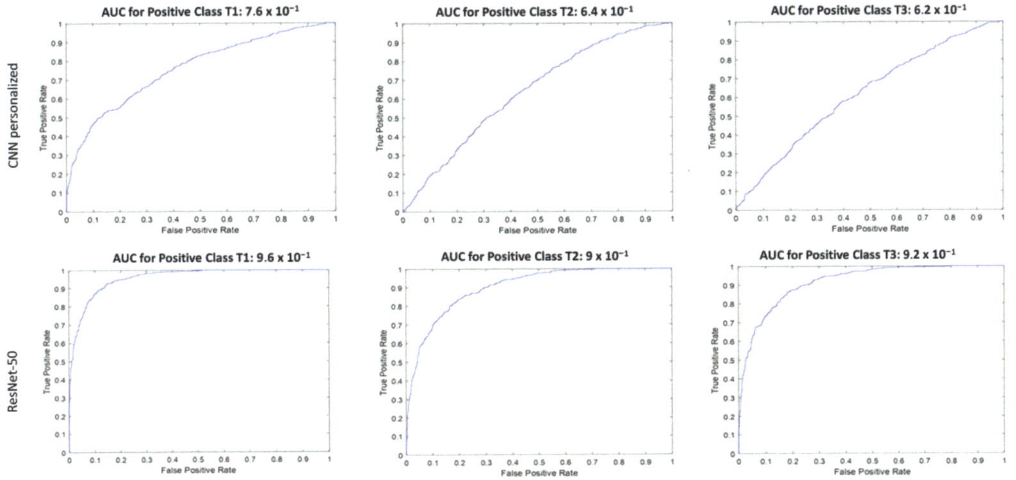

Figure 4. The AUC curves show the comparisons of the performance of the models as to personalized CNN and ResNet-50.

The analysis of the metrics highlighted that ResNet-50 had higher values in all metrics, especially in specificity, ranging from 64% to 69%, while personalized CNN had lower specificity, ranging from 41% to 44%.

The comparison between the architectures revealed that ResNet-50 achieved superior accuracy, standing out for its ability to identify distinct patterns, in contrast to the custom CNN, which had lower performance due to its lower capacity for learning and generalization. These results corroborate the findings of previous studies, such as those of Costa et al. [42], which achieved 86.6% recall when applying ResNet-50 for detection of defects in tomatoes, reinforcing the efficiency of this architecture in predicting adequate fertilization rates in strawberry plants.

The superior classification performance of the model using ResNet-50 may be related to the characteristics presented in this convnet. In this architecture, there are blocks with convolutional layers in sequence and a separate parallel identity layer [43], which can result in a more precise extraction of color or texture features from plant leaf images [44]. In the ResNet-50 used, the technique of transfer learning, combined with pre-training on large datasets, was applied, which conferred an essential generalization capacity for classifying different nitrogen concentrations in strawberry leaves [45]. In addition, the application of residuals to the convolutions, a differential of ResNet-50, allows the network to learn residual differences in the image features, optimizing the model's training.

4. Conclusions

The study focused on comparing the performance of convolutional neural network architectures in classifying the nitrogen status of strawberry leaves. The results highlighted the better performance of the Resnet-50 architecture compared to the custom CNN in the specific task of classifying nitrogen status in strawberry leaves by handling RGB images. Evaluating the two main performance analysis metrics of the models, it was found that Resnet-50, with total accuracy = 78% and AUC T1 = 0.76, T2 = 0.63 and T3 = 0.61, outperformed the custom CNN with total accuracy = 48% and AUC T1 = 0.96, T2 = 0.90 and T3 = 0.91. An increase of 38.5% was achieved with ResNet-50 when compared to the custom CNN. By offering a more accurate and efficient approach to nutritional status detection, Resnet-50 demonstrated its potential for improving monitoring processes, standing out in this work as an important tool in precision agriculture. Additionally, it is suggested that further studies explore image preprocessing techniques and fine-tuning, such as adjustments to model hyperparameters, regularization and data augmentation techniques, to further optimize the accuracy and effectiveness of the classification model. These initiatives have the potential to elevate Resnet-50 to a new level of performance, consolidating its role as an essential tool in precision agriculture. Thus, this research offers important practical knowledge for the efficient management of agricultural crops, promoting a more sustainable approach to modern agriculture. Finally, further field research involving other crops is recommended for the validation of this approach.

Author Contributions: Conceptualization, J.R.R., T.L.d.S. and E.J.d.S.S.; methodology, J.R.R., T.L.d.S., T.M.G., C.G.F. and E.J.d.S.S.; software, J.R.R., T.L.d.S., M.S.T. and E.J.d.S.S.; validation, J.R.R., T.L.d.S. and M.M.B.; formal analysis, J.R.R., T.L.d.S., M.S.T. and E.J.d.S.S.; resources, J.R.R.; data curation, J.R.R., T.L.d.S., M.S.T. and E.J.d.S.S.; writing—original draft preparation, J.R.R., T.L.d.S., M.S.T. and A.R.B.T.; writing—review and editing, J.R.R., T.L.d.S., M.S.T., E.J.d.S.S., T.M.G. and M.M.B.; visualization, J.R.R., J.L.C., C.G.F. and A.R.B.T.; supervision, J.R.R. and M.M.B. All authors have read and agreed to the published version of the manuscript.

Funding: This research was funded by the National Council for Scientific and Technological Development (Conselho Nacional de Desenvolvimento Científico e Tecnológico—CNPQ)—Funding Code 001, by Coordination for the Improvement of Higher Education Personnel (Coordenação de Aperfeiçoamento de Pessoal de Nível Superior—CAPES)—Brazil (Funding Code 001) and the Luiz de Queiroz Agricultural Studies Foundation (Fundação de Estudos Agrários Luiz de Queiroz—FEALQ).

Data Availability Statement: The raw data supporting the conclusions of this article will be made available by the authors on request.

Conflicts of Interest: The authors declare no conflicts of interest.

References

1. FAO. *World's Area Harvested and Production Quantity (2021)*; Faostat—Food and Agriculture Organization of the United Nations: Roma, Italy, 2021.
2. Hernández-Martínez, N.R.; Blanchard, C.; Wells, D.; Salazar-Gutiérrez, M.R. Current State and Future Perspectives of Commercial Strawberry Production: A Review. *Sci. Hortic.* **2023**, *312*, 111893. [CrossRef]
3. Zhou, Z.; Struik, P.C.; Gu, J.; van der Putten, P.E.L.; Wang, Z.; Yin, X.; Yang, J. Enhancing Leaf Photosynthesis from Altered Chlorophyll Content Requires Optimal Partitioning of Nitrogen. *Crop Environ.* **2023**, *2*, 24–36. [CrossRef]
4. Jamali, M.; Bakhshandeh, E.; Emadi, M. Effects of Water Source and Technology on Energy Use and Environmental Impacts of Rice Production in Northern Iran. *Water Resour. Res.* **2022**, *58*, e2021WR031546. [CrossRef]
5. Shah, S.H.; Angel, Y.; Houborg, R.; Ali, S.; McCabe, M.F. A Random Forest Machine Learning Approach for the Retrieval of Leaf Chlorophyll Content in Wheat. *Remote Sens.* **2019**, *11*, 920. [CrossRef]
6. Taha, M.F.; Mao, H.; Wang, Y.; ElManawy, A.I.; Elmasry, G.; Wu, L.; Memon, M.S.; Niu, Z.; Huang, T.; Qiu, Z. High-Throughput Analysis of Leaf Chlorophyll Content in Aquaponically Grown Lettuce Using Hyperspectral Reflectance and RGB Images. *Plants* **2024**, *13*, 392. [CrossRef] [PubMed]

7. Bakhoum, G.S.; Tawfik, M.M.; Kabesh, M.O.; Sadak, M.S. Potential Role of Algae Extract as a Natural Stimulating for Wheat Production under Reduced Nitrogen Fertilizer Rates and Water Deficit. *Biocatal. Agric. Biotechnol.* **2023**, *51*, 102794. [CrossRef]
8. Bakhoum, G.; Sadak, M.; Badr ELSayed, E. Influence of Boron and/or Potassium Accompanied by Two Irrigation Systems on Chickpea Growth, Yield and Quality under Sandy Soil Conditions. *Egypt. J. Chem.* **2022**, *165*, 103–117. [CrossRef]
9. Ali, M.M.; Al-Ani, A.; Eamus, D.; Tan, D.K.Y. Leaf Nitrogen Determination Using Non-Destructive Techniques–A Review. *J. Plant Nutr.* **2017**, *40*, 928–953. [CrossRef]
10. Bi, K.; Niu, Z.; Xiao, S.; Bai, J.; Sun, G.; Wang, J.; Han, Z.; Gao, S. Non-Destructive Monitoring of Maize Nitrogen Concentration Using a Hyperspectral LiDAR: An Evaluation from Leaf-Level to Plant-Level. *Remote Sens.* **2021**, *13*, 5025. [CrossRef]
11. Li, R.; Wang, D.; Zhu, B.; Liu, T.; Sun, C.; Zhang, Z. Estimation of Nitrogen Content in Wheat Using Indices Der-ived from RGB and Thermal Infrared Imaging. *Field Crops Res.* **2022**, *289*, 108735. [CrossRef]
12. Zhou, X.; Kono, Y.; Win, A.; Matsui, T.; Tanaka, T.S.T. Predicting Within-Field Variability in Grain Yield and Protein Content of Winter Wheat Using UAV-Based Multispectral Imagery and Machine Learning Approaches. *Plant Prod. Sci.* **2021**, *24*, 137–151. [CrossRef]
13. Soilueang, P.; Jaikrasen, K.; Chromkaew, Y.; Buachun, S.; Yimyam, N.; Sanjunthong, W.; Kullachonphuri, S.; Wicharuck, S.; Mawan, N.; Khongdee, N. Dynamics of Soil Nitrogen Availability Following Conversion of Natural Forests to Various Coffee Cropping Systems in Northern Thailand. *Heliyon* **2023**, *9*, e22988. [CrossRef] [PubMed]
14. Nyamangara, J.; Kodzwa, J.; Masvaya, E.N.; Soropa, G. The Role of Synthetic Fertilizers in Enhancing Ecosystem Services in Crop Production Systems in Developing Countries. In *The Role of Ecosystem Services in Sustainable Food Systems*; Academic Press: New York, NY, USA, 2020; pp. 95–117. [CrossRef]
15. Cheng, W.; Sun, D.W.; Pu, H.; Wei, Q. Chemical Spoilage Extent Traceability of Two Kinds of Processed Pork Meats Using One Multispectral System Developed by Hyperspectral Imaging Combined with Effective Variable Selection Methods. *Food Chem.* **2017**, *221*, 1989–1996. [CrossRef] [PubMed]
16. Liu, Y.; Li, J. Comparing the Effectiveness of Two Convolutional Neural Networks Methods on Fault Diagnosis. In Proceedings of the 2020 IEEE 9th Data Driven Control and Learning Systems Conference (DDCLS), Liuzhou, China, 19–21 June 2020; pp. 350–354. [CrossRef]
17. Podar, D.; Maathuis, F.J.M. Primary Nutrient Sensors in Plants. *iScience* **2022**, *25*, 104029. [CrossRef] [PubMed]
18. Wavrek, M.T.; Carr, E.; Jean-Philippe, S.; McKinney, M.L. Drone Remote Sensing in Urban Forest Management: A Case Study. *Urban For. Urban Green.* **2023**, *86*, 127978. [CrossRef]
19. Ferentinos, K.P. Deep Learning Models for Plant Disease Detection and Diagnosis. *Comput. Electron. Agric.* **2018**, *145*, 311–318. [CrossRef]
20. Teng, J.; Zhang, D.; Lee, D.J.; Chou, Y. Recognition of Chinese Food Using Convolutional Neural Network. *Multimed. Tools Appl.* **2019**, *78*, 11155–11172. [CrossRef]
21. Shin, J.; Chang, Y.K.; Heung, B.; Nguyen-Quang, T.; Price, G.W.; Al-Mallahi, A. A Deep Learning Approach for RGB Image-Based Powdery Mildew Disease Detection on Strawberry Leaves. *Comput. Electron. Agric.* **2021**, *183*, 106042. [CrossRef]
22. Passos, F.; Trani, P. *Calagem e Adubação do Morangueiro*; Instituto Agronômico (IAC): Botafogo, Brazil, 2013.
23. Jia, Z.; Zhang, J.; Jiang, W.; Wei, M.; Zhao, L.; Li, G. Different Nitrogen Concentrations Affect Strawberry Seedlings Nitrogen Form Preferences through Nitrogen Assimilation and Metabolic Pathways. *Sci. Hortic.* **2024**, *332*, 113236. [CrossRef]
24. Li, Q.; Liu, L.; Zhao, P.; Zhao, Q.; Wu, M.; Liu, J.; Cheng, C.; Li, L. Insights into the Promoting Effects of Water-Soluble Amino Acid Fertilizers on Strawberry Fruit Quality under Nitrogen Reduction Treatment. *Sci. Hortic.* **2024**, *329*, 112978. [CrossRef]
25. da Silva, F.C. *Manual de Análises Químicas de Solos, Plantas e Fertilizantes*; EMBRAPA: Rio de Janeiro, Brazil, 2009. Available online: https://www.infoteca.cnptia.embrapa.br/infoteca/bitstream/doc/330496/1/Manual-de-analises-quimicas-de-solos-plantas-e-fertilizantes-ed02-reimpressao-2014.pdf (accessed on 18 April 2024).
26. Zaitoun, N.M.; Aqel, M.J. Survey on Image Segmentation Techniques. In *Procedia Computer Science*; Elsevier: Amsterdam, The Netherlands, 2015; Volume 65, pp. 797–806. [CrossRef]
27. Shaukat, K.; Luo, S.; Varadharajan, V. A Novel Deep Learning-Based Approach for Malware Detection. *Eng. Appl. Artif. Intell.* **2023**, *122*, 106030. [CrossRef]
28. Taye, M.M. Theoretical Understanding of Convolutional Neural Network: Concepts, Architectures, Applications, Future Directions. *Computation* **2023**, *11*, 52. [CrossRef]
29. Yamashita, R.; Nishio, M.; Do, R.K.G.; Togashi, K. Convolutional Neural Networks: An Overview and Application in Radiology. *Insights Imaging* **2018**, *9*, 611–629. [CrossRef]
30. Ioffe, S.; Szegedy, C. Batch Normalization: Accelerating Deep Network Training by Reducing Internal Covariate Shift. *PMLR* **2015**, *37*, 448–456.
31. He, K.; Zhang, X.; Ren, S.; Sun, J. Deep Residual Learning for Image Recognition. In Proceedings of the IEEE Conference on Computer Vision and Pattern Recognition, Boston, MA, USA, 7–12 June 2015.
32. Alzubaidi, L.; Zhang, J.; Humaidi, A.J.; Al-Dujaili, A.; Duan, Y.; Al-Shamma, O.; Santamaría, J.; Fadhel, M.A.; Al-Amidie, M.; Farhan, L. Review of Deep Learning: Concepts, CNN Architectures, Challenges, Applications, Future Directions. *J. Big Data* **2021**, *8*, 53. [CrossRef]
33. Arabameri, A.; Blaschke, T.; Pradhan, B.; Pourghasemi, H.R.; Tiefenbacher, J.P.; Bui, D.T. Evaluation of Recent Advanced Soft Computing Techniques for Gully Erosion Susceptibility Mapping: A Comparative Study. *Sensors* **2020**, *20*, 335. [CrossRef]

34. Choi, J.H.; Kim, D.; Ko, M.S.; Lee, D.E.; Wi, K.; Lee, H.S. Compressive Strength Prediction of Ternary-Blended Concrete Using Deep Neural Network with Tuned Hyperparameters. *J. Build. Eng.* **2023**, *75*, 107004. [CrossRef]
35. Too, E.C.; Yujian, L.; Njuki, S.; Yingchun, L. A Comparative Study of Fine-Tuning Deep Learning Models for Plant Disease Identification. *Comput. Electron. Agric.* **2019**, *161*, 272–279. [CrossRef]
36. Harrison, M. *Machine Learning—Guia de Referência Rápida: Trabalhando com Dados Estruturados em Python*; Novatec Editora: São Paulo, Brazil, 2019; Volume 1.
37. Nawaz, M.; Nazir, T.; Javed, A.; Tawfik Amin, S.; Jeribi, F.; Tahir, A. CoffeeNet: A Deep Learning Approach for Coffee Plant Leaves Diseases Recognition. *Expert. Syst. Appl.* **2024**, *237*, 121481. [CrossRef]
38. Kaya, Y.; Gürsoy, E. A Novel Multi-Head CNN Design to Identify Plant Diseases Using the Fusion of RGB Images. *Ecol. Inform.* **2023**, *75*, 101998. [CrossRef]
39. Smith, S.; Elsen, E.; De, S. On the Generalization Benefit of Noise in Stochastic Gradient Descent. In Proceedings of the 37th International Conference on Machine Learning, Virtual, 13–18 July 2020; Singh, A.H.D., III, Ed.; Proceedings of Machine Learning Research. PMLR: New York, NY, USA, 2020; Volume 119, pp. 9058–9067.
40. Reddy, S.R.G.; Varma, G.P.S.; Davuluri, R.L. Resnet-Based Modified Red Deer Optimization with DLCNN Classifier for Plant Disease Identification and Classification. *Comput. Electr. Eng.* **2023**, *105*, 108492. [CrossRef]
41. Asad, M.A.U.; Guan, X.; Zhou, L.; Qian, Z.; Yan, Z.; Cheng, F. Involvement of Plant Signaling Network and Cell Metabolic Homeostasis in Nitrogen Deficiency-Induced Early Leaf Senescence. *Plant Sci.* **2023**, *336*, 111855. [CrossRef] [PubMed]
42. da Costa, A.Z.; Figueroa, H.E.H.; Fracarolli, J.A. Computer Vision Based Detection of External Defects on Tomatoes Using Deep Learning. *Biosyst. Eng.* **2020**, *190*, 131–144. [CrossRef]
43. Ramos-Ospina, M.; Gomez, L.; Trujillo, C.; Marulanda-Tobón, A. Deep Transfer Learning for Image Classification of Phosphorus Nutrition States in Individual Maize Leaves. *Electronics* **2023**, *13*, 16. [CrossRef]
44. Wan, S.; Chang, S.-H.; Peng, C.-T.; Chen, Y.-K. A Novel Study of Artificial Bee Colony with Clustering Technique on Paddy Rice Image Classification. *Arab. J. Geosci.* **2017**, *10*, 215. [CrossRef]
45. Ibarra-Pérez, T.; Jaramillo-Martínez, R.; Correa-Aguado, H.C.; Ndjatchi, C.; Martínez-Blanco, M.d.R.; Guerrero-Osuna, H.A.; Mirelez-Delgado, F.D.; Casas-Flores, J.I.; Reveles-Martínez, R.; Hernández-González, U.A. A Performance Comparison of CNN Models for Bean Phenology Classification Using Transfer Learning Techniques. *AgriEngineering* **2024**, *6*, 841–857. [CrossRef]

AgriEngineering

Article

A Performance Comparison of CNN Models for Bean Phenology Classification Using Transfer Learning Techniques

Teodoro Ibarra-Pérez [1], Ramón Jaramillo-Martínez [1,*], Hans C. Correa-Aguado [1], Christophe Ndjatchi [1], Ma. del Rosario Martínez-Blanco [2], Héctor A. Guerrero-Osuna [3], Flabio D. Mirelez-Delgado [1], José I. Casas-Flores [4], Rafael Reveles-Martínez [1] and Umanel A. Hernández-González [1]

1 Instituto Politécnico Nacional, Unidad Profesional Interdisciplinaria de Ingeniería Campus Zacatecas (UPIIZ), Zacatecas 98160, Mexico; tibarrap@ipn.mx (T.I.-P.); hcorreaa@ipn.mx (H.C.C.-A.); mndjatchi@ipn.mx (C.N.); fmirelezd@ipn.mx (F.D.M.-D.); rrevelesm@ipn.mx (R.R.-M.); uahernandez@ipn.mx (U.A.H.-G.)

2 Laboratorio de Inteligencia Artificial Avanzada (LIAA), Universidad Autónoma de Zacatecas, Zacatecas 98000, Mexico; mrosariomb@uaz.edu.mx

3 Posgrado en Ingeniería y Tecnología Aplicada, Unidad Académica de Ingeniería Eléctrica, Universidad Autónoma de Zacatecas, Zacatecas 98000, Mexico; hectorguerreroo@uaz.edu.mx

4 Instituto Nacional de Investigaciones Forestales, Agrícolas y Pecuarias, Campo Experimental Zacatecas (INIFAP), Zacatecas 98500, Mexico; casas.israel@inifap.gob.mx

* Correspondence: rjaramillom@ipn.mx

Citation: Ibarra-Pérez, T.; Jaramillo-Martínez, R.; Correa-Aguado, H.C.; Ndjatchi, C.; Martínez-Blanco, M.d.R.; Guerrero-Osuna, H.A.; Mirelez-Delgado, F.D.; Casas-Flores, J.I.; Reveles-Martínez, R.; Hernández-González, U.A. A Performance Comparison of CNN Models for Bean Phenology Classification Using Transfer Learning Techniques. *AgriEngineering* **2024**, *6*, 841–857. https://doi.org/10.3390/agriengineering6010048

Academic Editors: Ray E. Sheriff and Chiew Foong Kwong

Received: 4 January 2024
Revised: 1 March 2024
Accepted: 1 March 2024
Published: 18 March 2024

Abstract: The early and precise identification of the different phenological stages of the bean (*Phaseolus vulgaris* L.) allows for the determination of critical and timely moments for the implementation of certain agricultural activities that contribute in a significant manner to the output and quality of the harvest, as well as the necessary actions to prevent and control possible damage caused by plagues and diseases. Overall, the standard procedure for phenological identification is conducted by the farmer. This can lead to the possibility of overlooking important findings during the phenological development of the plant, which could result in the appearance of plagues and diseases. In recent years, deep learning (DL) methods have been used to analyze crop behavior and minimize risk in agricultural decision making. One of the most used DL methods in image processing is the convolutional neural network (CNN) due to its high capacity for learning relevant features and recognizing objects in images. In this article, a transfer learning approach and a data augmentation method were applied. A station equipped with RGB cameras was used to gather data from images during the complete phenological cycle of the bean. The information gathered was used to create a set of data to evaluate the performance of each of the four proposed network models: AlexNet, VGG19, SqueezeNet, and GoogleNet. The metrics used were accuracy, precision, sensitivity, specificity, and F1-Score. The results of the best architecture obtained in the validation were those of GoogleNet, which obtained 96.71% accuracy, 96.81% precision, 95.77% sensitivity, 98.73% specificity, and 96.25% F1-Score.

Keywords: deep learning; convolutional neural network; bean phenology; food security; transfer learning

1. Introduction

According to the Food and Agriculture Organization of the United Nations (FAO), plant plagues are among the main causes of the loss of over 40 percent of food crops worldwide, exceeding losses of up to USD 220 billion each year [1]. In Mexico, bean production contributes MXN 5927 million to the annual income. However, in 2021, the registered loss was more than MXN 222 million, mainly due to diseases caused by viruses transmitted in seeds, aphids, white flies, and other similar insects [2]. Several factors interfere with food security, such as climate change [3–5], the lack of pollinators [6,7], plagues and plant diseases [8], the result of the COVID-19 pandemic, and the present war between Russia and Ukraine [9], among others.

Insect plagues, diseases, and other organisms significantly affect the quality and production of crops. These organisms feed off plants and transmit diseases that can cause severe disruption in the growth and development of plants, causing a major impact on food security, the economy, and the environment, thereby decreasing the availability of food, increasing production costs, and affecting the growth of rural areas and developing countries [10].

It is important to mention the strategies used to mitigate the effects produced by plagues and diseases around the world, such as the selection of resistant varieties, crop rotation, the use of natural enemies of plagues, and the rational use of chemical products, among others. More efforts need to be made to implement mechanisms and innovative strategies to reduce loss in food crops and sustainably contribute to food security [9,11,12].

In recent years, the use of artificial intelligence (AI) in applications has increased exponentially. Proof of this is the appearance of works related to image recognition, especially in the field of agriculture, where various approaches to using deep learning (DL) methods to classify the phenology of different food crops around the world have been presented. This allows us to have knowledge of the record of critical moments in the life cycle of the plant to program treatments, effectively and timely apply pesticides or fungicides, and prevent and control plagues and diseases; this offers great advantages in precision agriculture in a nonharmful manner, and it helps minimize damage to crops [13–17].

DL methods are used to identify the different phenological stages of crops, and there exists a diversity of approaches to address the classification of topics related to agricultural decision making that mainly influence the estimation of agricultural production. In this regard, the present work proposes a comparative study of the performance of four models of convolutional neural networks (CNNs), AlexNet, VGG19, SqueezeNet, and GoogleNet, in classifying the phenological stages of bean crops; the performance of each of the models is compared through the following metrics: accuracy, precision, sensitivity, specificity, and F-1 Score. The results are used to choose the architecture that best models the classification problem in bean phenology.

The goal of analyzing the different CNN architectures is to identify the best-performing one and, in the future, to embed networks in compact systems so that farmers can identify the phenological stages of plants, allowing them to take preventive measures.

The organization of the present work is structured as follows: Section 2 describes the most relevant works on transfer learning, related concepts, and a description of the CNN models used in the present work. Section 3 describes the methodology of the investigation work. Section 4 contains the obtained results and their discussion. Section 5 presents the conclusions, and finally, future work is described.

2. Related Work

DL has been used to obtain high-quality maps. In this regard, Ge et al. [18] mapped crops from different regions in the period from planting to vigorous growth and compared the maps obtained by using conventional methods with those obtained with DL, where the latter reached 87% accuracy.

On the other hand, Yang et al. [17] proposed the identification of the different phenological phases of rice from RGB images captured by a drone with a CNN model that incorporated techniques such as spatial pyramidal sampling, knowledge transfer, and external data, which are essential for timely estimation and output, in comparison to previous approaches where data on the vegetation index in temporal series and diverse methods based on thresholds were used. The obtained results show that the approach has high precision in the identification of the phenology of rice, with 83.9% precision and a mean absolute error (MAE) of 0.18.

The approach of learning by transfer is used in various applications, such as the prediction of the performance of numerous crops worldwide, where Wang et al. [19] used remote sensors with satellite images to estimate the outcome of soy crops through algorithms and deep learning, offering an inexpensive and efficient alternative in comparison to conventional techniques that are generally expensive and difficult to expand in regions with limited access to data.

Identifying the phenology in diverse crops allows for determining critical moments for timely agricultural activities. In this regard, Reeb et al. [20] implemented a pre-trained CNN model named ResNet18 to classify the phenology of *A. petiolata*, comparing the results obtained to the classification conducted by a group of non-expert humans. During the validation stage of the model, 86.4% of the results obtained from a set of 2448 images were classified correctly by the proposed model. Subsequently, an evaluation of the precision of the model was made and compared to human precision, where the model correctly classified 81.7% of a total of 241 images. In contrast, the non-expert humans achieved a precision of 78.6%.

On the other hand, Datt et al. [15] used the CNN to recognize eight phenological stages in apple crops. The knowledge transfer model used was Inception-v3, which trained a set of images from the Srinagar region in India. The number of images captured in the area was 1200, extending to 7000 images through data augmentation techniques, obtaining results in comparison to other models such as Xception, Xception-v3, ResNet50, VGG16, and VGG19 where the metric used was for F1-Score and achieved values of 0.97, 0.96, 0.66, 0.96, and 0.95, respectively.

Through the use of characteristic descriptors, Yalcin et al. [16] proposed the implementation of automatic learning algorithms to compare learning algorithms based on a CNN to recognize and classify phenological stages in various types of crops such as wheat, corn, barley, lentil, cotton, and pepper through images collected by cameras at farming stations located in parts of the territory of Turkey. The AlexNet model used for the classification of phenology significantly exceeded automatic learning algorithms during the performance evaluation.

The combination of conventional techniques with DL methods can offer alternatives to the solution of classification problems, such as the case of the application of hybrid methods, to get to know the estimation of agricultural production in the work carried out by Zhao et al. [21], in which they used the knowledge transfer technique to learn from an existing model based on the combination of biomass algorithms of wheat crops and the model of simple performance. The results show a precise estimation of the wheat harvest with both models since they reveal a good correlation of $R^2 = 0.83$ and a root mean squared error (RMSE) of 1.91 t ha^{-1}.

The use of temporal series and other techniques for phenological classification is usually relevant in the work carried out by Taylor et al. [22], who propose a model for creating temporal series of the phenological cycle. They use a hidden Markov post-process model to address the temporal correlation between images, reaching F1-Scores of 0.86 to 0.91. The results show the temporal progression of the crops from the emergency to the harvest, providing the daily phenological stages on a temporal scale.

DL techniques are used to classify the different phenological stages in different types of crops, including bean crops. The diversity of approaches and techniques used to classify images varies the obtained results compared to techniques and strategies used in diverse studies during the last decade. Thereupon, CNN produces trusted results, has a grand capacity for generalization in the classification of images, and has a high capacity for extracting features related to the phenology in different crops.

2.1. *Convolutional Neural Networks (CNNs)*

CNNs basically consist of three blocks: the first in a layer of convolution that allows the extraction of features of an image; the second is a block that consists of a layer of maximum grouping to execute a subsample of pixels and reduce the dimensionality, allowing the reduction of computational costs; and finally, the third block involves fully connected layers to provide the network with the capacity of classification [23–25]. The general description of the CNN architecture is shown in Figure 1, where features of the images are identified, extracted, and classified.

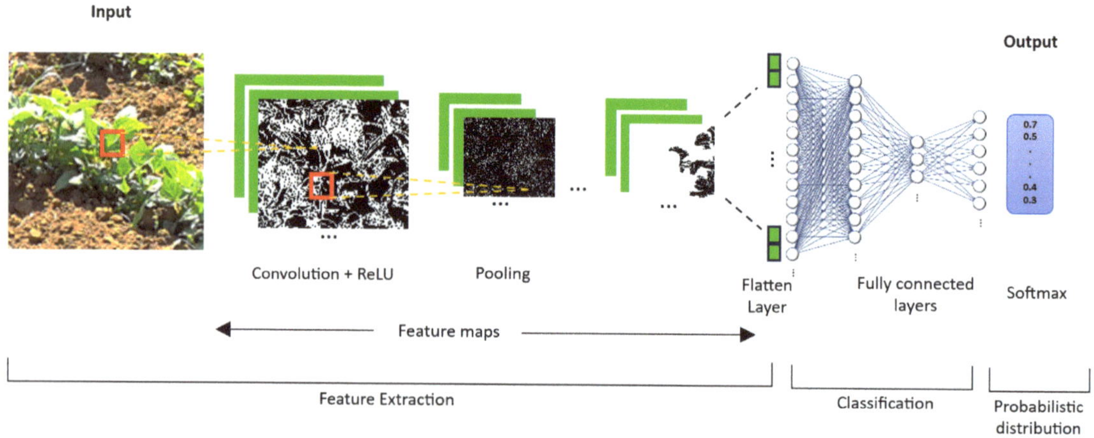

Figure 1. The architecture of convolutional network.

2.2. Transfer Learning

In general, CNN results are better if trained by more extensive data sets than small ones. However, many applications do not have large data sets, and transfer learning can be helpful in those applications where the data set is smaller in ImageNet [26]. For this reason, a re-trained model from large data sets can be used to learn new features from a comparatively smaller data set [27]. Figure 2 describes a block diagram of the transfer learning approach used in this study.

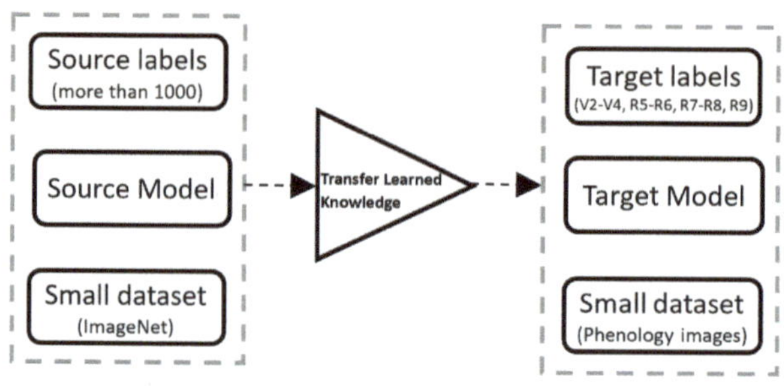

Figure 2. Block diagram of the concept of transfer learning.

Recently, the scientific community has taken a particular interest in the transfer learning approach in diverse fields, such as medicine and agriculture, among others [28–31]. This approach allows previously acquired knowledge and avoids training with large quantities of data when training new deep architecture models [28,32,33].

2.3. Re-Trained Neural Networks

Using a re-trained CNN model has significant advantages in comparison to the design of models from zero, which require large sets of data and training that can take considerable time, including weeks, translating into high computational costs. On the other hand, a re-trained model can have a high capacity for generalization and accelerate convergence [34].

In this study, four models of re-trained CNN were used to evaluate the performance in the image classification identifying phenological phases in bean crops: AlexNet, VGG19,

SqueezeNet, and GoogleNet. A brief introduction of each of the re-trained models is included below.

2.3.1. AlexNet Model

The algorithms for object detection and image classification were evaluated with the AlexNet model developed by Krizhevsky et al. [35] and the ImageNet Large-Scale Visual Recognition Challenge (LSVRC) for the model training [36]. The architecture is characterized by using a new activating function, the Rectified Linear Unit (ReLU), to add non-linearity, solve the gradient evanescent problem, and accelerate network training. CNN consists of eight layers in total: the first five layers are of convolution, some of which are followed by max-pooling layers, and the following three layers are fully connected, followed by an exit layer of 1000 neurons for the SoftMax for multiclass classification. The AlexNet model was trained with over one million images from the ImageNet database created by Deng et al. [26]. The entry size of the images is 227 × 227, a total of 60 million parameters and 650 thousand neurons.

2.3.2. VGG19

The VGG19 network is a nineteen-layer deep convolutional neural network developed by Simonyan et al. [37]. This model uses small filters of 3 × 3 in each of the sixteen convolution layers. Next, it uses three fully connected layers to classify images into 1000 categories. The ImageNet database developed by Deng et al. [26] was used to train the VGG19 model. The layers used for the extraction of features are separated into five groups where a layer of max-pooling follows each group, and it is required to insert and image the size of 224 × 224 to generate the label corresponding to the exit.

2.3.3. SqueezeNet

The SqueezeNet architecture uses compression techniques to reduce the model size without compromising its performance with a fire module that, instead of using convolutional layers followed by grouping layers, uses a combination of filters that combine convolutions of 1 × 1 and 3 × 3 to reduce the number of parameters. The SqueezeNet model proposed by Iandola et al. [38] contains fifty times fewer parameters than the AlexNet model, has eighteen layers of profundity, and requires a size of 227 × 227.

2.3.4. GoogleNet

This model, also known as Inception v1, was developed by Szegedy et al. [32]. It consists of twenty-two layers of profundity with entry images the size of 224 × 224. GoogleNet uses average pooling after the last convolutional layer instead of fully connected layers. The convolution modules called "Inception modules" are composed of multiple convolutions of different sizes (1 × 1, 3 × 3, and 5 × 5), allowing the network to capture the features on different spatial layers, facilitating the representation of fine details and complex patterns.

Table 1 summarizes the main features in terms of the parameters used, profundity, and size of the different architectures of CNN networks proposed in this study.

Table 1. Summary of presented architectures.

CNN Architecture	Year	Developer	Profundity (Number of Layers)	Size (MB)	Number of Parameters
AlexNet	2012	Krizhevsky et al. [35]	8	240	60 million
VGG19	2014	Simonyan et al. [37]	19	550	138 million
SqueezeNet	2016	Iandola et al. [38]	18	5	1.2 million
GoogleNet	2014	Google	22	50	4 million

3. Materials and Methods

The methodology used in this study consists of three phases, as shown in Figure 3. The first phase describes the data acquisition procedure and the construction and general features of the obtained images. The second phase describes the transfer learning of CNN architectures used in this study and the configuration of hyperparameters, such as the learning rate, the lot size by iteration, the number of epochs, and the optimizer. The third phase describes the evaluation of the proposed models to measure the performance by applying different metrics.

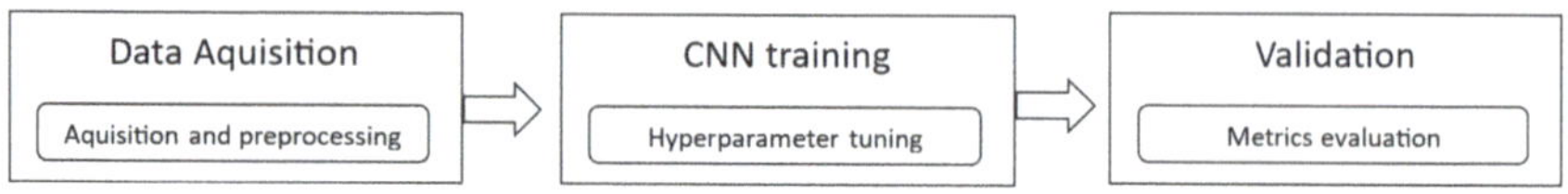

Figure 3. Diagram of proposed methodology.

3.1. *Acquisition of Data*

The two selected bean parcels are located in the municipality of Calera de Víctor Rosales in the state of Zacatecas, Mexico (22°54′14.6″ N 102°39′32.5″ W). The variety of beans used was pinto Saltillo, and the data were collected between 12 May and 15 August of the year 2023. The camera model used was HC-801Pro, with 4G technology, a range of optical vision of 120°, IP65 protection, and a resolution of 30 megapixels to acquire high-quality images.

Two cameras were installed to capture the images, as shown in Figure 4. To determine the number of images for the training and testing data set, according to Tylor et al. [22], the average time for bean harvest after its emergence is from 65 to 85 days approximately, which is why an average of eight to ten images were captured per day since the emergence of the plant.

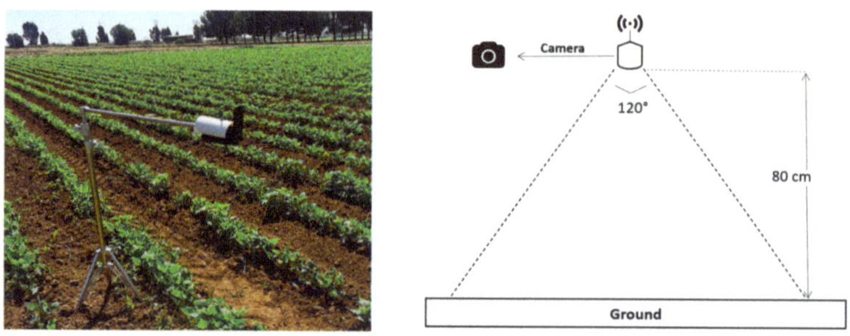

(a) (b)

Figure 4. Installation of the GSM camera station in the open field for the capture of images: (**a**) camera station for the capture of images; (**b**) schematic diagram for the acquisition of images.

The shooting method used was for two samples per sequence for intervals of time between 8:00, 10:00, 12:00, 16:00, and 18:00 h, obtaining a total of 814 images, allowing the experimental data to include the bean growing cycle in the vegetative phase and the reproductive phase, from the germination phase of the plant (V0) through the emergence phase of the plant (V1), primary leaves (V2), the first trifoliate leaf (V3), the third trifoliate leaf (V4), prefloration (R5), floration (R6), pod formation (R8), pod filling (R8), and maturation (R9).

Generally, the bean's phenology is classified into ten classes and divided into two main categories: the vegetative and the production phases. However, for this investigation, only four classes were selected according to the most significant number of examples per class since they tend to be the most representative, according to Etemadi et al. [39].

An example of each class of the obtained data set can be observed in Figure 5, labeling the vegetative phase in the stage of primary leaves, first and third trifoliate leaves (V2–V4), reproductive phase in the stage of prefloration and floration (R5–R6), reproductive phase in the stage of formation and filling of pods (R7–R8), and reproductive phase in the stage of maturation (R9).

Figure 5. Descriptive stages of the phenology of the bean: (**a**) vegetative phase in primary leaves, first and third trifoliate leaves; (**b**) reproductive phase in prefloration and floration; (**c**) reproductive stage in the formation and filling of pods; (**d**) reproductive phase in maturation.

The training data set and tests used per class can be observed in Figure 6. Most images have a resolution of 5120 × 3840 pixels. However, the images were re-dimensioned to adjust the size according to the entry specifications for each proposed model [34].

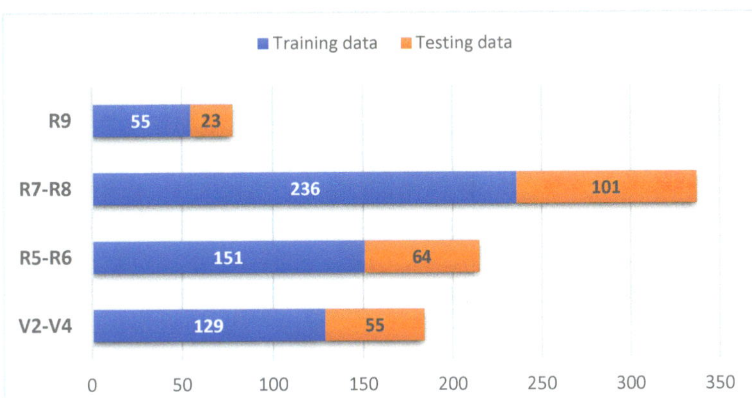

Figure 6. Images for training and tests per class.

3.2. Data Augmentation

The data augmentation contributes to avoiding overfitting the network and memorizing the exact details of the images during training. This is a common problem when the CNN model is exposed to small data sets where the learned patterns are not generalized into new data [40,41].

Currently, there is a tendency in deep learning training algorithms to allow for the increase of the initial data set through data augmentation techniques, obtaining results that can improve the precision performance in deep learning algorithms [42]. A series of aleatory transformations increased data to exploit the few examples of images and increase the precision of the proposed CNN models. The strategies of data augmentation used were rotation, translation, reflection, and scaling. Examples are shown in Figure 7.

(a) (b) (c) (d) (e)

Figure 7. Data augmentation in an image of the phenology of the bean: (**a**) original image without data increase; (**b**) image after rotation; (**c**) image after translation; (**d**) image after reflection; (**e**) image after scaling.

3.3. *Training of the Models*

For the training data set and tests, the images were divided randomly in a partition of 70% of the training data set and 30% for the test set. Table 2 shows the configuration of the experimental equipment used in this investigation. The four models previously selected were trained by the ImageNet database, which contains more than 15 million images [26].

Table 2. Configuration of experimental equipment.

Name of Configuration	Parameters
Processor	AMD Ryzen 7, 5700U, 8 cores at 1.8 GHz
Graphic Card	Radeon Graphics up to 4.3 GHz
RAM	16 GB
Operative system	Windows 11
Programing platform	MATLAB (2022a)

The hyperparameters selected in this study from the revised literature and the previously mentioned hardware capacity are described in Table 3. The selection of hyperparameters significantly affects the performance of CNN models, which is why a good selection is crucial. The hyperparameters were standardized for each model to compare the performance of the proposed models [31,43,44].

Table 3. Training hyperparameters of pre-trained models.

Re-Trained CNN Model	Image Size	Optimizer	Momentum	Epochs	Batch Size	Learning Rate
AlexNet VGG19 SqueezeNet GoogleNet	227 × 227 224 × 224 227 × 227 224 × 224	Stochastic Gradient Descent with Momentum	0.9	15	128	0.001

The optimizer used is the Stochastic Gradient Descent with Momentum (SGDM) method, which combines stochastic gradient descent and momentum techniques. Each iteration calculates the gradient using a random sample from the training set. Then, the weight is updated considering the previous update, allowing convergence acceleration and keeping it at a local minimum.

Momentum is employed to improve the precision and velocity of the training by adding a fraction from the previous step to the present step in the weight update. This allows it to overcome the obstacle of local minimums and maintain a constant impulse in the direction of the gradient.

The epochs refer to the number of iterations carried out regarding the correlation of forward and reverse propagation to reduce loss. The size of the epochs describes the number of examples used in each iteration of the training algorithm. The learning rate is defined by the velocity size in which the optimization function performs a search to converge [45].

3.4. Performance Evaluation

At present, an extended variety of metrics are used to evaluate the performance of CNN models, where information is given about the aspects and characteristics that allow the evaluation of the performance of the models. A number of true positives (TP), true negatives (TN), false positives (FP), and false negatives (FN) are needed to calculate the performance of the models. These cases represent the combinations of true and predicted classes in classification problems. Therefore, TP + TN + FP + FN equals the total number of samples and is described in a confusion matrix [46].

When trained by transfer learning, the biased distributions appear naturally, producing an intrinsic unbalance. This is why it is necessary to employ metrics that evaluate the global performance of each model. In this regard, the metrics employed [16,20,22,46,47] were used to compare the performance of the models without setting aside the different characteristics of the training and validation data used in this study.

TP is the true positive, which means the prediction is positive. FP is the false positive, which means the prediction is negative. However, the prediction is positive. FN is the false negative, which means a positive prediction, but the result is negative. TN is the true negative, which means a negative result prediction.

The confusion matrix is a tool that allows the visualization of a model's performance when classifying and containing the previously defined elements. The rows in the matrix represent the true class, and the columns, the predicted class, and the primary diagonal cells describe the correctly classified observations. In contrast, the lateral diagonals correspond to the incorrectly classified observations.

In this study, five metrics were used to evaluate the performance of the proposed models: accuracy, precision, sensitivity, specificity, and F1-Score [27,48]. Accuracy is the relation between the number of correct predictions and the total number of made predictions, as calculated by Equation (1).

$$\text{Accuracy} = \frac{\text{TP} + \text{TN}}{\text{TP} + \text{TN} + \text{FP} + \text{FN}} \tag{1}$$

Precision measures the proportion of correct predictions made by the model, in other words, the number of correctly classified elements as positives out of a total of elements identified as positive. The mathematical representation is described in Equation (2).

$$\text{Precision} = \frac{\text{TP}}{\text{TP} + \text{FP}} \tag{2}$$

Sensitivity is also known as recall; it calculates the proportion of correctly identified cases as positive from a total of true positives, as described in Equation (3).

$$\text{Sensitivity} = \frac{\text{TP}}{\text{TP} + \text{FN}} \tag{3}$$

Specificity is the opposite of sensitivity or recall and calculates the portion of cases identified as negatives. It is calculated by Equation (4).

$$\text{Specificity} = \frac{\text{TN}}{\text{TN} + \text{FP}} \tag{4}$$

F1-Score allows the combination of precision and sensitivity or recall, where the value of one indicates a good balance between precision and sensitivity in the classification model. Its mathematical representation is described in Equation (5).

$$\text{F1-score} = 2 \times \frac{\text{precision} \times \text{recall}}{\text{precision} + \text{recall}} \quad (5)$$

4. Results and Discussion

Table 4 provides a detailed calculation for each of the different metrics of AlexNet's architecture. A sensitivity of 100% was obtained for the prediction of classes R9 and V2–V4, which correspond to the reproductive phenological phase in the stage of maturation and the vegetative phase in the stage of primary leaves and first and third trifoliate leaves, respectively.

Table 4. Calculation of metric for the AlexNet model.

Name	Classes				Averaging
True Positive	62	93	23	55	58.25
False Positive	3	2	4	1	2.5
False Negative	2	8	0	0	2.5
True Negative	176	140	216	187	179.75
Accuracy	0.958847	0.958847	0.958847	0.958847	0.958847
Precision	0.953846	0.978947	0.851851	0.982142	0.941697
Sensitivity	0.968750	0.920792	1	1	0.972385
Specificity	0.983240	0.985915	0.981818	0.994680	0.986413
F1-Score	0.961240	0.948979	0.920000	0.990990	0.955302

The averages obtained for the accuracy, precision, sensitivity, specificity, and F1-Score were 95.8%, 94.1%, 97.2%, 98.6%, and 95.5% of the predicted classes during validation, respectively.

Table 5 shows the different metrics in the VGG19 model. In classes R5-R6, corresponding to the reproductive phenological phase in the stages of prefloration and floration, a precision of 100% was reached.

Table 5. Calculation of metrics for VGG19 model.

Name	Classes				Averaging
True Positive	60	97	23	55	58.75
False Positive	0	3	4	1	2
False Negative	4	4	0	0	2
True Negative	179	139	216	187	180.25
Accuracy	0.967078	0.967078	0.967078	0.967078	0.967078
Precision	1	0.970000	0.851851	0.982142	0.950998
Sensitivity	0.937500	0.960396	1	1	0.974474
Specificity	1	0.978873	0.981818	0.994680	0.988843
F1-Score	0.967741	0.965174	0.920000	0.990990	0.960976

A 95% precision average for all classes can be observed. On the other hand, maximum sensitivity was also obtained for classes R9 and V2–V4, corresponding to the phenological phase of reproduction in the maturation stage and vegetative phase of primary leaves and first and third trifoliate leaves, respectively. An average of 97.4% of sensitivity is observed in each class. In addition, averages reached for specificity are 98.8% and 96% F1-Score in all classes, achieving the best scores compared to the other architectures.

A detailed calculation for each metric used in the architecture SqueezeNet can be shown in Table 6. For the classes V2–V4, which corresponds to the vegetative phenological phase in the stage of primary leaves and first and third trifoliate leaves, a sensitivity of 100%

was obtained. On the other hand, averages in accuracy, precision, sensitivity, and F1-Score of 95.8%, 93.4%, 95.9%, 98.6%, and 94.4%, respectively, are observed for all predicted classes during validation.

Table 6. Calculation of metrics for SqueezeNet model.

Name	Classes				Averaging
True Positive	59	97	22	55	58.25
False Positive	0	3	6	1	2.50
False Negative	5	4	1	0	2.50
True Negative	179	139	214	187	179.75
Accuracy	0.958847	0.958847	0.958847	0.958847	0.958847
Precision	1	0.970000	0.785714	0.982142	0.934464
Sensitivity	0.921875	0.960396	0.956521	1	0.959698
Specificity	1	0.978873	0.972727	0.994680	0.986570
F1-Score	0.959349	0.965174	0.862745	0.990990	0.944564

The different metrics of the GoogleNet model can be observed in Table 7, where the average obtained for precision is 96.8% in all predicted classes, and there is a maximum sensitivity in the prediction of classes V2–V4 that corresponds to the vegetative phenological phase in the stage of primary leaves and first and third trifoliate leaves. The averages observed in the metrics of accuracy, precision, sensitivity, specificity, and F1-Score are 96.7%, 96.8%, 95.7%, 98.7%, and 96.2%, respectively, for all predicted classes during validation, which concur with the metric of precision obtained for the VGG19 model.

Table 7. Calculation of metrics for GoogleNet model.

Name	Classes				Averaging
True Positive	60	99	21	55	58.75
False Positive	1	5	1	1	2
False Negative	4	2	2	0	2
True Negative	178	137	219	187	180.25
Accuracy	0.967078	0.967078	0.967078	0.967078	0.967078
Precision	0.983606	0.951923	0.954545	0.982142	0.968054
Sensitivity	0.937500	0.980198	0.913043	1	0.957685
Specificity	0.994413	0.964788	0.995454	0.994680	0.987334
F1-Score	0.960000	0.965853	0.933333	0.990990	0.962544

Table 8 shows the results obtained in each metric, with the best values obtained during the architecture's evaluation highlighted in bold. It shows that the architecture of VGG19 and GoogleNet obtained the best performance, and both concur in accuracy. On the other hand, the architecture with the lowest performance observed is SqueezeNet due to the values obtained that are generally lower than those obtained in other architectures. However, SqueezeNet required the least training time in comparison to the others.

Table 8. Comparison of results for each of the architectures.

Metrics	AlexNet	VGG19	SqueezeNet	GoogleNet
Accuracy	0.9588	0.9671	0.9588	0.9671
Precision	0.9417	0.9510	0.9345	0.9681
Sensitivity	0.9724	0.9745	0.9597	0.9577
Specificity	0.9864	0.9888	0.9866	0.9873
F1-Score	0.9553	0.9610	0.9446	0.9625
Time [minutes]	17.31	168.76	20.19	25.48

A comparison of the accuracy obtained in each of the models during the training and validation stages is shown in Figure 8, where the architecture AlexNet reaches the highest accuracy percentage compared to the other models during the training stage. However, in the validation stage, VGG19 reached the highest accuracy percentage. This stage shows that the GoogleNet model obtained the lowest accuracy percentage compared to the other models. However, it is observed that this model reached the highest accuracy percentage during the validation stage, just like VGG19; in other words, they obtained a greater capacity.

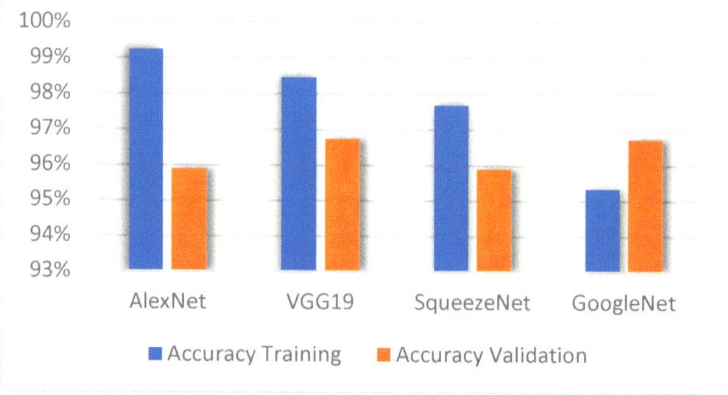

Figure 8. Accuracy of models during training and validation.

The architecture performance summary with measurements obtained in each metric calculated during the validation stage can be observed in Figure 9; the accuracy metric obtained during the training stage is included for each model. It is observed that the GoogleNet architecture maintained the best balance in the projection of all metrics. On the other hand, it is observed that the architecture SqueezeNet obtained the best performance compared to the rest of the architectures, except in precision, where the result is the same as the AlexNet architecture, and sensitivity is less than the obtained with the GoogleNet architecture.

Based on the obtained results, the GoogleNet architecture has a higher performance during the validation compared to its performance during the training process; the cause of this behavior could be due to the limited amount of training data, the adequate selection of hyperparameters, and possibly an over-adjustment. However, the difference between the precision of training and validation is 1.4% compared to the models AlexNet, VGG19, and SqueezeNet, which present a difference of 1.7%, 3.3%, and 1.7%, respectively—considering that increased training data will give an outcome with a tendency to decrease the performance during the validation of each architecture.

According to the behavior during the validation, the architectures AlexNet and SqueezeNet presented a low-balanced tendency in the metrics, obtaining low results for precision and F1-Score. On the other hand, the architecture VGG19 registered the same level of performance as GoogleNet but with lower precision and F1-Score, giving a reason to consider the architecture GoogleNet as having the best global performance.

Figure 10 shows the confusion matrix of the four proposed CNN models. It also provides a detailed analysis of instance numbers correctly classified by each proposed architecture. Compared to other architectures, the AlexNet architecture presented problems in correctly classifying the class R9, which corresponds to the reproductive phenological phase in the stage of maturation, achieving the classification of only 85.2% of instances.

The confusion matrix of the VGG19 architecture shows its high capacity to classify instances correctly. The diagonal shows the correctly classified instances; however, class R9 presented the most difficulty. The SqueezeNet architecture, like the previous architectures, shows more difficulties in correctly classifying the class R9; however, for the GoogleNet architecture, the class presents minimal difficulties.

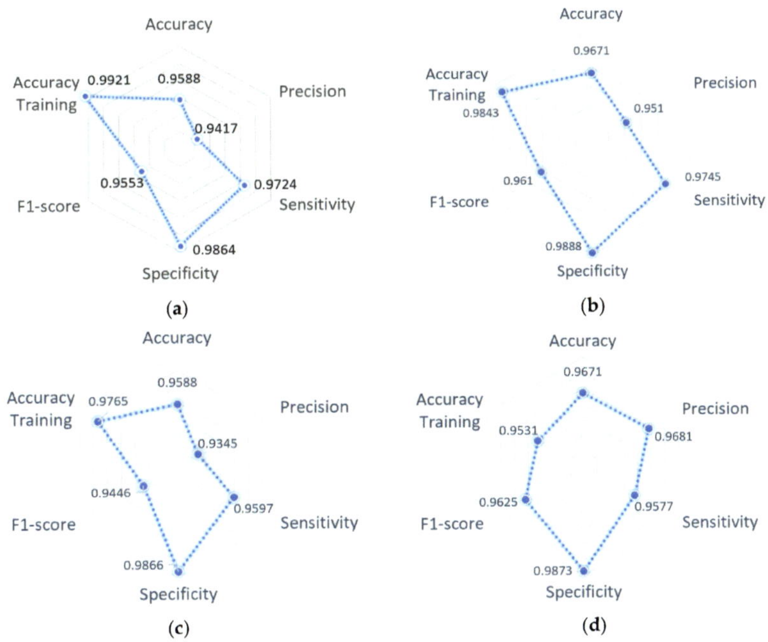

Figure 9. Summary of model performance: (a) AlexNet model; (b) VGG19 model; (c) SqueezeNet model; (d) GoogleNet model.

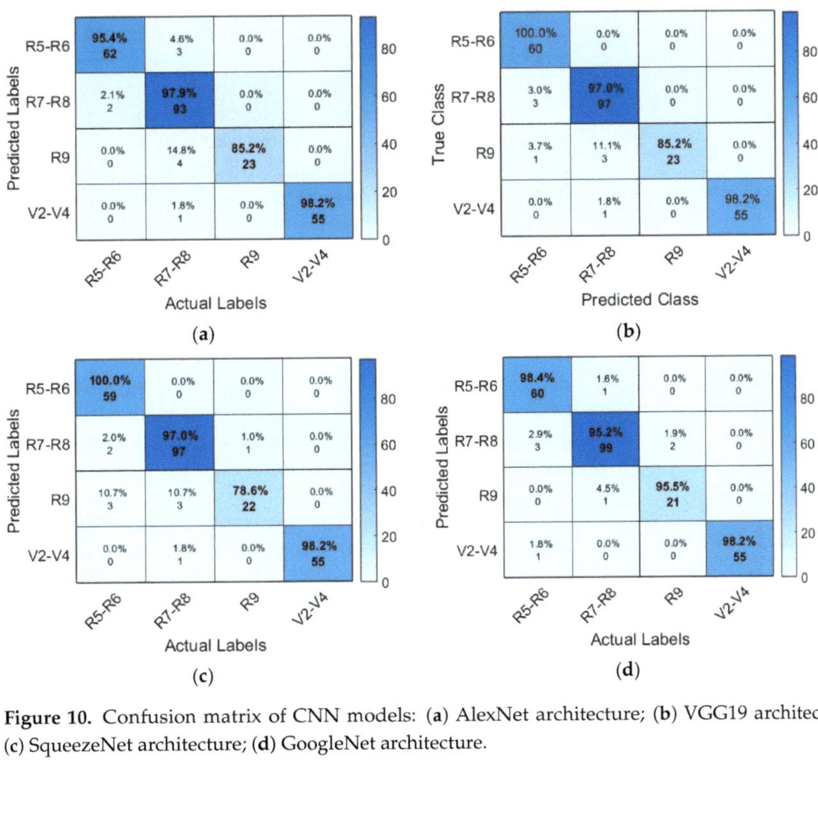

Figure 10. Confusion matrix of CNN models: (a) AlexNet architecture; (b) VGG19 architecture; (c) SqueezeNet architecture; (d) GoogleNet architecture.

Table 9 describes a comparison summary of the performance results obtained by other authors concerning the techniques and metrics used for this work. It breaks down the values obtained for accuracy, precision, sensitivity, and F1-Score for each proposed architecture.

Table 9. Comparison of similar work recently published.

Author	Phenology	Technique	Data Augmentation	Accuracy [%]	Precision [%]	Sensitivity [%]	F1-Score [%]
Yalcin [16]	Wheat, barley, lentil, cotton, and pepper	Re-trained CNN AlexNet model	Not	87.14	88.12	87.24	87.28
Qin et al. [46]	Rice	Re-trained CNN ResNet-50 model	Yes	87.33	-	-	-
Reeb et al. [20]	*Alliaria petiolata*	Re-trained CNN ResNet-18 model	Not	86.4	-	-	-
Taylor et al. [22]	Corn, wheat, barley, soybean, and alfalfa	Re-trained CNN VGG16 model and temporal model of Markov	Not	-	92	90	91
Han et al. [47]	Rice	CNN model with support vector machine (SVM)	Yes	91.3	-	-	93
This study	Bean	Re-trained CNN GoogleNet model	Yes	96.71	95.1	97.45	96.1

5. Conclusions

The proposed methodology shows that the proposed CNN models allow the correct classification of more than 90% of the samples, even when working with an unbalanced and relatively minor data set. In addition, each analyzed architecture has different characteristics, such as the number of layers and used filters. However, it is crucial to highlight a suitable selection of metrics to discriminate one architecture from the other.

Evaluating different CNN topologies is significant for future work since the architectures can present bias due to being trained with numerous images from which many are not part of the final classification. In this regard, evaluating the performance by transfer with new data lays the foundation for new work, such as identifying nutrients or plagues for this species. The joint evaluation of the metrics accuracy, precision specificity, sensitivity, and F1-Score allows the obtention of a multifaceted analysis, resulting in a higher performance GoogleNet architecture. Even though the global performance of each model is acceptable, data augmentation can modify the performance of all architectures.

One of the main limitations in the implementation of CNN models is the lack of data for certain classes, for which the main contribution of this study is to be able to distinguish the performance obtained through a reduced data set, where the application of data augmentation techniques other than reducing the overfitting in training helps improve the capacity of generalization in the network in comparison to other studies where augmentation techniques were not applied in the same way as the performance results of the models in Table 9.

On the other hand, through a methodological analysis, the performance was compared and evaluated by applying five metrics to four CNN models. The GoogleNet architecture obtained the best performance, showing the best results in most metrics, obtaining 96.71% accuracy, 96.81% precision, 95.77% sensitivity, 98.73% specificity, and 96.25% F1-Score.

6. Future Work

This study will open other alternatives that could be applied using the same transfer-by-learning approach for the controlled prevention of plagues and diseases in bean crops through timely intervention and automated computerized image classification.

Author Contributions: Conceptualization, T.I.-P. and U.A.H.-G.; methodology, R.J.-M., M.d.R.M.-B. and J.I.C.-F.; software, U.A.H.-G. and H.A.G.-O.; validation, R.R.-M. and T.I.-P.; formal analysis, H.C.C.-A., U.A.H.-G. and R.R.-M.; investigation, C.N., H.C.C.-A. and R.J.-M.; resources, U.A.H.-G. and H.A.G.-O.; data curation, C.N., M.d.R.M.-B. and J.I.C.-F.; writing—original draft preparation, T.I.-P. and J.I.C.-F.; writing—review and editing, T.I.-P. and R.J.-M.; visualization, R.R.-M. and H.A.G.-O.; supervision, C.N. and M.d.R.M.-B.; project administration, T.I.-P. and F.D.M.-D.; funding acquisition, H.C.C.-A., F.D.M.-D. and T.I.-P. All authors have read and agreed to the published version of the manuscript.

Funding: This research was supported by the Instituto Politécnico Nacional (IPN) under grant number SIP/20230388.

Data Availability Statement: The data presented in this study are available upon request from the corresponding author.

Acknowledgments: We want to deeply thank Instituto Nacional de Investigaciones Forestales, Agrícolas y Pecuarias Campo Experimental Zacatecas (INIFAP) for providing us with the experimental field for this research and the Consejo Zacatecano de Ciencia, Tecnología e Innovación (COZCyT). We sincerely thank the people who provided support and advice for this paper.

Conflicts of Interest: The authors declare no conflicts of interest.

References

1. FAO. Food and Agriculture Organization of the United Nations International Year of Plant Health. Available online: https://www.fao.org/plant-health-2020/about/en/ (accessed on 11 December 2023).
2. Velia, A.; Garay, A.; Alberto, J.; Gallegos, A.; Muro, L.R. *El Cultivo Del Frijol Presente y Futuro Para México*; INIFAP: Celaya, Gto., México, 2021; Volume 1, ISBN 978-607-37-1318-4.
3. Gregory, P.J.; Ingram, J.S.I.; Brklacich, M. Climate Change and Food Security. *Philos. Trans. R. Soc. B Biol. Sci.* **2005**, *360*, 2139–2148. [CrossRef]
4. Chakraborty, S.; Newton, A.C. Climate Change, Plant Diseases and Food Security: An Overview. *Plant Pathol.* **2011**, *60*, 2–14. [CrossRef]
5. Mutengwa, C.S.; Mnkeni, P.; Kondwakwenda, A. Climate-Smart Agriculture and Food Security in Southern Africa: A Review of the Vulnerability of Smallholder Agriculture and Food Security to Climate Change. *Sustainability* **2023**, *15*, 2882. [CrossRef]
6. Bailes, E.J.; Ollerton, J.; Pattrick, J.G.; Glover, B.J. How Can an Understanding of Plant-Pollinator Interactions Contribute to Global Food Security? *Curr. Opin. Plant Biol.* **2015**, *26*, 72–79. [CrossRef]
7. Saha, H.; Chatterjee, S.; Paul, A. Role of Pollinators in Plant Reproduction and Food Security: A Concise Review. *Res. J. Agric. Sci.* **2023**, *14*, 72–79.
8. Trebicki, P.; Finlay, K. *Pests and Diseases under Climate Change; Its Threat to Food Security*; John Wiley & Sons Ltd.: Chichester, UK, 2019; Volume 1, ISBN 9781119180654.
9. Alam, F.B.; Tushar, S.R.; Zaman, S.M.; Gonzalez, E.D.S.; Bari, A.M.; Karmaker, C.L. Analysis of the Drivers of Agriculture 4.0 Implementation in the Emerging Economies: Implications towards Sustainability and Food Security. *Green Technol. Sustain.* **2023**, *1*, 100021. [CrossRef]
10. Mcbeath, J.H.; Mcbeath, J. *Environmental Change and Food Security in China*; Beniston, M., Ed.; Springer: Fairbanks, AK, USA, 2010; Volume 35, ISBN 978-1-4020-9179-7.
11. Calicioglu, O.; Flammini, A.; Bracco, S.; Bellù, L.; Sims, R. The Future Challenges of Food and Agriculture: An Integrated Analysis of Trends and Solutions. *Sustainability* **2019**, *11*, 222. [CrossRef]
12. He, J.; Chen, K.; Pan, X.; Zhai, J.; Lin, X. Advanced Biosensing Technologies for Monitoring of Agriculture Pests and Diseases: A Review. *J. Semicond.* **2023**, *44*, 23104. [CrossRef]
13. Mallick, M.D.T.; Biswas, S.; Das, A.K.; Saha, H.N.; Chakrabarti, A.; Deb, N. Deep Learning Based Automated Disease Detection and Pest Classification in Indian Mung Bean. *Multimed. Tools Appl.* **2023**, *82*, 12017–12041. [CrossRef]
14. Hadipour-Rokni, R.; Asli-Ardeh, E.A.; Jahanbakhshi, A.; Paeen-Afrakoti, I.E.; Sabzi, S. Intelligent Detection of Citrus Fruit Pests Using Machine Vision System and Convolutional Neural Network through Transfer Learning Technique. *Comput. Biol. Med.* **2023**, *155*, 106611. [CrossRef]
15. Datt, R.M.; Kukreja, V. Phenological Stage Recognition Model for Apple Crops Using Transfer Learning. In Proceedings of the 2022 2nd International Conference on Advance Computing and Innovative Technologies in Engineering (ICACITE), Greater Noida, India, 28–29 April 2022; pp. 1537–1542.
16. Yalcin, H. Phenology Recognition Using Deep Learning. In Proceedings of the 2018 Electric Electronics, Computer Science, Biomedical Engineerings' Meeting (EBBT), Istanbul, Turkey, 18–19 April 2018; pp. 1–5.
17. Yang, Q.; Shi, L.; Han, J.; Yu, J.; Huang, K. A near Real-Time Deep Learning Approach for Detecting Rice Phenology Based on UAV Images. *Agric. For. Meteorol.* **2020**, *287*, 107938. [CrossRef]

18. Ge, S.; Zhang, J.; Pan, Y.; Yang, Z.; Zhu, S. Transferable Deep Learning Model Based on the Phenological Matching Principle for Mapping Crop Extent. *Int. J. Appl. Earth Obs. Geoinf.* **2021**, *102*, 102451. [CrossRef]
19. Wang, A.X.; Tran, C.; Desai, N.; Lobell, D.; Ermon, S. Deep Transfer Learning for Crop Yield Prediction with Remote Sensing Data. In Proceedings of the 1st ACM SIGCAS Conference on Computing and Sustainable Societies, COMPASS, Menlo Park and San Jose, CA, USA, 20–22 June 2018.
20. Reeb, R.A.; Aziz, N.; Lapp, S.M.; Kitzes, J.; Heberling, J.M.; Kuebbing, S.E. Using Convolutional Neural Networks to Efficiently Extract Immense Phenological Data From Community Science Images. *Front. Plant Sci.* **2022**, *12*, 787407. [CrossRef]
21. Zhao, Y.; Han, S.; Meng, Y.; Feng, H.; Li, Z.; Chen, J.; Song, X.; Zhu, Y.; Yang, G. Transfer-Learning-Based Approach for Yield Prediction of Winter Wheat from Planet Data and SAFY Model. *Remote Sens.* **2022**, *14*, 5474. [CrossRef]
22. Taylor, S.D.; Browning, D.M. Classification of Daily Crop Phenology in PhenoCams Using Deep Learning and Hidden Markov Models. *Remote Sens.* **2022**, *14*, 286. [CrossRef]
23. Bailer, C.; Habtegebrial, T.; Varanasi, K.; Stricker, D. Fast Feature Extraction with CNNs with Pooling Layers. *arXiv* **2018**, arXiv:1805.03096. [CrossRef]
24. Paymode, A.S.; Malode, V.B. Transfer Learning for Multi-Crop Leaf Disease Image Classification Using Convolutional Neural Network VGG. *Artif. Intell. Agric.* **2022**, *6*, 23–33. [CrossRef]
25. Bosilj, P.; Aptoula, E.; Duckett, T.; Cielniak, G. Transfer Learning between Crop Types for Semantic Segmentation of Crops versus Weeds in Precision Agriculture. *J. Field Robot.* **2020**, *37*, 7–19. [CrossRef]
26. Deng, J.; Dong, W.; Socher, R.; Li, L.-J.; Li, K.; Fei-Fei, L. ImageNet: A Large-Scale Hierarchical Image Database. In Proceedings of the 2009 IEEE Conference on Computer Vision and Pattern Recognition, Miami, FL, USA, 20–25 June 2009; pp. 248–255.
27. Rahman, T.; Chowdhury, M.E.H.; Khandakar, A.; Islam, K.R.; Islam, K.F.; Mahbub, Z.B.; Kadir, M.A.; Kashem, S. Transfer Learning with Deep Convolutional Neural Network (CNN) for Pneumonia Detection Using Chest X-ray. *Appl. Sci.* **2020**, *10*, 3233. [CrossRef]
28. Barbhuiya, A.A.; Karsh, R.K.; Jain, R. CNN Based Feature Extraction and Classification for Sign Language. *Multimed Tools Appl.* **2021**, *80*, 3051–3069. [CrossRef]
29. Salehi, A.W.; Khan, S.; Gupta, G.; Alabduallah, B.I.; Almjally, A.; Alsolai, H.; Siddiqui, T.; Mellit, A. A Study of CNN and Transfer Learning in Medical Imaging: Advantages, Challenges, Future Scope. *Sustainability* **2023**, *15*, 5930. [CrossRef]
30. Kim, H.E.; Cosa-Linan, A.; Santhanam, N.; Jannesari, M.; Maros, M.E.; Ganslandt, T. Comparison of Three Dimensional Reconstruction and Conventional Computer Tomography Angiography in Patients Undergoing Zero-Ischemia Laparoscopic Partial Nephrectomy. *BMC Med. Imaging* **2022**, *22*, 47. [CrossRef]
31. Narvekar, C.; Rao, M. Flower Classification Using CNN and Transfer Learning in CNN-Agriculture Perspective. In Proceedings of the 2020 3rd International Conference on Intelligent Sustainable Systems (ICISS), Thoothukudi, India, 3–5 December 2020; pp. 660–664.
32. Szegedy, C.; Liu, W.; Jia, Y.; Sermanet, P.; Reed, S.; Anguelov, D.; Erhan, D.; Vanhoucke, V.; Rabinovich, A. Going Deeper with Convolutions. In Proceedings of the 2015 IEEE Conference on Computer Vision and Pattern Recognition (CVPR), Boston, MA, USA, 7–12 June 2015; pp. 1–9.
33. Jogin, M.; Mohana, M.; Madhulika, M.; Divya, G.; Meghana, R.; Apoorva, S. Feature Extraction Using Convolution Neural Networks (CNN) and Deep Learning. In Proceedings of the 2018 3rd IEEE International Conference on Recent Trends in Electronics, Information & Communication Technology (RTEICT), Bangalore, India, 18–19 May 2018; pp. 2319–2323.
34. Cuevas-Rodriguez, E.O.; Galvan-Tejada, C.E.; Maeda-Gutiérrez, V.; Moreno-Chávez, G.; Galván-Tejada, J.I.; Gamboa-Rosales, H.; Luna-García, H.; Moreno-Baez, A.; Celaya-Padilla, J.M. Comparative Study of Convolutional Neural Network Architectures for Gastrointestinal Lesions Classification. *PeerJ* **2023**, *11*, e14806. [CrossRef]
35. Krizhevsky, A.; Sutskever, I.; Hinton, G.E. ImageNet Classification with Deep Convolutional Neural Networks. *Commun. ACM* **2017**, *60*, 84–90. [CrossRef]
36. Russakovsky, O.; Deng, J.; Su, H.; Krause, J.; Satheesh, S.; Ma, S.; Huang, Z.; Karpathy, A.; Khosla, A.; Bernstein, M.; et al. ImageNet Large Scale Visual Recognition Challenge. *Int. J. Comput. Vis.* **2015**, *115*, 211–252. [CrossRef]
37. Simonyan, K.; Zisserman, A. Very Deep Convolutional Networks for Large-Scale Image Recognition. *arXiv* **2014**, arXiv:1409.1556.
38. Iandola, F.N.; Han, S.; Moskewicz, M.W.; Ashraf, K.; Dally, W.J.; Keutzer, K. SqueezeNet: AlexNet-Level Accuracy with 50× Fewer Parameters and <0.5 MB Model Size. *arXiv* **2016**, arXiv:1602.07360.
39. Etemadi, F.; Hashemi, M.; Zandvakili, O.R.; Mangan, F.X. Phenology, Yield and Growth Pattern of Faba Bean Varieties. *Int. J. Plant Prod.* **2018**, *12*, 243–250. [CrossRef]
40. Kolar, Z.; Chen, H.; Luo, X. Transfer Learning and Deep Convolutional Neural Networks for Safety Guardrail Detection in 2D Images. *Autom. Constr.* **2018**, *89*, 58–70. [CrossRef]
41. Lopez, A.; Giro-I-Nieto, X.; Burdick, J.; Marques, O. Skin Lesion Classification from Dermoscopic Images Using Deep Learning Techniques. In Proceedings of the 2017 13th IASTED International Conference on Biomedical Engineering (BioMed), Innsbruck, Austria, 20–21 February 2017; pp. 49–54.
42. Kurek, J.; Antoniuk, I.; Górski, J.; Jegorowa, A.; Świderski, B.; Kruk, M.; Wieczorek, G.; Pach, J.; Orłowski, A.; Aleksiejuk-Gawron, J. Data Augmentation Techniques for Transfer Learning Improvement in Drill Wear Classification Using Convolutional Neural Network. *Mach. Graph. Vis.* **2019**, *28*, 3–12. [CrossRef]

43. Hassan, S.M.; Maji, A.K.; Jasiński, M.; Leonowicz, Z.; Jasińska, E. Identification of Plant-Leaf Diseases Using Cnn and Transfer-Learning Approach. *Electronics* **2021**, *10*, 1388. [CrossRef]
44. Thenmozhi, K.; Reddy, U.S. Crop Pest Classification Based on Deep Convolutional Neural Network and Transfer Learning. *Comput. Electron. Agric.* **2019**, *164*, 104906. [CrossRef]
45. Alzubaidi, L.; Zhang, J.; Humaidi, A.J.; Al-Dujaili, A.; Duan, Y.; Al-Shamma, O.; Santamaría, J.; Fadhel, M.A.; Al-Amidie, M.; Farhan, L. Review of Deep Learning: Concepts, CNN Architectures, Challenges, Applications, Future Directions. *J. Big Data* **2021**, *8*, 53. [CrossRef] [PubMed]
46. Qin, J.; Hu, T.; Yuan, J.; Liu, Q.; Wang, W.; Liu, J.; Guo, L.; Song, G. Deep-Learning-Based Rice Phenological Stage Recognition. *Remote Sens.* **2023**, *15*, 2891. [CrossRef]
47. Han, J.; Shi, L.; Yang, Q.; Huang, K.; Zha, Y.; Yu, J. Real-Time Detection of Rice Phenology through Convolutional Neural Network Using Handheld Camera Images. *Precis. Agric.* **2021**, *22*, 154–178. [CrossRef]
48. Johnson, J.M.; Khoshgoftaar, T.M. Survey on Deep Learning with Class Imbalance. *J. Big Data* **2019**, *6*, 27. [CrossRef]

AgriEngineering

Article

Predicting the Power Requirement of Agricultural Machinery Using ANN and Regression Models and the Optimization of Parameters Using an ANN–PSO Technique

Ganesh Upadhyay [1,*], Neeraj Kumar [2], Hifjur Raheman [3] and Rashmi Dubey [4]

1 Department of Farm Machinery and Power Engineering, College of Agricultural Engineering and Technology, CCS Haryana Agricultural University, Hisar 125004, Haryana, India
2 ICAR—Indian Institute of Wheat and Barley Research, Karnal 132001, Haryana, India; neeraj.kumar2@icar.gov.in
3 Department of Agricultural and Food Engineering, Indian Institute of Technology Kharagpur, Kharagpur 721302, West Bengal, India; hifjur@agfe.iitkgp.ac.in
4 School of Engineering, University of Petroleum and Energy Studies, Dehradun 248007, Uttarakhand, India; rashmi.dubey@ddn.upes.ac.in
* Correspondence: ganesh.upadhyay0@hau.ac.in

Abstract: Optimizing the design and operational parameters for tillage tools is crucial for improved performance. Recently, artificial intelligence approaches, like ANN with learning capabilities, have gained attention for cost-effective and timely problem solving. Soil-bin experiments were conducted and data were used to develop ANN and regression models using gang angle, velocity ratio, soil CI, and depth as input parameters, while tractor equivalent PTO (PTO_{eq}) power was used as an output. Both models were trained with a randomly selected 90% of the data, reserving 10% for testing purposes. In regression, models were iteratively fitted using nonlinear least-squares optimization. The ANN model utilized a multilayer feed-forward network with a backpropagation algorithm. The comparative performance of both models was evaluated in terms of R^2 and mean square error (MSE). The ANN model outperformed the regression model in the training, testing, and validation phases. A well-trained ANN model was integrated with the particle-swarm optimization (PSO) technique for optimization of the operational parameters. The optimized configuration featured a 36.6° gang angle, 0.50 MPa CI, 100 mm depth, and 3.90 velocity ratio for a predicted tractor PTO_{eq} power of 3.36 kW against an actual value of 3.45 kW. ANN–PSO predicted the optimal parameters with a variation between the predicted and the actual tractor PTO_{eq} power within ±6.85%.

Keywords: artificial neural network; particle-swarm optimization; specific draft; specific torque; equivalent PTO power

Citation: Upadhyay, G.; Kumar, N.; Raheman, H.; Dubey, R. Predicting the Power Requirement of Agricultural Machinery Using ANN and Regression Models and the Optimization of Parameters Using an ANN–PSO Technique. *AgriEngineering* **2024**, *6*, 185–204. https://doi.org/10.3390/agriengineering6010012

Academic Editors: Ray E. Sheriff and Chiew Foong Kwong

Received: 6 December 2023
Revised: 15 January 2024
Accepted: 15 January 2024
Published: 18 January 2024

1. Introduction

The careful selection of energy-efficient agricultural machinery plays a pivotal role in reducing labor-intensive tasks, increasing cropping frequency, and minimizing field-preparation time. Traditional tillage implements often necessitate multiple passes to achieve the desired seedbed, leading to soil compaction that results from repetitive tractor passes. Combining active and passive tools strategically allows the forward thrust produced by the active tool to contribute to the power requirements of the passive tool. This reduces draft and specific energy requirements for tillage tasks, increases field productivity, and minimizes slip due to fewer passes [1–6]. Within an active–passive tillage implement, the performance and power requirement are affected by operational parameters [7,8]. The optimization of the design and operational parameters of agricultural machinery is a significant approach for improving tool performance, efficiency in production, and quality of cultivation while cutting down the energy expenditure and its harmful impact on the environment. The field operation of tillage implements under optimal parameters can fulfill the task of improving tilth quality with a minimum expenditure of energy [9,10]. Numerous research studies have demonstrated that the use of tillage equipment in field operations often demands a substantial amount of energy. Effectively managing this energy can lead to reduced fuel consumption and cost savings.

Predictive models help researchers predict the draft and power demand of tillage tools with the use of limited data, thereby eliminating the need to perform both costly and time-intensive field trials every time. It also helps research workers and manufacturers improve the design of tools through the comparison and analysis of multiple factors that affect the draft requirement of tools. Soft computing techniques have gained significant popularity over the past twenty years due to their better accuracy and fast process. In recent times, artificial intelligence-based techniques like artificial neural networks (ANN) have been used effectively for predictive modeling in various fields. One advantageous aspect of ANN is its capability to establish correlations within extensive and intricate datasets, devoid of any prior understanding of the interconnections. ANN models have seen a growing application in the realm of agricultural engineering due to their aptitude for addressing a wide range of challenging issues. This includes tasks such as pattern classification, prediction, and their ability to maintain performance even when dealing with noisy, incomplete, or inconsistent data [11,12]. ANNs comprise interconnected nonlinear processing elements, each featuring numerous inputs and outputs. These networks operate through three key phases: training, validation and testing, and application. Given the constraints of linear regression techniques for approximating functions, artificial neural networks (ANNs) can serve as a valuable method for predicting the necessary energy for tillage by considering various soil parameters and speed data. The nonlinear and stochastic characteristics of interactions between soil and tools, coupled with the constraints of linear regression techniques for function approximation attract the application of ANNs for the predictive modeling of soil-working machines due to their learning nature and ability to solve complex problems by transforming them into a differentiable function [13,14].

Researchers used various tools, viz. regression model, response surface methodology, ANN, genetic algorithm (GA), and genetic particle swarm optimization (GAPSO), for modeling and the optimization of operational parameters of tillage and seeding machinery [15–17]. The integrated approach of modeling tools and optimization techniques has been successfully applied by researchers in various fields [18,19]. The integration of ANN with global optimization techniques like GA and particle-swarm optimization (PSO) might be helpful for modeling and the optimization of parameters of a tillage implement. However, the ability of the PSO technique to work in discrete as well as analog systems, along with better computational efficiency, makes it advantageous over other techniques [20,21].

Hence, in light of these considerations, this study was undertaken to develop ANN and regression models for predicting the power requirement of active–passive tillage machinery. To achieve this, soil-bin experiments were conducted at various gang angles, u/v ratio, soil cone index, and operative depth, and the equivalent PTO (PTO_{eq}) power was recorded. The data were used to develop ANN and regression models using gang angle, velocity ratio, soil CI, and operating depth as input parameters, while PTO_{eq} power was used as an output parameter. The well-trained ANN model was also integrated with PSO, and different combinations of optimal parameters were predicted by ANN–PSO.

2. Materials and Methods

2.1. Prototype Active–Passive Disc Harrow and Its Associated Test Rig for Soil-Bin Investigations

Experiments were conducted using a test rig of a combined active–passive disc harrow (APDH). To aid in its development, SolidWorks 2013 software (version SP1.0) was utilized to construct a 3D model of the APDH test rig. The isometric view of the developed laboratory prototype APDH and its associated test rig, along with the nomenclature of different parts, are shown in Figure 1. The developed APDH comprised 03 notched discs (13) in the front active gang and 03 plain discs (16) in the rear passive gang. Each of these discs featured an edge diameter of 510 mm, with a spacing of 225 mm between adjacent discs. These discs were mounted on two separate 25 × 25 mm square gang axles (15), and these gangs were affixed to the primary frame constructed from MS angle iron (2) using suitable bearings (17). In order to supply power to the front gang axle during operation, a 7.5 kW three-phase electric motor running at 1425 rpm (5) was securely positioned on top of the primary frame, employing a chain and sprocket drive mechanism. Additionally, for adjusting and controlling its operating depth, a double-acting hydraulic cylinder (19) was positioned (vertically) at the central point of the primary frame.

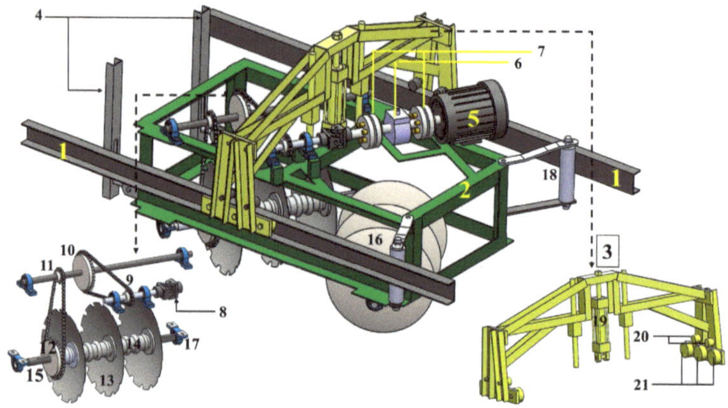

Figure 1. Test rig of the laboratory prototype APDH. (**1**) Side rail; (**2**) primary frame; (**3**) guide frame; (**4**) C-bar linkage having vertical slot; (**5**) motor; (**6**) torque transducer; (**7**) bellow coupling; (**8**) universal joint; (**9**) sprocket 1; (**10**) sprocket 2; (**11**) sprocket 3; (**12**) sprocket 4; (**13**) notched disc; (**14**) spool bearing; (**15**) gang axle; (**16**) plain disc; (**17**) pillow block bearing; (**18**) side roller; (**19**) double-acting hydraulic cylinder; (**20**) roller on top of the side rail; (**21**) roller on C-section of the side rail.

A guide frame (3) was integrated to ease the mobility of the test rig inside the soil bin. It was accomplished by a firm and secure linkage between the primary frame of the test rig and the bin's side rails (1) by employing appropriate rollers. Four rollers (20) were affixed to the guide frame and positioned atop the side rails to guide and maintain the intended path of the test rig inside the soil bin. Additionally, six rollers (21) were mounted on the guide frame within the C-section of the side rails (1) to counterbalance the upward force and uphold the desired operating depth of the test rig while it moved on the carriage. These rollers also served to prevent the guide frame from moving upward during the hydraulic depth adjustment of the test rig at the beginning of each experiment, as they were secured to the side rails. The primary frame was additionally outfitted with two cylindrical side rollers (18) that were pressed against the rails. These rollers played a crucial role in mitigating the lateral force generated while the disc was in operation. This ensured the primary frame's stability and prevented it from slipping out of the rail. The developed harrow had an operating width of 630 mm, and its operating depth could be adjusted up to 200 mm. Furthermore, an arrangement existed for altering the gang angle of the front active set (α) by manipulating the angles of the shafts responsible for transmitting motion via the universal joints (8) and by using a set of holes in the primary frame, along with adjusting the positions of the pillow block bearings (17). The primary frame was equipped with two C-channels (4), which served as a means to attach the primary frame to the intermediary carriage and allow for a free vertical movement of the test rig during the adjustment of its operating depth. Table 1 outlines the test rig's specifications.

Table 1. APDH test rig's details.

Particulars	Value
Type of disc harrow	Offset
Type of discs	Spherical with notched concave surfaces at the front and smooth discs at the rear
Diameter of discs, mm	510
Concavity of discs, mm	60
Spacing between discs, mm	225
Gang axle, mm × mm	25 × 25
Working width, mm	630
Gang angle adjustment, degrees	25 to 40
Overall dimensions (l × w × h), mm × mm × mm	1965 × 1620 × 1237

During the soil-bin study, the forward speed was set at a constant 3.2 km h^{-1}, which was the maximum allowable speed within the soil bin. The circumferential speed of the discs was manipulated by varying the disc's rpm within a range of 80 to 150. This manipulation resulted in achieving different u/v ratios, specifically 2.40, 3.00, 3.60, and 4.60, for the APDH.

2.2. Experimental Plan for Tests in the Soil Bin

The design of active tillage machinery is primarily influenced by critical factors such as the gang angle, rotational direction of the tillage tool, operating depth, and the speed ratio (which is the ratio of the circumferential velocity of the tool (u) to the forward velocity of the implement (v)). These parameters collectively impact the draft and torque requirements of powered discs [22–25]. The rotation of the discs in the same direction as the travel is referred to as concurrent rotation while rotating in the opposite direction to travel is termed as non-concurrent or reverse rotation. To conduct experiments in the non-concurrent mode for the front-powered discs, the rotational direction of the discs was altered using a motor controller switch. However, after an initial assessment of the front active set in the non-concurrent mode, it became evident that soil accumulation in front of the discs led to a significant and undesirable increase in the draft and torque requirements. Consequently, experiments were not pursued in the non-concurrent mode for the front set.

The speed ratio (u/v) plays a pivotal role in determining soil tilth/pulverization at the expense of power consumption. An increased u/v ratio leads to an unwarranted rise in power consumption, while a lower u/v ratio leads to inadequate soil pulverization and increased fluctuations in cutting resistance [2,24]. The performance of APDH was evaluated by operating the discs in the concurrent mode of the front gang axle at various u/v ratios, α, operating depth, and soil cone indices, as mentioned in Table 2. Each experiment was conducted three times.

Given the numerous possible combinations of α for both the front and rear gangs, it is impractical to test every conceivable combination. Consequently, a strategic approach was adopted, focusing on the key parameters that exert a significant influence on the outcomes to streamline the process. Specifically, the rear gang angle (β) was kept constant (30 degrees) throughout each test. In contrast, α was systematically adjusted, ranging from 25 to 40 degrees. This variation in α was implemented due to the front gang's operation in virgin soil, where it played a predominant role in contributing to the overall draft.

Table 2. Research plan for the soil-bin study of APDH.

Variables	Levels	Values
General factors		
Soil texture	1	Sandy clay loam
Soil moisture content (MC), % (db)	1	10 ± 1
Working width, mm	1	630
Forward speed, km h^{-1}	1	3.2
Rear gang angle (β), degrees	1	30
Independent factors		
Front gang angle (α), degrees	4	25, 30, 35, and 40
Operating depth, mm	3	100, 120, and 140
u/v ratio	4	2.40, 3.00, 3.60, and 4.60
Soil CI, MPa (bulk density, g cm^{-3})	3	0.50 ± 0.03 (1.43), 0.08 ± 0.03 (1.58), and 1.10 ± 0.03 (1.73)
Dependent factors		
Draft, kN		
Torque, kN-m		

2.3. Instrumentation and Measurements for Tests in the Soil Bin

The effects of α, forward speed, operating depth, and u/v ratio on the performance of APDH were studied under uniform soil conditions. The experiments were conducted in the soil bin of the AgFE Department, IIT Kharagpur, India, during the months of June to October. In order to assess the consistency of the test bed, measurements of soil CI, bulk density, and moisture content (m.c.) were taken before commencing each experiment. The MC content of the soil was determined using a rapid infrared moisture meter, and this measurement process was repeated three times for each set of experiments to ensure accuracy and reliability. For the assessment of soil CI, a hydraulically operated soil cone penetrometer was employed. The procedures followed for these measurements adhered to the guidelines specified in the ASABE Standards S313.3 [26], ensuring consistency and standardization in the data-collection process. Prior to commencing each experiment, CI readings were systematically recorded at six specific locations spaced approximately 1 m apart along the prepared test bed. This vigilant monitoring process allowed us to assess the uniformity of the soil conditions. In instances where CI readings deviated substantially from the intended values, indicating non-uniformity, corrective action was taken. The test bed was intentionally disturbed and reprepared to ensure that the soil conditions met the required standards before proceeding with the experiments. A core sampler (150 mm length × 50 mm dia.) was inserted vertically into the soil to extract soil samples for the purpose of determining the soil's bulk density. This sampling process was repeated three times for each set of experiments, ensuring the collection of representative soil samples [27].

The test rig's vertical positioning was facilitated by a hydraulic system, and the operating depth was precisely measured using a combination of a potentiometer (rotary type) and a rack-and-pinion mechanism. To determine the forward speed accurately, a magnetic proximity sensor was affixed to the soil-processing carriage, allowing for precise measurement of the carriage's movement speed. For assessing the draft requirement, a calibrated S-type load cell with a 2000 kg capacity was horizontally mounted between the soil-processing carriage and the intermediary carriage, as depicted in Figure 2. This load cell provided accurate measurements of the force or draft exerted during the experiments. The load cell was equipped with a nominal sensitivity of 2.0 millivolts per volt, exhibiting a composition error within the range of ±0.03%. It functions within an excitation voltage range of 9 to 12 volts (DC). For measuring the torque requirement of the front gang axle of the APDH, a high-precision HBM torque transducer with a capacity of 1000 N-m was employed. This torque transducer was coupled to the output shaft of a 7.5 kW three-phase induction motor, which was connected to the test rig using a bellow coupling. The outputs from all the sensors utilized in the experiment were directed to the Spider-8 data-acquisition system (HBM), which was configured to sample data at a frequency of 50 Hz. Furthermore, these data were saved to a PC for subsequent analysis and processing.

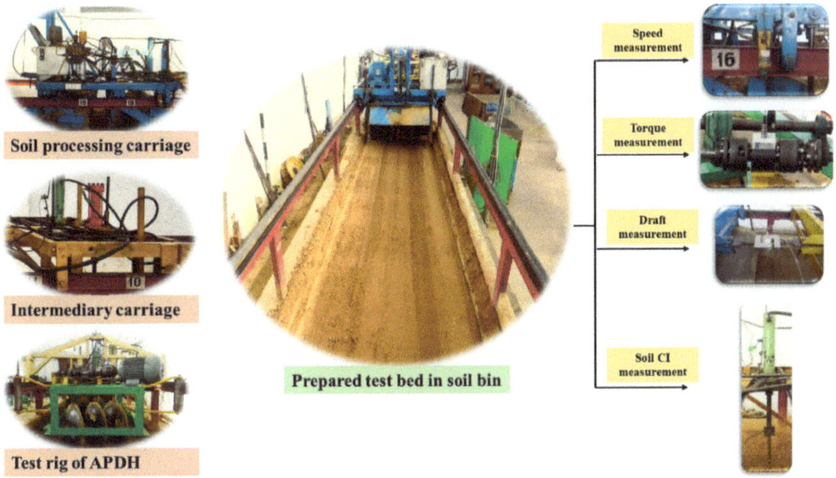

Figure 2. Experimental setup of the soil bin.

2.4. Experimental Procedure for Tests in the Soil Bin

The scope of this research was confined to a single soil texture, specifically sandy clay loam, maintaining an MC within a range of 10 ± 1% (dry basis). This deliberate limitation was imposed to ensure that the experiments remained manageable and focused. Prior to each test, soil preparation was carried out using the soil-processing carriage. This process involved the utilization of a rotary tiller to alleviate stresses within the soil. Subsequently, water was applied to bring the MC within the desired range. The soil within the soil bin was leveled using a leveling blade and compacted through a hydraulic roller operation to achieve the targeted CI. Experiments were conducted only after confirming the uniformity of the test bed, which was assessed by measuring the MC, CI, and bulk density of the soil. If a substantial difference ($p \leq 0.01$) was detected in the recorded CI and bulk density compared to the targeted values, the test bed was intentionally disturbed and reprepared. This rigorous procedure was consistently followed to attain the desired soil conditions before each experiment.

Before conducting each test, the data on soil condition were collected using a core sampler and the hydraulically operated cone penetrometer. Experiments were conducted over a test span of seven meters within the central section of the established test bed with each test replicated thrice. The test apparatus was moved through the soil after configuring the α to the desired settings, choosing an appropriate gear to set the forward speed at different levels, utilizing a suitable chain and sprocket drive system to attain the desired front gang axle rotation speed, and adjusting the operating depth as needed. A calibrated S-type load cell, positioned between the soil-processing carriage and the intermediary carriage, along with a torque transducer and proximity switch, continuously collected data on draft, torque, and operating speed. These data were consistently gathered and monitored by the data-acquisition (DAQ) system. Upon completing each test, the soil bed was intentionally disturbed and subsequently prepared again, ensuring consistency and uniformity in the soil conditions for conducting subsequent tests.

2.5. Estimation of Equivalent PTO (PTO_{eq}) Power of the Tractor

The tractor's necessary equivalent PTO (PTO_{eq}) power for performing harrowing in real field conditions was calculated using the collected draft and torque data obtained from the soil-bin testing following Equation (1), as given below:

$$PTO_{eq}\ power = \frac{D_f \times v_a}{3.6 \times \eta_{PTO\ to\ DB}} + \frac{2 \times \pi \times N_{FGA} \times T_{FGA}}{60 \times \eta_{PTO\ to\ FGA}} \quad (1)$$

Equation (1) for PTO_{eq} power could be rewritten in terms of specific values as follows (Equation (2)):

$$PTO_{eq}\ power = \left(\frac{SD \times v_a}{3.6 \times \eta_{PTO\ to\ DB}} + \frac{2 \times \pi \times N_{FGA} \times ST}{60 \times \eta_{PTO\ to\ FGA}} \right) \times A_{sd} \quad (2)$$

where PTO_{eq} power is the equivalent PTO power of the tractor in kW; D_f is the draft in kN; SD is the specific draft in kN m^{-2}; v_a is the forward speed in km h^{-1}; T_{FGA} is the torque required at front gang axle in kN-m; ST is the specific torque requirement in kN-m m^{-2}; A_{sd} is the area of soil disturbance in m^2; N_{FGA} is the rpm of front gang axle; $\eta_{PTO\ to\ DB}$ is the transmission efficiency from PTO to drawbar, and $\eta_{PTO\ to\ FGA}$ is the transmission efficiency from PTO to the front gang axle = 0.98 × 0.98, taking into account dual speed reduction.

In accordance with ASABE standards D497.5 [28], the transmission efficiency from PTO to drawbar ($\eta_{PTO\ to\ DB}$) was taken into account for agricultural 2WD tractors. Specifically, it was set at 0.55 for soft soil conditions (0.50 ± 0.03 kPa), 0.67 for tilled soil conditions (0.80 ± 0.03 kPa), and 0.72 for firm soil conditions (1.10 ± 0.03 kPa) for calculating the PTO_{eq} power of the tractor.

2.6. Development of ANN and Regression Models

2.6.1. ANN Model

In the present study, a multilayer feedforward network utilizing the backpropagation algorithm was considered due to its learning capability for any complex problem [18]. The independent parameters, viz. α, cone index (MPa), depth (mm), and u/v ratio, were used as the input while the dependent parameter, namely PTO_{eq} power, was used as the output in the ANN model, as presented in Figure 3. The model underwent training using a randomly chosen 90% of the complete dataset obtained from the soil-bin experiments. The trained model was then tested with the remaining 10% of the data. The model was trained using two hidden layers with the 'tansig' transfer activation function and the 'purelin' transfer function in the output layer. The ANN coding was run multiple times using the trial-and-error method in MATLAB to reach the convergence criteria of the minimum mean square error (MSE). A good-performing ANN model with low MSE and high R^2 was then combined with particle-swarm optimization (PSO) to optimize different parameters.

2.6.2. Regression Model

In this study, multiple regression analyses were conducted, employing curve-fitting techniques, to investigate the relationships between individual independent factors (namely, CI, u/v ratio, depth, and α) and the dependent factor (PTO_{eq} power). The analyses were carried out using SPSS Statistics 22.0 from IBM Corporation in New York, NY, USA. These individual relationships were subsequently integrated into the final regression model for estimating the PTO_{eq} power requirement of APDH. The regression model was fitted iteratively to the test data utilizing nonlinear least-square optimization employing the Levenberg–Marquardt algorithm proposed by [29]. Interactions between independent parameters were omitted during model development to prevent unnecessary complexity, as they did not significantly enhance model accuracy.

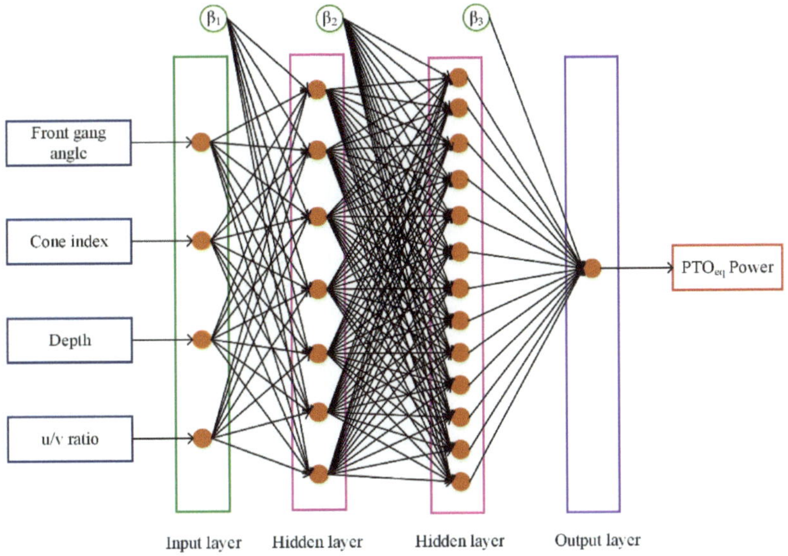

Figure 3. Pattern of ANN architecture used in the present study.

The effectiveness of both the ANN and regression models was evaluated using a range of statistical criteria, including performance indices such as RMSE, R^2, and MAPE, as detailed below:

$$\text{RMSE} = \sqrt[2]{\left\{\frac{\sum_{i=1}^{n}(E_i - M_i)^2}{n}\right\}} \tag{3}$$

$$R^2 = 1 - \frac{\sum_{i=1}^{n}(M_i - E_i)^2}{\sum_{i=1}^{n}(M_i - A_i)^2} \tag{4}$$

$$\text{MAPE} = \frac{\sum_{i=1}^{n}|(M_i - E_i)\ /\ M_i|}{n} \times 100 \tag{5}$$

where, E_i, M_i, and A are the values of predicted, actual, and average PTO_{eq} power, and n is the sample size. RMSE represents the standard deviation of the residuals and serves as an indicator of the predictive capability of the model. An RMSE of 0 would indicate an outstanding model. R^2 is another statistical criterion commonly utilized in soft computing models. Mean absolute percentage error (MAPE), which is typically expressed as a percentage, was also employed in the evaluation as a performance measure of the model.

2.7. Optimization Using Particle Swarm Optimization (PSO)

The particle-swarm optimization (PSO) technique is inspired by the social behavior of fish and birds. It is a powerful meta-heuristic optimization algorithm with faster convergence and the ability to search the wide space for the global minimization or maximization of parameters of both a discrete and a continuous nature, which makes it advantageous over other techniques like genetic algorithm [20,21,30]. The working strategy of the PSO algorithm is given in Figure 4. The algorithm starts with a random solution, i.e., the unplanned allocation of position and velocity to each particle. The particles are speedily moved towards their personal best locations. After attaining the personal best locations, all particles are converged to a global best location. During the entire procedure, the algorithm keeps track of the positions and velocities of all particles. The information of existing position, existing velocity, distance between existing position and pbest, and distance between existing position and gbest is used for changing the position and velocity of particles, as given in Equations (6) and (7).

$$v_m^{n+1} = wv_m^n + c_1 r_1 (pb_m - x_m^n) + c_2 r_2 (gb_m - x_m^n) \tag{6}$$

$$x_m^{n+1} = x_m^n + v_m^{n+1} \tag{7}$$

where v_m^n denotes the velocity of the mth particle at iteration n, w denotes the inertia weight factor, c_1 and c_2 are individual and social learning parameters, respectively, pb_m represents the position of the mth particle at its personal best value, gb_m denotes the position of the mth group at its global best value, x_m^n indicates the position of the mth particle at iteration n, and r_1 and r_2 are uniformly distributed random numbers.

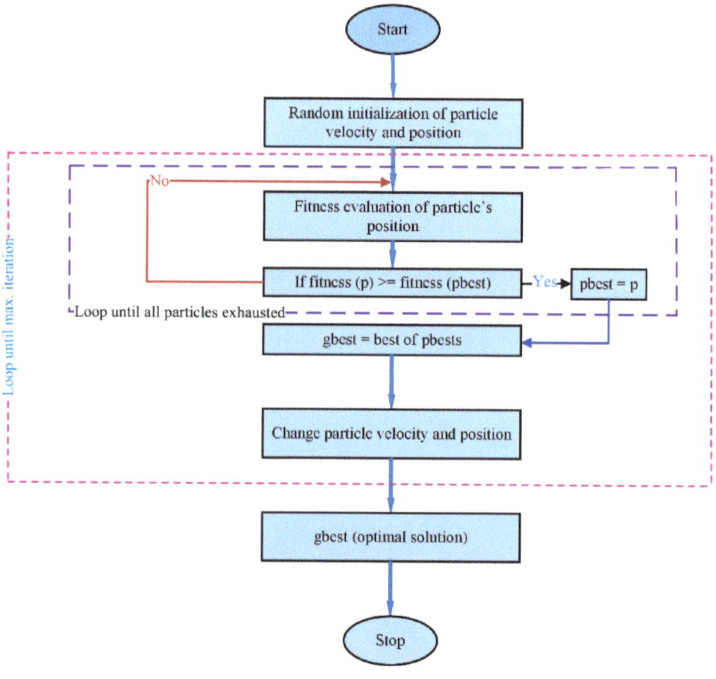

Figure 4. Working strategy of PSO algorithm.

In the present study, the design problem was of a minimization nature, involving α, cone index, depth, and u/v ratio, as represented by X_1, X_2, X_3, and X_4, respectively. Therefore, this problem can be represented mathematically in the following form:

$$\min_{X \in \mathbf{X}^T} f(X)$$

$$25 \leq X_1 \leq 40$$

$$0.5 \leq X_2 \leq 1.1$$

$$100 \leq X_3 \leq 140$$

$$2.40 \leq X_4 \leq 4.60$$

where $X = [X_1, X_2, X_3, X_4]^T$ indicates the decision solution vector.

3. Results and Discussion

As per the experimental plan outlined in Table 2, the test rig for the APDH underwent testing within the soil bin under varying conditions. These conditions included different values for α at 25°, 30°, 35°, and 40°, various u/v ratios at 2.40, 3.00, 3.60, and 4.60, cone index (CI) values within the ranges of 0.50 ± 0.03, 0.80 ± 0.03, and 1.10 ± 0.03 kPa, and different working depths set at 100, 120, and 140 mm. It is noteworthy that the rear gang angle (β) remained fixed at 30° throughout the experiments. The research design adopted a fully randomized approach, and 03 replications were conducted for each soil condition and operational factor. To assess the significance of the various independent factors and their interactions, an ANOVA was performed. Furthermore, to examine the presence of significant differences between treatment means, multiple comparison procedures were carried out using Duncan's multiple range (DMR) tests. The detailed findings of the experiments are presented in Table 3.

Table 3. Performance results of the soil-bin tests.

Variable	Levels	Specific Draft, kN m^{-2}	Specific Torque, kN m m^{-2}	Equivalent PTO Power Pe, kW
Front gang angle (α), degrees	25	13.23 ^{a}	5.15 ^{a}	5.73 ^{a}
	30	14.71 ^{b}	4.50 ^{b}	5.50 ^{b}
	35	17.04 ^{c}	3.69 ^{c}	5.19 ^{c}
	40	23.12 ^{d}	3.34 ^{d}	5.63 ^{a}
u/v ratio	2.40	20.57 ^{a}	5.38 ^{a}	5.67 ^{a}
	3.00	18.32 ^{b}	4.19 ^{b}	5.34 ^{b}
	3.60	14.99 ^{c}	3.57 ^{c}	5.08 ^{c}
	4.60	14.22 ^{d}	3.54 ^{c}	5.95 ^{d}
CI, MPa	0.50	13.47 ^{a}	3.06 ^{a}	4.40 ^{a}
	0.80	17.08 ^{b}	4.29 ^{b}	5.58 ^{b}
	1.10	20.52 ^{c}	5.16 ^{c}	6.55 ^{c}
Operating depth, mm	100	18.00 ^{a}	4.48 ^{a}	4.90 ^{a}
	120	16.85 ^{b}	4.14 ^{b}	5.50 ^{b}
	140	16.23 ^{c}	3.89 ^{c}	6.13 ^{c}

According to DMR tests, average values that are denoted by distinct lowercase letters within the same column for a particular variable are considered to be significantly different at a confidence level of 95%.

3.1. *Effect of Operational Parameters on Specific Draft of APDH*

Table 3 illustrates a noteworthy trend in the specific draft of the APDH. It consistently decreased as the u/v ratio increased across all settings of α and various operating conditions. What is particularly interesting is that this reduction in specific drafts became more noticeable up to a u/v ratio of 3.60. Beyond this point, any further increase in the u/v ratio resulted in a negligible reduction in the specific draft. This observation suggests that there is a point of diminishing returns when it comes to increasing the rpm of the discs. Beyond a certain threshold (a u/v ratio of 3.60 in this instance), the soil's shear strength appears to have reached a minimum, indicating thorough soil pulverization. Consequently, further intensifying the no. of bites per unit time by increasing the disc's rpm seems to have no significant effect on reducing specific draft. This reduction in the draft, leading to a decrease in the specific draft with an increase in the u/v ratio, is consistent with findings from previous experiments conducted with powered discs [22,24,25].

When examining the impact of α on the draft, two critical factors come into play. The first factor is the area of contact between the convex surface of the disc and the furrow wall, which we refer to as the rear-side bearing area. The second factor involves the amount of soil that the discs interact with during their forward movement. When the α increases, it leads to a larger area of contact between the convex surface of the disc and the furrow wall. This increased contact area results in greater scrubbing of the discs, which, in turn, demands higher energy to surmount the friction forces when the discs are in free-rolling mode. However, this extended rear-side bearing area also contributes to generating forward thrust or reducing the draft requirement when the discs are actively rotating. This reduction in draft occurs because the furrow wall provides bearing support to the discs. Therefore, the intrinsic result of the increase in α is to lower the draft when the discs are in free rolling mode, and increase it when they are in powered mode. This change in the draft is primarily due to the alteration in the rear-side bearing area. However, as α increases further, it also results in the discs handling a larger quantity of soil. This increase in the quantity of soil being processed leads to an elevated draft requirement for both the active and free-rolling modes of the discs. Consequently, the overall variation in draft force with changes in α is determined by the cumulative impact of these parameters. Additionally, it is important to note that, as α continues to increase, the rear-side bearing area of the discs eventually reduces to zero at a specific gang angle. This point is known as the critical gang angle. Beyond this critical angle, any further increase in α only serves to increase the amount of soil manipulated by the discs as also documented by [31].

From the data presented in Table 3, it is evident that the specific draft of the APDH exhibited a moderate increase, with an increase in α up to 35°, regardless of the u/v ratio and soil condition. However, beyond this point, there was a sharp and notable increase in the specific draft. This increase in specific draft with a higher α can be attributed to two primary factors. (1) As α increases, there is a reduction in the rear bearing area. This decrease in bearing area results in less support for the discs from the furrow wall, leading to an increase in the specific draft. (2) Additionally, an increase in α corresponds to a larger volume of soil manipulated by the discs. The greater quantity of soil being processed contributes to an elevated draft requirement. At higher α values, especially beyond 35°, the rear bearing area becomes nearly insignificant, and the substantial increase in the amount of soil being manipulated intensifies the specific draft. The rise in draft associated with an augmentation in disc angle plays a substantial role in driving up the specific draft of the APDH. Similar observations regarding draft have been reported in experiments conducted with powered discs in Bangkok clay and sandy loam soils by researchers such as [23–25,32]. These findings collectively support the observed trend of increased specific draft with higher α in the APDH.

The specific draft of APDH was augmented with an increase in soil CI and operating depth at all tested conditions (Table 3) which is a direct consequence of the heightened soil resistance associated with these conditions.

3.2. *Effect of Operational Parameters on Specific Torque of APDH*

The data presented in Table 3 reveals a consistent pattern: The specific torque requirement of the APDH decreased as the u/v ratio (rpm of the front gang axle) increased across all tested conditions. This reduction in specific torque requirement was particularly notable up to a u/v ratio of 3.60, beyond which any further elevation in the u/v ratio resulted in only a negligible reduction. This is because, at higher rotational speeds of the discs (higher u/v ratios), there is a decrease in the force necessary for cutting. Essentially, the resistance to cutting is reduced at these higher speeds. Furthermore, at elevated u/v ratio settings, the powered discs operated in the same area of tilled soil for an extended period of time and repeatedly interacted with the soil that had already been tilled. This repeated interaction contributed to a decrease in the average torque required. The minimal decrease in specific torque beyond the u/v ratio of 3.60 may be due to a balance between factors. While there may be an increased torque requirement to overcome air resistance and the increased soil displacement at a higher rpm of the discs, this increase is counteracted by the reduced torque required for cutting the pulverized soil. This equilibrium results in no significant change in the total torque requirement when the u/v ratio is increased from 3.60 to 4.60.

Similar observations regarding the decrease in torque demand with an increase in the u/v ratio have been reported in tests carried out utilizing powered discs in Bangkok clay and sandy loam soils by researchers such as [23–25]. These findings corroborate the observed trend of reduced torque demand with higher u/v ratios in the APDH.

Furthermore, Table 3 reveals a noticeable trend. The specific torque requirement of the APDH decreased as α increased under all tested conditions. This reduction was particularly pronounced from 25° to 35°. However, when α increased further from 35° to 40°, this reduction became almost negligible. The potential explanation for this decrease in specific torque requirement lies in the lower friction force generated because of the diminished contact area between the rear side of the disc and the furrow wall as α increases. Essentially, the greater α leads to less contact between the discs and the furrow wall, resulting in reduced friction and, consequently, a decreased torque requirement. The substantial decrease in the specific torque requirement observed from 25° to 35° is a direct consequence of the increased volume of soil being handled. However, as α continues to increase from 35° to 40°, the effect of this reduced rear-side bearing area on the torque requirement is partially offset by the substantial increase in the amount of soil being processed. This balancing act leads to a lesser reduction in the torque demand of the APDH within this range of α.

It is worth noting that this phenomenon of reduced torque, responsible for the decline in specific torque requirement with an increase in disc angle, has also been reported in tests conducted with powered discs in Bangkok clay and sandy loam soils by researchers such as [23–25]. These research findings align with the observed trend of the decreased specific torque requirement with an increased α in the APDH.

The specific torque demand of the APDH increased as both the soil CI and operating depth increased across all tested conditions (Table 3). This increase in specific torque is because of the increased soil resistance linked with greater CI values and operating at greater depths.

3.3. *Effect of Operational Parameters on PTO_{eq} Power of APDH*

The results of the ANOVA for the PTO_{eq} power for APDH are given in Table 4. From this table, it is evident that the effects of all considered factors were significant for the PTO_{eq} power for APDH at a 99% confidence interval. No significant effect was found for their interactions except the interactions of depth with α and soil CI. The F values given in Table 4 demonstrate that the soil CI had the utmost influence on the PTO_{eq} power followed by the depth, u/v ratio, and α in that order. The individual effects of these parameters on the PTO_{eq} power for APDH are discussed separately in the following sections.

Table 4. Results of ANOVA for the estimated PTO_{eq} for APDH.

Dependent Variable: PTO_{eq} Power, kW			
Source	**df**	**Mean Square**	**Computed F Value**
Model	144	94.75	751.95 **
α	3	5.99	47.56 **
CI	2	167.44	1328.83 **
depth	2	54.89	435.61 **
u/v ratio	3	15.55	123.40 **
$\alpha \times$ CI	6	0.10	0.82
$\alpha \times$ depth	6	0.29	2.33 **
$\alpha \times$ u/v ratio	9	0.21	1.68
CI $\times$ depth	4	1.44	11.47 **
CI $\times$ u/v ratio	6	0.33	2.65
depth $\times$ u/v ratio	6	0.04	0.35
$\alpha \times$ CI $\times$ depth	12	0.04	0.33
$\alpha \times$ CI $\times$ u/v ratio	18	0.04	0.29
$\alpha \times$ depth $\times$ u/v ratio	18	0.01	0.09
CI $\times$ depth $\times$ u/v ratio	12	0.02	0.13
$\alpha \times$ CI $\times$ depth $\times$ u/v ratio	36	0.01	0.08
Error	288	0.13	

** significant at 1% probability level, $R^2 = 0.99$, adjusted $R^2 = 0.99$, df = degrees of freedom.

Table 3 highlights an interesting trend; PTO_{eq} power for the APDH decreased as the u/v ratio increased from 2.40 to 3.60. However, beyond this point, there was a sharp increase in PTO_{eq} power when the u/v ratio reached 4.6 across all investigated combinations of soil and working parameters. This pattern clearly indicates the presence of an optimal u/v ratio for actively rotating discs in terms of energy consumption. The initial decline in PTO_{eq} power from a u/v ratio of 2.40 to 3.60 can be explained by the significant reduction in the draft and torque requirements of the APDH as the u/v ratio increased. This reduction effectively counteracted the higher power requirement resulting from an increase in the rpm of the front gang. Notably, there was only a marginal reduction in the draft and torque demand of the APDH when the u/v ratio was increased from 3.60 to 4.60. However, as the u/v ratio exceeded 3.60, the disc rpm continued to increase, leading to a sharp rise in PTO_{eq} power (according to Equation (2)). In essence, beyond a u/v ratio of 3.60, the increase in disc rpm surpassed the decrease in the draft and torque requirements of the APDH. Consequently, increasing the u/v ratio beyond 3.60 provided no advantage in terms of draft and torque reduction; instead, it resulted in unnecessary power consumption to overcome air resistance due to higher disc rotation, greater soil-slice displacement, and the additional task of further soil pulverization.

The mean PTO_{eq} powers for APDH for the u/v ratios of 2.40, 3.00, 3.60, and 4.60 was found to be 5.67, 5.34, 5.08, and 5.95 kW, respectively, for the entire test range of α, depth, and soil CI. Significant differences existed between the average values of PTO_{eq} power for APDH recorded at various levels of u/v ratio at a 95% confidence interval.

As shown in Table 3, the PTO_{eq} power for the APDH displayed a particular pattern in response to changes in α (gang angle) across all tested soil and working parameter combinations. Initially, PTO_{eq} power decreased as α increased from 25° to 35°. However, beyond 35°, there was a sharp increase in PTO_{eq} power when α reached 40°, indicating the presence of an optimal gang angle setting for the APDH near 35°. The initial decline in PTO_{eq} power with the increase in α from 25° to 35° was primarily attributed to a significant reduction in torque requirements. However, as α continued to increase beyond 35°, PTO_{eq} power increased due to a sharp rise in the draft, while the torque requirement exhibited almost negligible reduction.

The mean PTO_{eq} powers for α of 25°, 30°, 35°, and 40° were found to be 5.73, 5.50, 5.19, and 5.63 kW, respectively, for the test range of u/v ratio, depth, and soil CI. No substantial difference was found to exist among the mean PTO_{eq} power at α of 25° and 40°, whereas

PTO_{eq} power at α of 30° and 35° was found to be significantly different from each other as well as from PTO_{eq} power at α of 25° and 40°.

In general, the PTO_{eq} power for APDH was found to increase with an increase in operating depth at all tested levels of α, u/v ratio, and soil CI. The obvious explanation for the increase in PTO_{eq} power for APDH with an increase in depth was the higher draft and torque requirement of APDH at higher depths of operation. Furthermore, the PTO_{eq} power for APDH enhanced with an increase in soil CI at all tested conditions because of the higher penetration resistance linked with more CI values.

3.4. Multiple Regression Model

To develop the regression model, the Levenberg–Marquardt estimation technique [29] was followed with the negligence of interactions between parameters. This was because it added complexity to the model without yielding substantial improvement in accuracy. Multicollinearity was checked by taking into account the tolerance and variance inflation factor (IF_{var}) as measures of any possible correlations between the factors. The results indicated no evidence of multicollinearity, as the IF_{var} value was one.

The average draft and torque data obtained from the soil-bin tests were at three levels each of operating depth and soil CI; four levels each of u/v ratio and α were analyzed using SPSS 22.0 software and were used to develop the PTO_{eq} power model for APDH using a multiple regression technique. Soil CI and operating depth were found to have a linear relationship with PTO_{eq} power requirement. However, α and u/v ratio had a quadratic relationship with PTO_{eq} power requirement. The formulated model for estimating the PTO_{eq} power requirement of APDH is as follows (Equation (8)):

$$PTO_{eq}\text{ power} = C_0 + C_1 \times \alpha + C_2 \times \alpha^2 + C_3 \times \left(\frac{u}{v}\right) + C_4 \times \left(\frac{u}{v}\right)^2 + C_5 \times CI + C_6 \times d \quad (8)$$

where, PTO_{eq} power is in kW; α is the gang angle of the front set in degrees; CI is the cone index of the soil before tillage in MPa; d is the operating depth in mm; u/v is the speed ratio; and C_i = regression coefficients, the values of which are provided in Table 5 along with the standard error values, i = 0, 1, 2, 3, 4, 5.

Table 5. Regression coefficients and their respective standard errors for the PTO_{eq} power prediction model.

Regression Coefficients	PTO_{eq} Power Model	
	Estimate	Standard Error
C_0	12.30	0.903
C_1	−0.452	0.049
C_2	0.007	0.001
C_3	−3.76	0.247
C_4	0.55	0.035
C_5	3.59	0.077
C_6	0.03	0.001

The results of regression analysis for the PTO_{eq} power prediction model of APDH are given in Table 6. The high R^2 affirmed the precision of the formulated model and suggested that the variables under consideration could account for a significant portion of the variance observed in the experimental data.

Table 6. Results of regression analysis for the formulated regression model.

Source	PTO_{eq} Power Model (R^2 = 0.95)		
	SS	df	MS
Regression	4541.18	7	648.74
Residual	7.10	137	0.05
UT	4548.28	144	-
CT	174.44	143	-

SS = sum of squares (Type III); df = degrees of freedom; MS = mean square; UT = uncorrected total; CT = corrected total.

3.5. Performance of ANN and Regression Models during the Training and Testing Phases

It is evident from the regression plot presented in Figure 5 that the formulated ANN model predicted PTO_{eq} power reasonably well, with values close to the actual one during the training phase. In the training process of the model, mean square error (MSE) reached a minimum value within 10 epochs for validation, relating it to a faster convergence of the model. The comparative performances of the ANN and regression model during the training and testing phases are exhibited in Figure 6 and Table 7. Both models predicted the PTO_{eq} power data close to the actual one; however, the performance of ANN was superior to the regression model in terms of the R^2 value and MSE during the training as well as the testing phase.

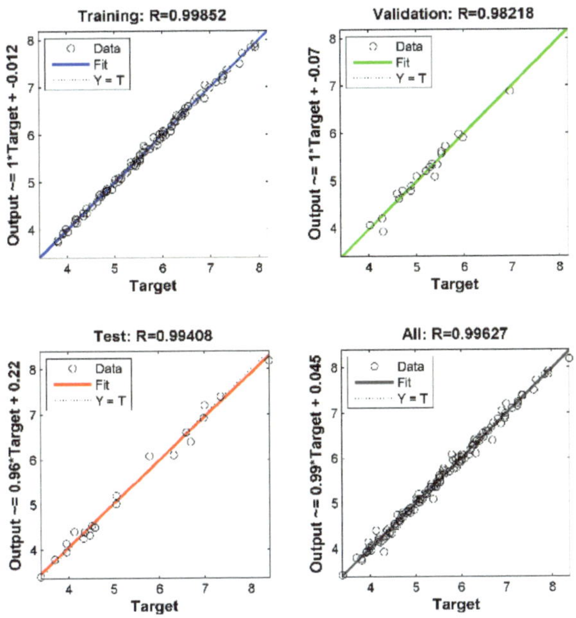

(a)

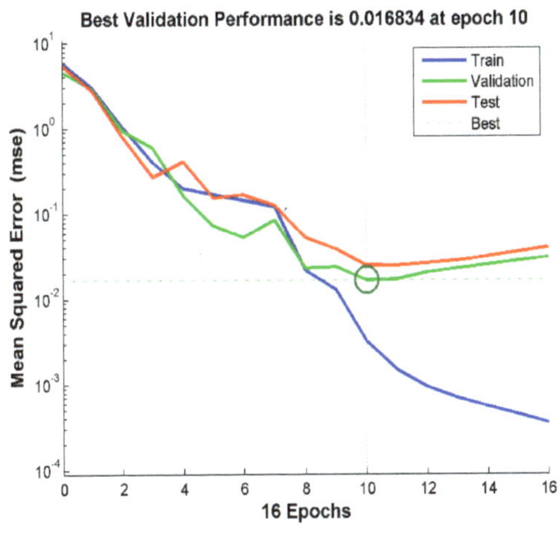

(b)

Figure 5. Regression (**a**) and MSE plots (**b**) during the training phase of the ANN model.

Table 7. Performance indices of ANN and regression models.

Model	RMSE		MAPE		R^2	
	Training	Testing	Training	Testing	Training	Testing
ANN	0.093	0.176	1.214	2.871	0.99	0.98
Regression	0.271	0.229	4.001	3.644	0.95	0.96

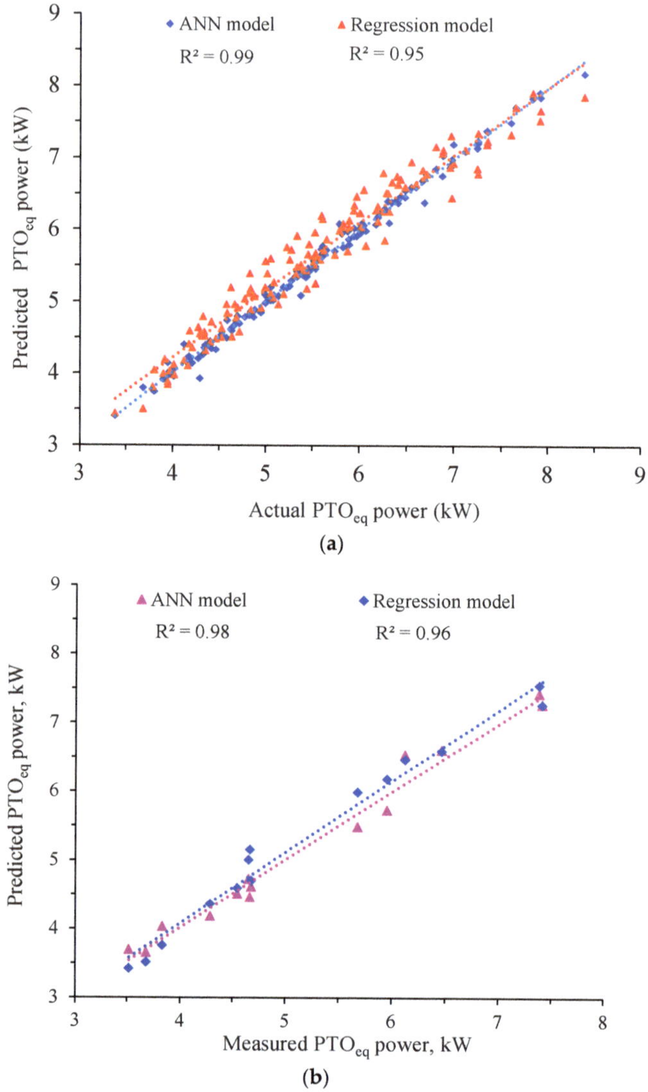

Figure 6. Comparative performance of the ANN and regression models during the (**a**) training and (**b**) testing phases.

3.6. Validation of Formulated Models with Independent Data

The formulated models were also validated by keeping the levels of front gang angles, cone index, and depth as used in the experiment but varying the u/v ratio now by changing the forward speed (2.5–3.2 km h^{-1}) while keeping the disc speed fixed as 100 rpm (Figure 7). During the validation of models with independent data also, ANN outperformed the regression model with a high R^2 value and low MSE. From the presented results, it is evident that, overall, the ANN model showed superior performance and precisely predicted PTO$_{eq}$ power close to the measured values.

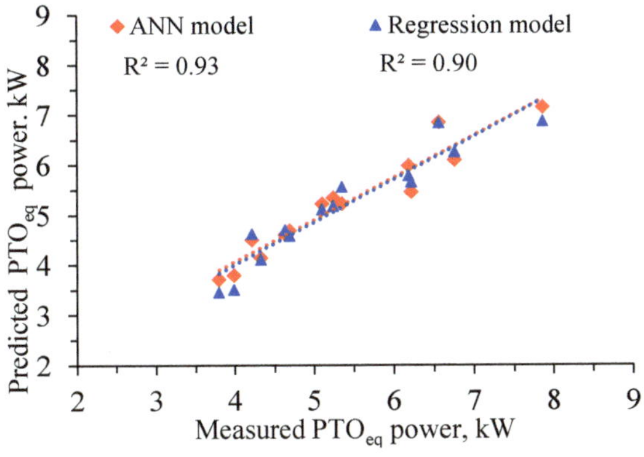

Figure 7. Performance of the ANN and regression models during the validation with independent data.

3.7. Prediction of Optimal Parameters Using ANN–PSO

The well-trained ANN model was coupled with PSO to predict the optimal operational settings. The program was run repeatedly to estimate different combinations of optimal parameters, as given in Table 8. The predicted PTO_{eq} power was compared with the actual PTO_{eq} power available either in the standard run or alternatively, through experimentation with the new sets of conditions. It is interesting to see that the PSO-trained ANN model indicated better performance than the regression model in terms of RMSE, MAPE, and R^2 value during the training phase of the model. However, in some instances, the performance of the regression model was better during the testing phase compared to the ANN–PSO technique (for instance, combinations given at serial numbers 1 and 2).

Table 8. Various sets of optimal parameters forecasted through the use of ANN–PSO.

S. No.	RMSE		MAPE		R^2		α, °	CI, MPa	Depth, mm	u/v Ratio	PTO_{eq} Power		Variation, %
	Training	Testing	Training	Testing	Training	Testing					Predicted	Actual	
1	0.154	0.302	1.509	4.018	0.98	0.95	35.0	0.5	100	3.60	3.36	3.59	−6.85
2	0.165	0.266	2.058	4.903	0.98	0.90	35.5	0.5	120	3.40	4.00	3.89	2.75
3	0.093	0.176	1.214	2.871	0.99	0.98	36.6	0.5	100	3.90	3.36	3.45	−2.67
4	0.125	0.111	1.521	1.773	0.99	0.99	35.7	0.5	140	3.90	4.20	4.02	4.29

One interesting thing to be observed in Table 8 is that all combinations exhibited an optimal front gang angle of 35 ± 2° and a cone index of 0.5 MPa to achieve the minimum PTO_{eq} power. Among various combinations, optimal parameters having a 36.6° front gang angle, 0.5 MPa cone index, 100 mm depth, and 3.90 u/v ratio were considered according to the performance indices of the model, i.e., low RMSE and high R^2 value along with lesser variability in the predicted and actual PTO_{eq} power. The results given in Table 8 indicate that the ANN–PSO model predicted the optimal parameters with reasonably good accuracy and low variation (±6.85%) between the predicted and actual PTO_{eq} power. The learning ability of ANN and the optimization of network parameters by the PSO technique makes ANN–PSO a faster and more accurate technique for handling complex and nonlinear problems. The convergence plot of the fitness function of PSO is presented in Figure 8. It is evident from Figure 8 that all particles converged rapidly towards the global minimum within 20 iterations, which makes it faster and advantageous over other techniques [30].

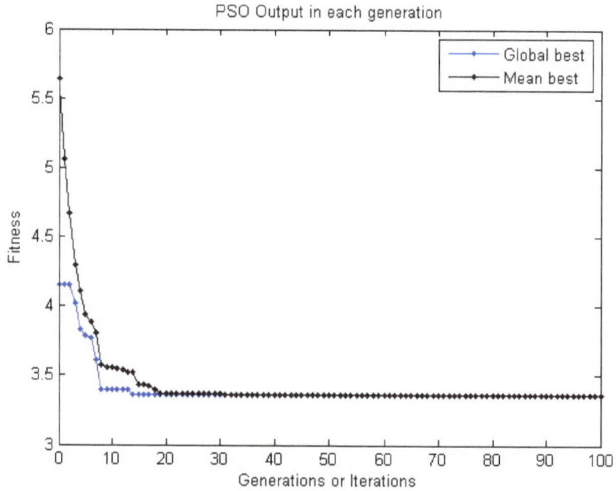

Figure 8. Convergence plot of PSO fitness function.

4. Conclusions

Predicting the power requirements of agricultural machinery is crucial for optimizing performance, reducing energy consumption, and improving overall efficiency. In this study, predictive models for the power requirement of tillage machinery, particularly an active–passive tillage machinery, were formulated by means of artificial neural networks (ANNs) and regression techniques using data from soil-bin experiments conducted at various gang angles, u/v ratios, soil cone indexes, and operative depths. The following conclusions were drawn from the study:

1. Both the ANN and regression models performed well in predicting the power requirement; however, the ANN model had superior performance over the regression model, as evidenced by high R^2 values and low MSE during the training, testing, and validation phases.
2. The well-trained ANN model was then integrated with PSO and different combinations of optimal parameters were predicted by ANN–PSO with good accuracy and lesser variation (±6.85%) between the predicted and actual PTO_{eq} power.
3. The combination of a 36.6° front gang angle, 0.50 MPa cone index, 100 mm depth, and 3.90 u/v ratio was found to be an optimal setting for the predicted PTO_{eq} power of 3.36 kW against 3.45 kW (actual).
4. It was observed that all combinations of soil and working parameters (i.e., CI and depth) exhibited an optimal front gang angle in the range of 35 ± 2° and a u/v ratio of 3.65 ± 0.25 to achieve the minimum PTO_{eq} power.

These findings can have practical implications in agriculture by aiding in the selection of optimal equipment settings to reduce energy consumption, minimize operational costs, and improve fieldwork efficiency. In conclusion, this study underscores the significant role of techniques like ANN and PSO in the precise prediction of performance characteristics and the optimization of operational parameters of tillage tools for improved efficiency and quality of work with minimal expenditure of energy.

Author Contributions: Conceptualization, G.U.; methodology, G.U. and H.R.; software, N.K. and R.D.; validation, G.U. and N.K.; formal analysis, G.U. and R.D.; investigation, G.U. and H.R.; resources, H.R.; data curation, G.U. and R.D.; writing—original draft preparation, G.U.; writing—review and editing, G.U. and N.K.; visualization, H.R.; supervision, H.R.; project administration, G.U. and H.R.; funding acquisition, H.R. All authors have read and agreed to the published version of the manuscript.

Funding: This research received funding from MHRD, Govt. of India.

Data Availability Statement: The data are contained in the article.

Acknowledgments: The funding and cooperation received from MHRD, Govt. of India, to carry out this research work is sincerely acknowledged.

Conflicts of Interest: The authors declare no conflicts of interest.

References

1. Wilkes, R.D.; Addai, S.H. The use of the 'Wye Double Digger' as an alternative to the plough to reduce energy requirement per hectare and soil damage. In Proceedings of the International Conference on Agricultural Engineering, Richmond, Australia, 25–30 September 1988; Paper (No. 88-190).
2. Shinners, K.J.; Wilkes, J.M.; England, T.D. Performance characteristics of a tillage machine with active-passive components. *J. Agric. Eng. Res.* **1993**, *55*, 277–297.
3. Manian, R.; Kathirvel, K. Development and evaluation of an active-passive tillage machine. *Agric. Mech. Asia Afr. Lat. Am.* **2001**, *32*, 9–18.
4. Mehta, C.R.; Singh, K.; Selvan, M.M. A decision support system for selection of tractor–implement system used on Indian farms. *J. Terramech.* **2011**, *48*, 65–73. [CrossRef]
5. Upadhyay, G.; Raheman, H. Effect of velocity ratio on performance characteristics of an active-passive combination tillage implement. *Biosyst. Eng.* **2020**, *191*, 1–12.
6. Sarkar, P.; Upadhyay, G.; Raheman, H. Active-passive and passive-passive configurations of combined tillage implements for improved tillage and tractive performance: A review. *Span. J. Agric. Res.* **2021**, *19*, e02R01. [CrossRef]
7. Ranjbarian, S.; Askari, M.; Jannatkhah, J. Performance of tractor and tillage implements in clay soil. *J. Saudi Soc. Agric. Sci.* **2017**, *16*, 154–162.
8. Nataraj, E.; Sarkar, P.; Raheman, H.; Upadhyay, G. Embedded digital display and warning system of velocity ratio and wheel slip for tractor operated active tillage implements. *J. Terramech.* **2021**, *97*, 35–43.
9. Damanauskas, V.; Velykis, A.; Satkus, A. Efficiency of disc harrow adjustment for stubble tillage quality and fuel consumption. *Soil Tillage Res.* **2019**, *194*, 104311. [CrossRef]
10. Fawzi, H.; Mostafa, S.A.; Ahmed, D.; Alduais, N.; Mohammed, M.A.; Elhoseny, M. TOQO: A new tillage operations quality optimization model based on parallel and dynamic decision support system. *J. Clean. Prod.* **2021**, *316*, 128263.
11. Sablani, S.S.; Ramaswamy, H.S.; Sreekanth, S.; Prasher, S.O. Neural network modeling of heat transfer to liquid particle mixtures in cans subjected to end-over-end processing. *Food Res. Int.* **1997**, *30*, 105–116.
12. Kumar, N.; Upadhyay, G.; Kumar, P. Comparative Performance of Multiple Linear Regression and Artificial Neural Network Based Models in Estimation of Evaporation. *Adv. Res.* **2017**, *11*, 1–11.
13. Kalogirou, S.A. Artificial neural networks in renewable energy systems applications: A review. *Renew. Sustain. Energy Rev.* **2001**, *5*, 373–401.
14. Kumar, M.; Yadav, N. Multilayer perceptrons and radial basis function neural network methods for the solution of differential equations: A survey. *Comput. Math. Appl.* **2011**, *62*, 3796–3811. [CrossRef]
15. Zeng, Z.; Chen, Y.; Zhang, X. Modelling the interaction of a deep tillage tool with heterogeneous soil. *Comput. Electron. Agric.* **2017**, *143*, 130–138. [CrossRef]
16. Askari, M.; Abbaspour-Gilandeh, Y.; Taghinezhad, E.; Hegazy, R.; Okasha, M. Prediction and optimizing the multiple responses of the overall energy efficiency (OEE) of a tractor-implement system using response surface methodology. *J. Terramech.* **2022**, *103*, 11–17. [CrossRef]
17. Wang, S.; Zhao, B.; Yi, S.; Zhou, Z.; Zhao, X. GAPSO-Optimized fuzzy PID controller for electric-driven seeding. *Sensors* **2022**, *22*, 6678. [CrossRef]
18. Alzoubi, I.; Delavar, M.R.; Mirzaei, F.; Nadjar Arrabi, B. Comparing ANFIS and integrating algorithm models (ICA-ANN, PSO-ANN, and GA-ANN) for prediction of energy consumption for irrigation land leveling. *Geosystem Eng.* **2018**, *21*, 81–94. [CrossRef]
19. Rabbani, A.; Samui, P.; Kumari, S. A novel hybrid model of augmented grey wolf optimizer and artificial neural network for predicting shear strength of soil. *Model. Earth Syst. Environ.* **2023**, *9*, 2327–2347. [CrossRef]
20. Hassan, R.; Cohanim, B.; De Weck, O.; Venter, G. A comparison of particle swarm optimization and the genetic algorithm. In Proceedings of the 46th AIAA/ASME/ASCE/AHS/ASC Structures, Structural Dynamics and Materials Conference, Austin, TX, USA, 18–21 April 2005; p. 1897.
21. Kundu, P.; Paul, V.; Kumar, V.; Mishra, I.M. Formulation development, modeling and optimization of emulsification process using evolving RSM coupled hybrid ANN-GA framework. *Chem. Eng. Res. Des.* **2015**, *104*, 773–790. [CrossRef]
22. Hoki, M.; Burkhardt, T.H.; Wilkinson, R.H.; Tanoue, T. Study of PTO driven powered disk tiller. *Trans. ASAE* **1988**, *31*, 1355–1360. [CrossRef]
23. Salokhe, V.M.; Quang, N.B. Dynamics of a powered disk in clay soil. *J. Terramech.* **1995**, *32*, 231–244. [CrossRef]
24. Hann, M.J.; Giessibl, J. Force measurements on driven discs. *J. Agric. Eng. Res.* **1998**, *69*, 149–157. [CrossRef]

25. Nalavade, P.P.; Salokhe, V.M.; Niyamapa, T.; Soni, P. Performance of free rolling and powered tillage discs. *Soil Tillage Res.* **2010**, *109*, 87–93.
26. *ASAE S313.3;* Soil Cone Penetrometer. ASABE Standards: St. Joseph, MI, USA, 2001.
27. *ASA 30-2.2-1965;* Methods of Soil Analysis. Bulk Density. ASA Standards: Madison, WI, USA, 1965.
28. *ASAE D497.5;* Agricultural Machinery Management Data. ASABE Standards: St. Joseph, MI, USA, 2006.
29. Marquardt, D.W. An algorithm for least-squares estimation of nonlinear parameters. *J. Soc. Ind. Appl. Math.* **1963**, *11*, 431–441.
30. Parrott, D.; Li, X. Locating and tracking multiple dynamic optima by a particle swarm model using speciation. *IEEE Trans. Evol. Comput.* **2006**, *10*, 440–458. [CrossRef]
31. O'Dogherty, M.J.; Godwin, R.J.; Hann, M.J.; Al-Ghazal, A.A. A geometrical analysis of inclined and tilted spherical plough discs. *J. Agric. Eng. Res.* **1996**, *63*, 205–217.
32. Islam, M.S.; Salokhe, V.M.; Gupta, C.P.; Hoki, M. Effects of PTO-powered disk tilling on some physical properties of Bangkok clay soil. *Soil Tillage Res.* **1994**, *32*, 93–104. [CrossRef]

AgriEngineering

Article

Dynamic Behavior Forecast of an Experimental Indirect Solar Dryer Using an Artificial Neural Network

Angel Tlatelpa Becerro [1,*], Ramiro Rico Martínez [2,*], Erick César López-Vidaña [3], Esteban Montiel Palacios [4], César Torres Segundo [4] and José Luis Gadea Pacheco [4]

1 Departamento de Ingeniería en Robótica y Manufactura Industrial, Escuela de Estudios Superiores de Yecapixtla, Universidad Autónoma del Estado de Morelos, Yecapixtla 62824, Mexico
2 Tecnológico Nacional de México, I.T. Celaya, Celaya 38000, Mexico
3 Consejo Nacional de Humanidades, Ciencias y Tecnología, Centro de Investigación en Materiales Avanzados S.C., Durango 34147, Mexico; erick.lopez@cimav.edu.mx
4 Escuela de Estudios Superiores de Xalostoc, Universidad Autónoma del Estado de Morelos, Ayala 62725, Mexico; esteban.montiel@uaem.mx (E.M.P.); cesar.torres@uaem.mx (C.T.S.); jose.gadea@uaem.mx (J.L.G.P.)
* Correspondence: angel.tlatelpa@uaem.mx (A.T.B.); ramiro@iqcelaya.itc.mx (R.R.M.)

Abstract: This research presents the prediction of temperatures in the chamber of a solar dryer using artificial neural networks (ANN). The dryer is a forced-flow type and indirect. Climatic conditions, temperatures, airflow, and geometric parameters were considered to build the ANN model. The model was a feed-forward network trained using a backpropagation algorithm and Levenberg–Marquardt optimization. The configuration of the optimal neural network to carry out the verification and validation processes was nine neurons in the input layer, one in the output layer, and two hidden layers of thirteen and twelve neurons each (9-13-12-1). The percentage error of the predictive model was below 1%. The predictive model has been successfully tested, achieving a predictor with good capabilities. This consistency is reflected in the relative error between the predicted and experimental temperatures. The error is below 0.25% for the model's verification and validation. Moreover, this model could be the basis for developing a powerful real-time operation optimization tool and the optimal design for indirect solar dryers to reduce cost and time in food-drying processes.

Keywords: solar dryer; thermal analysis; electronic instrumentation; artificial neural networks; feedforward propagation algorithm

Citation: Tlatelpa Becerro, A.; Rico Martínez, R.; López-Vidaña, E.C.; Montiel Palacios, E.; Torres Segundo, C.; Gadea Pacheco, J.L. Dynamic Behavior Forecast of an Experimental Indirect Solar Dryer Using an Artificial Neural Network. *AgriEngineering* **2023**, *5*, 2423–2438. https://doi.org/10.3390/agriengineering5040149

Academic Editors: Ray E. Sheriff and Chiew Foong Kwong

Received: 26 October 2023
Revised: 1 December 2023
Accepted: 8 December 2023
Published: 14 December 2023

1. Introduction

Developing dehydrated vegetable and meat products has gained economic importance worldwide to preserve food and increase its shelf life. Drying is one of the oldest and most used techniques for food preservation [1]. Its function is to eliminate the water that contains the food, which inhibits the proliferation of microorganisms and hinders putrefaction. However, many processes are based on artisan techniques, which provide poor quality and repeatability in the resulting products. Nevertheless, the globalization trend has aroused interest in increasing its production, controlling the quality of its products, and expanding its offerings by accelerating the drying processes, seeking to add value to "artisanal" products [2].

Furthermore, there is a vision of sustainable development for food preservation, which is sought to be supported by using renewable energies, such as solar energy, both for artisanal and industrial processes [3,4].

Based on this perspective, solar drying through solar dryers is an attractive process for food dehydration applications [5]. Some solar dryers are considered direct because the foods are exposed to radiation [1], and they use a black surface to increase the air temperature to dry the product. Indirect solar dryers, on the other hand, comprise the solar collector and the drying chamber, and the solar radiation does not affect the products' surface; this equipment's performance can be by natural or forced convection. For this configuration, the collector receives solar radiation to increase the airflow temperature supplied through it, and the chamber is where the air circulates to remove moisture from the product. This system is desirable as it does not require high temperatures, exhibits greater drying efficiency, and does not require energy storage because the temperature required for drying fruits and vegetables is between 30 and 70 °C [1] and because it manages to reach these temperatures. However, these devices operate in areas with abundant solar radiation and climatic conditions, such as those in tropical regions, under semi-cloudy and clear-sky conditions.

In this study, an experimental forced-flow indirect solar dryer is used for research purposes. This system is considered complex for its geometry and operating principle since its performance depends on weather conditions. It requires a system to supply airflow that guarantees adequate air circulation.

The design, development, and analysis of the thermal behavior of solar dryers have been widely studied theoretically with numerical analysis and experimental studies. For numerical analysis, computational fluid dynamics (CFD) was used to analyze heat and mass transfer to determine the behavior of temperatures, flows, and moisture loss in the dryers [6–8]. In experimental studies, dryers were instrumented to monitor the behavior of temperatures, airflows, and moisture and the data compared with computational numerical simulations to validate their models [3,9–11]. On the other hand, different methodologies have been applied to obtain mathematical models and numerically simulate dryers to describe their performance [12,13]. For these study cases, it has been found that moisture tends to decrease when the temperature increases inside the drying chamber. This improves the equipment's performance in drying food. However, each of these papers aims to determine a design that allows the ideal temperature inside the drying chamber to be obtained by evaluating thermal performance by applying mathematical models, numerical computational simulations, and experimental studies.

The ideal conditions in a solar dryer allow for calculating the drying kinetics of products such as vegetables, meats, and other materials [4,12–16]. Due to the system's complexity, it is difficult to determine the performance of the solar dryer. Moreover, the techniques used become tedious and computationally expensive. Looking for new practical alternatives to predict the air temperature inside the chamber of a solar dryer to improve the designs and optimize the operation is a challenge. One of the computational models that impact a predictive analysis tool in thermal systems is the artificial neural networks (ANN) technique.

This technique consists of models inspired by the biology of the human brain [17]. It offers an alternative to solving complex problems in real situations. Otherwise, it would be tough to characterize using analytical techniques and numerical simulation since they do not require detailed knowledge of the physical phenomena of thermal systems [18]. Recently, ANNs have been used in different engineering fields in pattern recognition, adaptive control systems, and complex thermal systems [19–21]. They have been applied to modeling and predicting thermal performance in heat pumps for cooling and heating [18,22,23] and performing modeling in reactors for storing sorption energy to predict the dynamics of charge and temperature state [24]. Likewise, neural networks have been applied to simulate solar collectors' thermal performance and calculate solar radiation [25–28].

Similarly, studies have been carried out in heat exchangers for the predictive analysis of heat transfer, pressure drop, and control [20,29–33]. Also, this technique has been used to predict temperatures and mass flow and calculate thermal behavior in heat exchangers [34–36]. It has been applied to describe fluidized bed dryers, predicting the drying characteristics of potatoes, garlic, and melon [37] and microwave-assisted green pea-drying time prediction [38]. Likewise, predictions have been made to calculate the moisture content in wheat [39]. Also, ANNs have worked on modeling the performance of a parabolic greenhouse dryer for drying lychee pulp [40].

This research presents the prediction of temperatures in the chamber of a solar dryer using artificial neural networks (ANN). The dryer under study is indirect and forced flow. It comprises a solar collector and a drying chamber with trays for food. The objective is to build a predictive model of the temperature of the drying chamber. For this purpose, solar radiation and ambient temperature are considered, as well as the temperatures of the pipes and the speed of the airflow inside the drying chamber. A model dryer was instrumented with temperature sensors, an airflow meter, and a meteorological station for climate conditions for experimental purposes. The latter were recorded with an Arduino-type microcontroller and LabVIEW Student Software Suite 2020, respectively. Also, the number and spacing of trays were considered. The proposed artificial neural network model was applied to model the system and validated with experimental data. This network model could be a powerful tool to assist in the design of solar dryers as well as the optimization of their operation.

2. Materials and Methods

The study of solar dryers remains a complex problem. Different innovative proposals in the literature combine theoretical, numerical, and experimental procedures for thermal analysis. However, only some of them are suitable for online applications. In some cases, the analysis is limited to the effect of the internal geometry of the dryer on its dynamic behavior. For this reason, this research paper proposes using artificial neural networks to predict the air temperature inside the drying chamber for different numbers and positions of trays inside the dryer chamber.

This section describes the experimental solar dryer in detail, then the electronic communication between a microcontroller and the LabVIEW Student Software Suite 2020 for applied electronic instrumentation and practical data acquisition. Finally, artificial neural networks are presented to predict the state of the solar dryer.

2.1. Experimental Setup of the Solar Dryer

This experimental work was developed in Yecapixtla, Morelos, Mexico, at 18°51′09.6″ north latitude and 98°52′11.0″ west longitude. The solar dryer is outdoors and faces south for maximum use of solar energy. The collector has a tilt angle of 24° and the following dimensions: length (L_1) 2.145 m, width (W_1) 0.95 m, and height (D_1) 0.13 m, and for the chamber, the following: height (L_2) 1.27 m, width (W_2) 1.0 m, and depth (D_2) 0.455 m (Figure 1a). The solar collector comprises three parts: a 20-gauge stainless steel casing; an arrangement of sixteen copper ducts with dimensions of 1.5 m in length, an external diameter of 0.0266 m, and a 0.001 cell for airflow passage; and lastly, a clear acrylic cover. The main component of the collector is the energy-absorbing pipes. These are painted matte black to absorb and take advantage of the most remarkable solar radiation available during the day.

The underside and the sides are insulated to minimize heat loss. The thermal insulation is a 3 cm thick polyurethane foam between the housing and a reflective aluminum sheet. The clear acrylic cover reduces convection heat loss from the absorber pipes. The air inlet to the collector is in the lower part (Color blue arrows), and the air outlet in the upper (Color yellow arrows) part connects to the drying chamber (Figure 1b). The airflow that enters the collector passes into the ducts, raises the temperature, and is transported to the drying chamber. The drying chamber is constructed of stainless steel and painted matte black to absorb radiant energy and increase indoor air temperature. The airflow that enters the chamber is in the lower part (Color yellow arrows), and the outlet (Color green arrows) is in the upper part (Figure 1c). The drying chamber contains spaces for a maximum of fifteen mobile trays. The trays are used to dehydrate or dry a product. For this study, the chamber was arranged for fifteen, twelve, ten, eight, and five trays equally spaced from each other. The trays were constructed with an anodized aluminum frame and nylon mesh. These are arranged horizontally, with a dimension length (L) of 0.88 m and a width (W) of 0.41 m (Figure 1d).

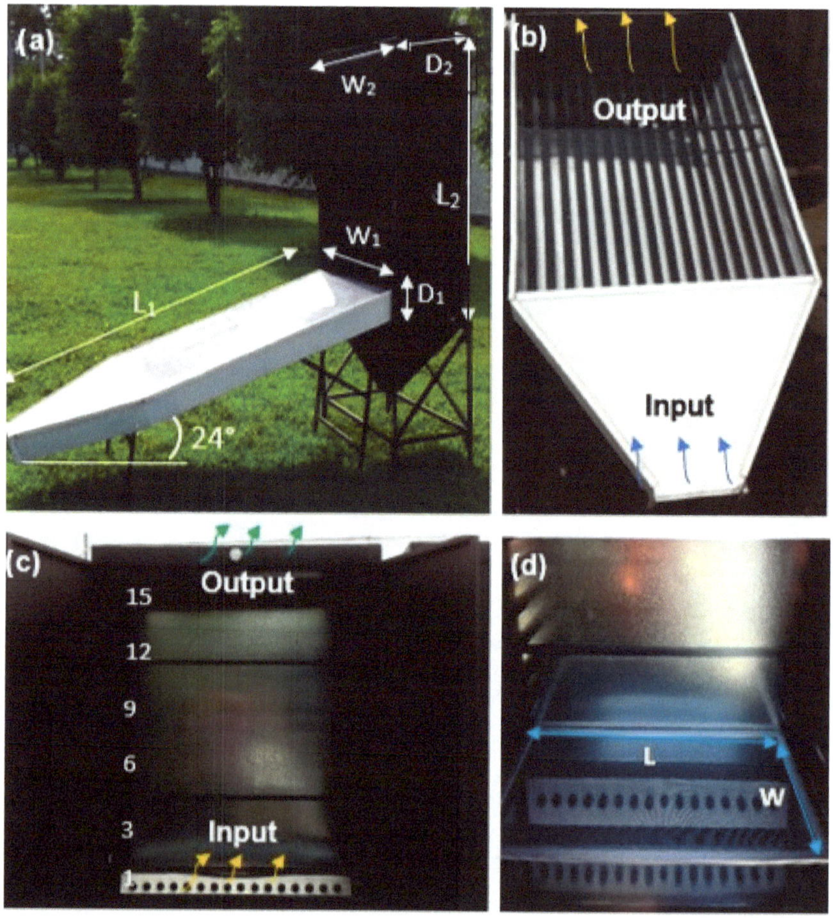

Figure 1. Experimental solar dryer facing south. (**a**) Exterior view. (**b**) Airflow ducts of the solar collector. (**c**) Drying chamber. (**d**) Trays for drying.

2.2. Data Acquisition

The solar dryer has sensors for the measurement of temperature and airflow velocity. Four of the solar collector's sixteen pipes were designed to contain type K thermocouples with their Max 6675 modules. The sensors are located and represented with S_1, S_2, S_3, and S_4 (Figure 2a). It has a measurement range between −100 °C to 1250 °C. The sensitivity of the sensor is approximately between 41 to 42 µV/°C with an accuracy of ±0.7 °C. Likewise, two fans of type NMB with 12 volts and 1.3 ampere operating characteristics were installed at the input of the solar collector (Figure 2a). These were used to supply the airflow to the solar collector and transport the air into the drying chamber. The flow rate was controlled with voltage modules FZ0430 and current modules ACS712. The flow velocity was measured at the input of the drying chamber (see Figure 1c) with a type UT363 anemometer. The measurement range is 0 to 30 $\frac{m}{s}$ with a precision of ±5% rdg + 0.5 (Figure 2a). Also, the drying chamber was fitted with a DHT22 (S_5) temperature sensor with a precision of ±0.1 °C (Figure 2b).

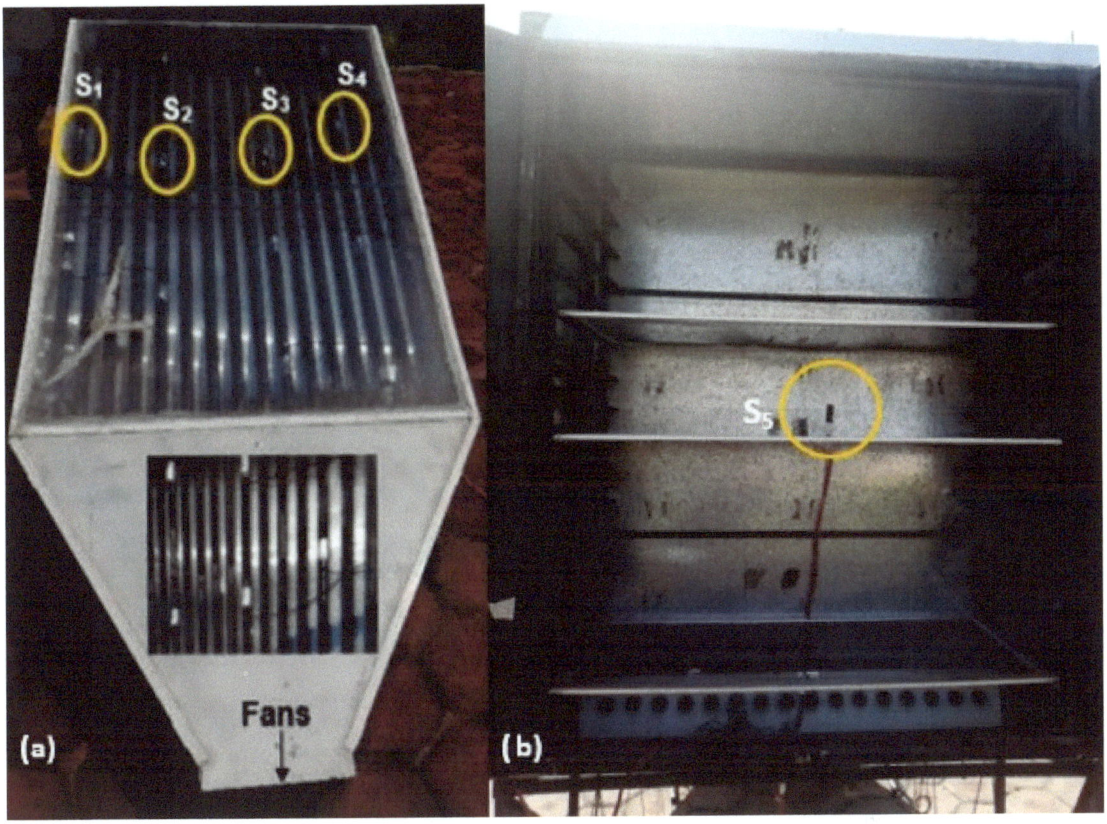

Figure 2. Arrangement of the solar dryer. (**a**) Location of fans, K-type thermocouples, and anemometers in the solar collector. (**b**) The DHT22 temperature sensors are inside the drying chamber.

On the other hand, a microcontroller (Arduino, Monza, Italy) was used to interconnect the temperature and humidity sensors and for data acquisition. The microcontroller is connected to a personal computer through a USB port to acquire data. LabVIEW software allows communication between the Arduino microcontrollers for data recording. LabVIEW was used to manage the intensity of the electrical current and, in turn, control the speed of the airflow supplied by the fans.

2.3. Weather Station

A weather station (Ambient Weather, Chandler, AZ, USA; model WS-2902A) was used to measure weather conditions (Figure 3a). Solar radiation and ambient temperature are taken into consideration for the construction of the predictive model. Temperature sensor specifications correspond to a range between −4.4 and 60 °C and solar radiation between 0 and 120 k Lux ($\frac{W}{mm^2}$, lux, fc). The station was installed at a height of 3.5 m and was located near the solar dryer to provide more accurate data (Figure 3b). The data were acquired and recorded remotely into a personal computer then monitored using a display console.

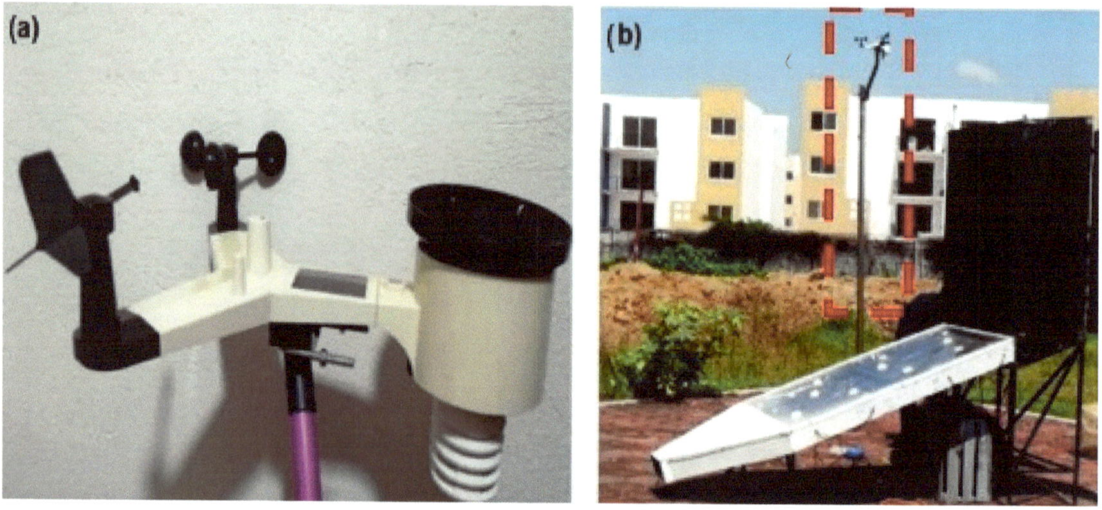

Figure 3. Weather station. (**a**) Meteorological station for measuring solar radiation and ambient temperature. (**b**) A 3.5 m high weather station located next to the solar dryer (Red-lined square).

2.4. *Artificial Neural Network Model*

The method applied is backpropagation, which is used in the training process of forwarding connection networks. Backpropagation is a supervised learning method with a descending gradient: an input pattern propagates through different network layers until it produces an output target. The target is compared to the desired output, and the output neurons calculate the error. The estimated error is transmitted backward from the output layer to the other neurons in the hidden layers. It is proportional to its contribution to the total error of the artificial neural network. With this, the mistakes of the synaptic weights of each neuron used are adjusted.

The input neurons correspond to the first layer, while the intermediate layer is the hidden layer. However, multilayer networks can use the tan-sigmoid (tansig) transfer function. For the last layer, output neurons with linear transfer functions (purelin) are used. Thus, these network outputs can take any value without being limited.

The input vectors normalize between 0.1 and 0.9 due to the tan-sigmoid and purelin transfer function constraints for the hidden and output layers. The transfer function [41] and the normalization equation [42] are given by Equations (1) and (2), respectively.

$$g(x) = \frac{[1 + tanh(x)]}{2} \tag{1}$$

$$X_i = 0.8\left(\frac{X_i - X_{min}}{X_{max} - X_{min}}\right) + 0.1 \tag{2}$$

X_i is the value to be normalized; X_{max} is the maximum value; X_{min} the minimum value of the data set.

In this work, the Levenberg–Marquardt (trainlm) algorithm of multilayer backpropagation was used to evaluate the performance of different configurations of the neural network architecture to determine the ideal. Once the backpropagation training algorithm of the multilayer network is known, the architecture is arranged for four layers. The first layer corresponds to the input neurons and a neuron in the output layer. However, due to the complexity and non-linearity of the system on its own, two hidden layers are considered. As there is no rigorous way to determine an optimal number of neurons in the hidden layer, the selection results from a compromise between the computational effort required for ANN training and an estimate of the minimum number of neurons necessary for capturing dynamic behavior. Therefore, the mean square error (MSE) is applied as an indicator of the performance of the training process, Equation (3) [42]. In this way, it is possible to obtain the ideal architecture for the verification and validation process of the predictive model.

$$SE = \frac{\sum_{i=1}^{i=N} \left(y_{i,pred} - y_{i,exp} \right)^2}{N} \quad (3)$$

where N is the total number of data points; $y_{i,pred}$ is the network prediction; $y_{i,exp}$ is the experimental response; and i is an index of data.

The predictive model is built considering nine input parameters and one output parameter defining the system. The neural network's inputs are the following parameters:

- Solar radiation (R_a)
- Airflow velocity (V_a)
- Ambient temperature (T_a)
- Solar collector pipe temperatures (T_{t1}, T_{t2}, T_{t3}, T_{t4})
- Number of trays (N_{tray})
- Separation between each of the trays (D_t)
- Drying chamber temperature (T_c).

The data set available (188), 75%, is used for the training process. At the same time, 23% is separated to validate the training.

3. Results

3.1. Climatic Data and Parameters of the Solar Dryer

This section presents the temperature variables, air velocity, and climatological data described in Section 2. The weather data recording was completed every 10 min, corresponding to a cloudy day on 7 August and a clear and non-cloudy day on 18 August 2020. The recording time on 7 August corresponds to approximately between the 12:10 and 2:00 pm hours of the day (central Mexico time). Meanwhile, for the 18th, it fits a schedule approximately between 11:30 am and 2:30 pm. These established schedules have been considered to capture solar radiation as much as possible with the solar dryer.

The following figures present the experimental time series. Figure 4 shows the time series of solar radiation (S_{R1} and S_{R2}) and air temperature (T_{a1} and T_{a2}). Meanwhile, Figure 5 shows the airflow velocity (V_{a1} and V_{a2}). However, the pipe and air temperatures inside the drying chamber are recorded each 1 min over the specified time. Figure 6 shows the temperatures of the pipes, T_{t1}, T_{t2}, T_{t1v}, and T_{t2v}, and the drying chamber, T_c and T_{cv}. Also, Figure 7 presents the temperatures of the pipes: T_{t3}, T_{t4}, T_{t3v}, and T_{t4v}. S_{R1}, T_{a1}, V_{a1}, T_{t1}, T_{t2}, T_{t3}, T_{t4}, and T_c correspond to the data recorded on 18 August. S_{R2}, T_{a2}, V_{a2}, T_{t1v}, T_{t2v}, T_{t3v}, T_{t4v}, and T_{cv} correspond to the data of 7 August. Each of the variables shows disturbances throughout the trajectory. These disturbances are associated with the presence of experimental noise. Figures 6 and 7 also offer the smooth course of each time series. Smoothing makes it possible to reveal meaningful patterns in our data while omitting the contribution of noise. Here, a moving average filter has been applied. Figure 6 shows pipe temperatures: T_{t1s}, T_{t2s}, T_{t1vs}, and T_{t2vs}, and the drying chamber: T_{cs} and T_{cvs}. Figure 7 presents pipe temperatures: T_{t3s}, T_{t4s}, T_{t3vs}, and T_{t4vs}.

Table 1 shows the essential parameters of the solar dryer geometry to construct the neural network model to predict the temperature inside the drying chamber. The predictive model can provide the chamber temperature for different numbers of symmetrically separated trays. For this, three arrays of the number of trays (N_{tray}) with their respective separation distances (D_t) are used in training and verifying the model. For the validation, five tray arrangements are considered; see Table 1.

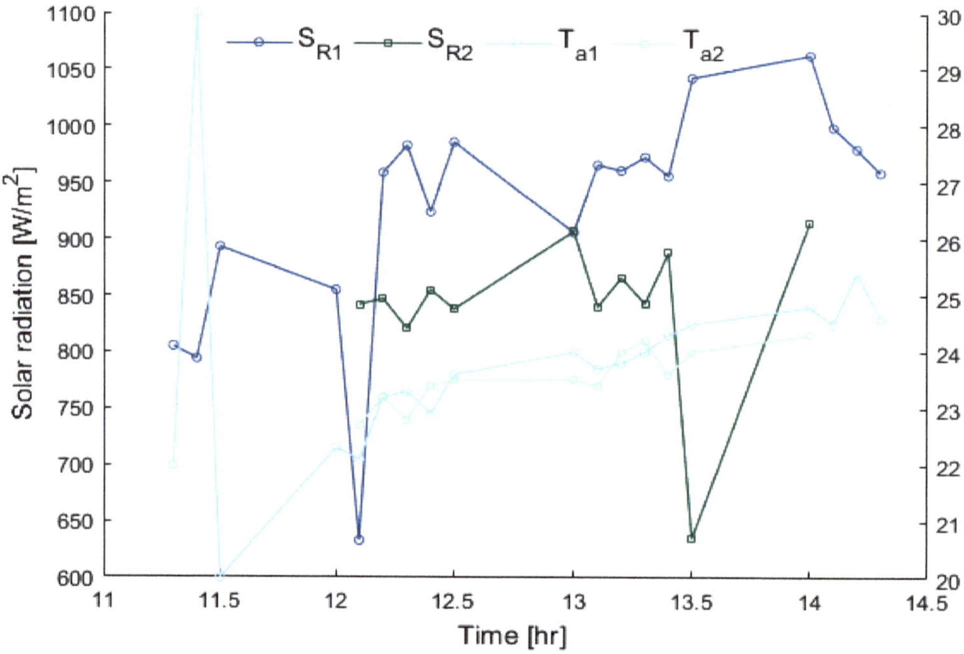

Figure 4. Solar radiation time series and air temperature time series. S_{R1} and T_{a1} are data recorded on 18 August. S_{R2} and T_{a2} are data recorded on 7 August.

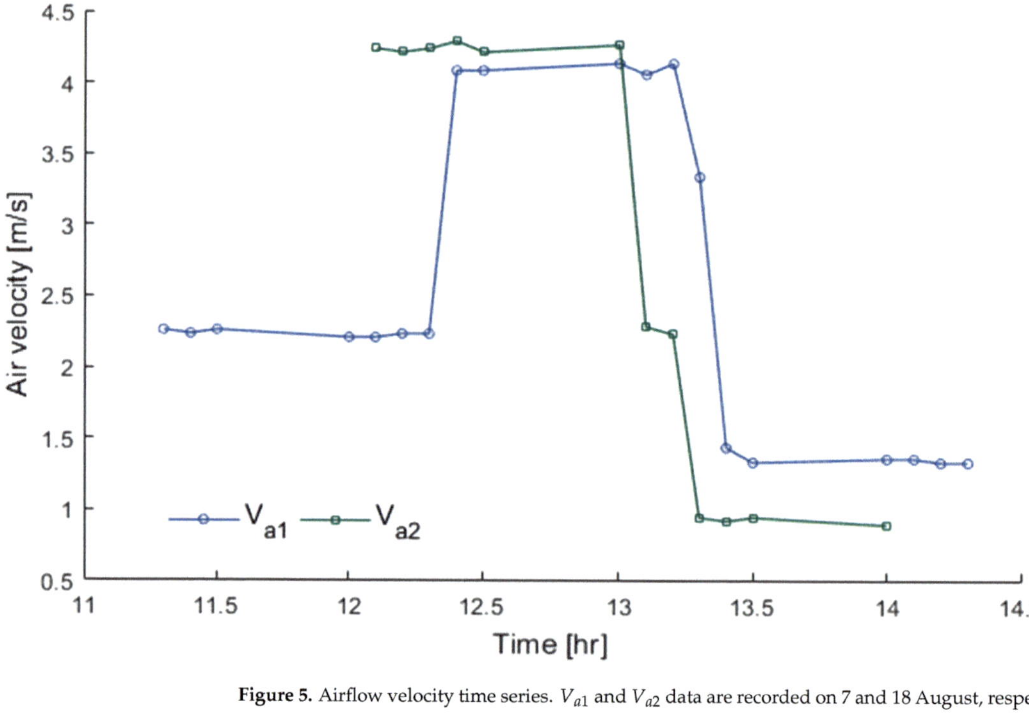

Figure 5. Airflow velocity time series. V_{a1} and V_{a2} data are recorded on 7 and 18 August, respectively.

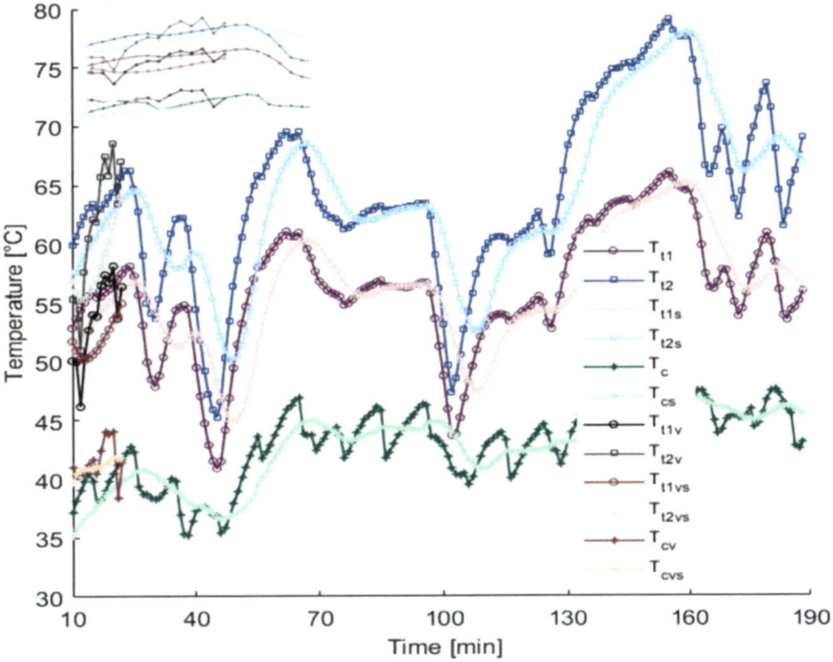

Figure 6. Pipe temperature time series. Experimental: T_{t1}, T_{t2}, T_{t1v}, T_{t2v}, T_c, and T_{cv} versus smoothed: T_{t1s}, T_{t2s}, T_{t1vs}, T_{t2vs}, T_{cs}, and T_{cvs}.

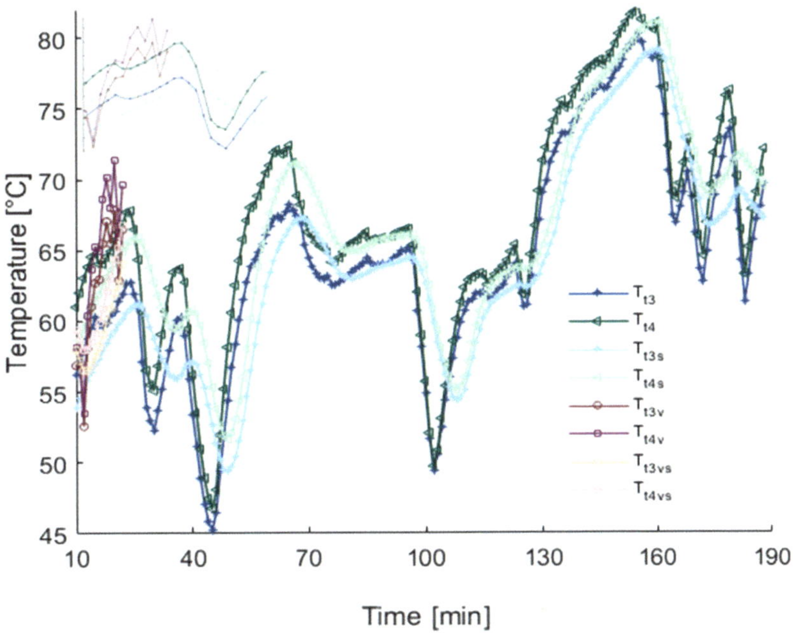

Figure 7. Pipe temperature time series. Experimental ones: T_{t3}, T_{t4}, T_{t3v}, and T_{t4v} vs. smoothed: T_{t3s}, T_{t4s}, T_{t3vs}, and T_{t4vs}.

Table 1. Drying chamber configuration.

Training–Verification		Validation	
(N_{tray})	D_t (m)	(N_{tray})	D_t (m)
15	0.07	15	0.07
10	0.105	12	0.087
5	0.21	10	0.105
		8	0.131
		5	0.21

3.2. Results of the Artificial Neural Network

According to the linear regression statistical modeling technique, the backpropagation algorithms were evaluated using the multiple determination coefficients (R^2) and the mean square error (MSE) criteria. The equation obtained with the linear regression minimized the distance between a fit line for experimental and predicted data points in the training process. This technique describes a continuous response variable as a function of one or more predictor variables [43–45]. The algorithm with the best performance was Levenberg–Marquardt (trainlm), with an R^2 of 0.94162 and an MSE of 0.9725. This last algorithm is used to conduct the training and prediction process and select the number of neurons within each hidden layer.

No technique or methodology exists to determine the optimal number of neurons within the hidden layers, as mentioned in Section 2.4. However, some authors recommend conducting a study for the calculation [46–48]. Therefore, the maximum number of neurons obtained is based on the trade-off between the computational effort required during the training process of the ANN and the estimate of the minimum number of neurons needed. On the other hand, an approximation is made through the geometric pyramid rule to start from an initial number of neurons in the hidden layers until the ideal is obtained. The trial-and-error phase is carried out from this process to find the best performance for the prediction process. In this training process, 75% of the data set (188) was used, while the rest was used for model verification. The criterion for obtaining the number of neurons in the hidden layers was based on the deviations used to obtain the architecture's configuration (MSE). According to performance analysis and the MSE estimation during the training process of the ANN, the best network configuration was 9-13-12-1, with an MSE below 1%. Therefore, the ANN configuration verifies and validates the model (see Figure 8).

3.3. Verification and Validation of Results

The verification and validation of the artificial neural network model for predicting the temperature in the drying chamber are shown in this section. Likewise, the relative error between the experimental and predicted data is also shown.

Forty-seven data points, from one hundred eighty-eight of 18 August 2020, were considered for verifying the predictive model. Figure 9 presents the results of the temperature prediction between the predicted (T_{ANN}) and experimental (T_{exp}) data from the drying chamber. Here, neural network results follow the trajectory throughout the trend, trying to resemble the experimental data. The relative error between the predicted and observed data is below 0.25% (see Figure 10). The error calculated makes the model reliable for making predictions with similar conditions to those proposed. For this, the neural network model is validated with a set of 24 data points.

However, for the validation process of the same model, 24 data points from 7 August 2020, of that same year, are considered. Figure 11 shows the validation results of the ANN model. Again, the neural network predicts data very close to the experimental values. The estimated relative error is below 0.25% (Figure 10).

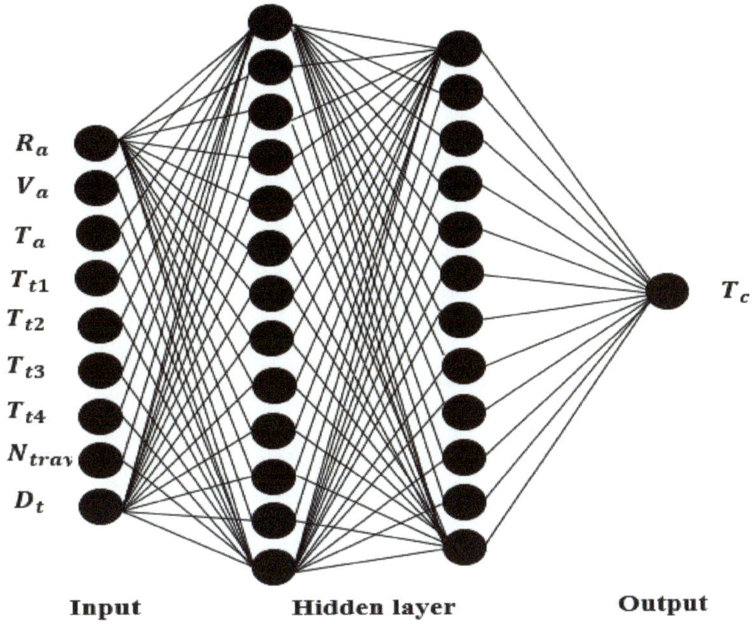

Figure 8. The architecture of the artificial neural network. The configuration is nine neurons in the input layer, two hidden layers with thirteen and twelve neurons, and one in the output layer (9-13-12-1). The parameters are solar radiation (R_a), airflow velocity (V_a), ambient temperature (T_a), temperatures in tube 1 (T_{t1}), temperatures in tube 2 (T_{t2}), temperatures in tube 3 (T_{t3}), temperatures in tube 4 (T_{t4}), number of trays (N_{tray}), the separation between each of the trays (D_t), and the temperature in the drying chamber (T_c).

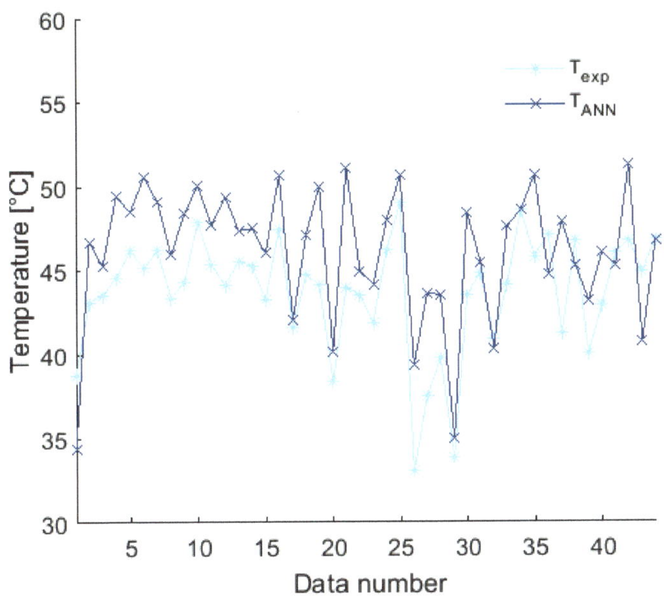

Figure 9. Predicted temperatures (T_{ANN}) vs. experimental temperatures (T_{exp}) of the model verification.

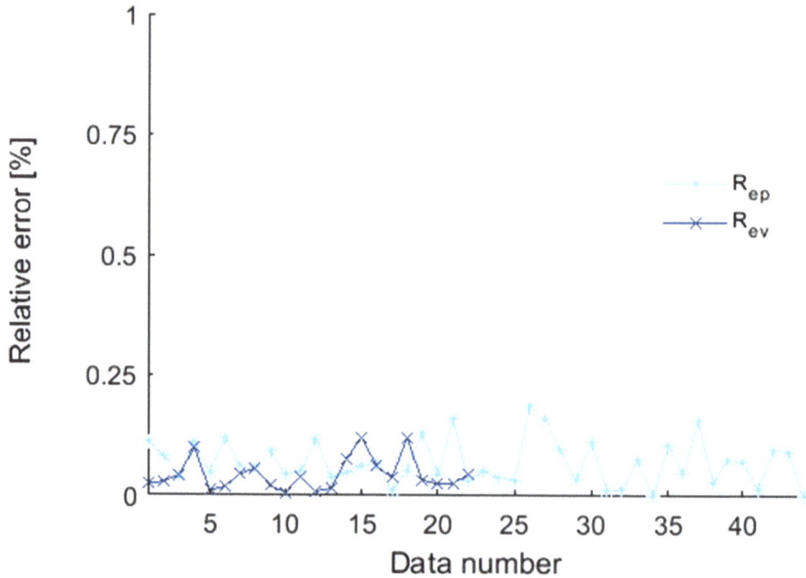

Figure 10. Relative error between predicted vs. experimental data. R_{ep} relative error for verification. R_{ev} relative error for validation.

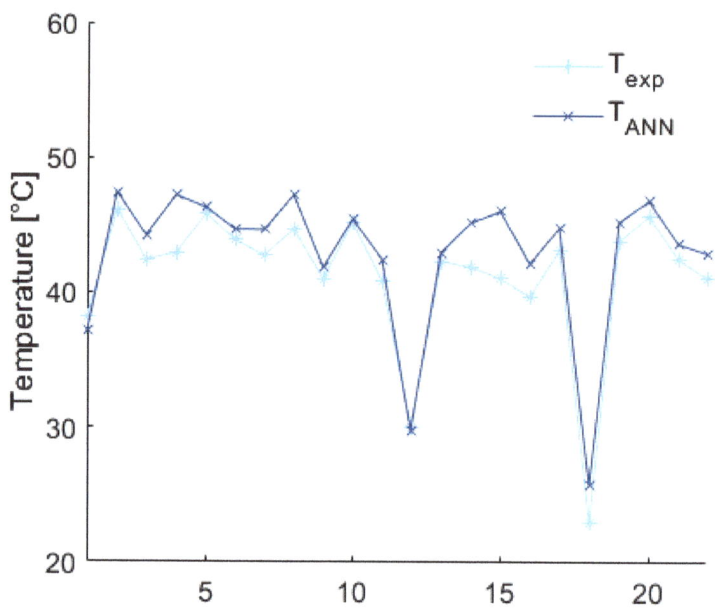

Figure 11. Predicted temperatures (T_{ANN}) vs. experimental temperatures (T_{exp}) of the model validation.

4. Discussion

The temperature inside the chamber plays a vital role in the drying process. The importance lies in the determining factor of temperature prediction, which correlates with how much heat the solar dryer can provide for drying. However, time is an important parameter that needs to be known; a future study should consider the type and quantity of products. Therefore, in this work, the drying chamber temperatures are predicted to help investigate a future model that involves the type and amount of product to be dried based on its moisture content. Likewise, raising the chamber's temperature allows the humidity of the air to be reduced through the ducts of the solar collector to be transported into the chamber. The ideal temperature for drying vegetables, fruits, and meats, among others, is between 30 and 70 °C [10,11,15,16,38,49]. Obtaining lower temperatures would cause the decomposition of the product subjected to drying due to the proliferation of microorganisms. When the temperature exceeds 70 °C, the product subjected to drying may exhibit degradation of its functional compounds [11,50]. Maintaining the temperature in solar dryers is a complex problem due to the variable weather conditions.

For our study, weather conditions, pipe temperatures, the air temperature in the drying chamber, and the speed of the airflow supplied to the solar collector were correctly recorded. Due to the noise present in each variable, these were smoothed through the moving average filter technique to create the neural network model. The smoothing of the data allowed the development of the predictive model of the network. Different neural network configurations were proposed, and the best structure determined was 9-13-12-1. The neural network model's capabilities were tested, and the verification was performed for an experimental data set. The model presented good predictive capabilities, obtaining relative errors below 0.25%. This type of error makes the model reliable with other weather conditions if they are within the range of conditions established in model training. The model was validated with experimental data for the same variables and input parameters for another specific day from 12:10 to 14:00. Again, the response was favorable as the prediction results closely followed the experimental trajectory, with an error like the verification.

The predictive model of the neural network has once again demonstrated its viability with similar conditions that have been applied for another time of year. It should be noted that this model can replace traditional models that use computational and experimental numerical models [6–13] to analyze thermal behavior in solar dryers, which become tedious due to the number of variables involved and the geometric shapes for their study. However, one of the essential advantages of this model is that it provides better predictive capabilities without considering the complexity of the system but considering the variables of interest that allow us to know its behavior. This model could be a basis for developing a real-time optimization tool for the operation and optimal design of indirect solar dryers involving the type, quantity, and moisture content of the product to be dried.

The solar dryer design for this case study uses radiant energy to raise the airflow temperature through the collector and transport it by a forced medium toward the chamber, reaching a temperature between 40 to 50 °C for different quantities of trays inside the drying chamber.

5. Conclusions

An indirect solar dryer has been presented to know the system's dynamics. Weather conditions, airflow velocities, ambient temperature, and solar collector piping have been successfully recorded. A configuration of five trays in the drying chamber was proposed to determine the temperature behavior.

A feed-forward neural network trained with the Levenberg–Marquardt (trainlm) algorithm showed good predictive capabilities. This algorithm was used to determine the ideal configuration of the network architecture. The network configuration was obtained for nine neurons in the input layer, one in the output layer, and two hidden layers of thirteen and twelve neurons each (9-13-12-1), with an MSE error of 0.9725. This configuration reduced the cost and time of the network learning process.

The predictive model was successfully tested, achieving a good predictor for forecasting the air temperature inside the drying chamber. It is reflected when the relative error between predicted and experimental temperatures is obtained. The relative error was below 0.25% for the model's verification and validation.

In addition, this model could serve as a basis for developing a real-time optimization tool for the operation and optimal design of indirect solar dryers. However, other factors must be considered, such as extreme weather conditions, low or high temperatures, and the type, quantity, and moisture content of the product to be dried.

Author Contributions: Conceptualization, A.T.B. and R.R.M.; methodology, A.T.B., R.R.M. and E.C.L.-V.; software, A.T.B. and C.T.S.; validation, A.T.B., R.R.M. and E.M.P.; formal analysis, A.T.B. and R.R.M.; investigation, A.T.B.; resources, A.T.B. and J.L.G.P.; data curation, A.T.B. and E.C.L.-V.; writing—original draft preparation, A.T.B. and R.R.M.; writing—review and editing, A.T.B., R.R.M. and C.T.S.; visualization, A.T.B. and E.M.P.; supervision, A.T.B., R.R.M. and J.L.G.P.; project administration, A.T.B. and R.R.M. All authors have read and agreed to the published version of the manuscript.

Funding: This research received no external funding.

Data Availability Statement: No new data were created or analyzed in this study. Data sharing is not applicable to this article.

Conflicts of Interest: The authors declare no conflict of interest.

References

1. Tlatelpa-Becerro, A.; Rico-Martínez, R.; Urquiza, G.; Calderón-Ramírez, M. Obtaining of Crataegus mexicana leaflets using an indirect solar dryer. *Rev. Mex. Ing. Quim.* **2020**, *2*, 669–676. [CrossRef]
2. Marwan, B.; Chawki, L.; Mirna, F.L.; Christy, L. Solar drying simulation of different products: Lebanese case. *Energy Rep.* **2020**, *6*, 548–564. [CrossRef]
3. Abhay, L.; Chandramohan, V.P.; Raju, V.R.K. Design, Development and Performance of Indirect Type Solar Dryer for Banana Drying. *Energy Procedia* **2017**, *109*, 409–416. [CrossRef]
4. Kokatea, D.H.; Kale, D.M.; Korpale, V.S.; Shinde, Y.H.; Panse, S.V.; Deshmukh, S.P.; Pandit, A.B. Energy Conservation Through Solar Energy Assisted Dryer for Plastic Processing Industry. *Energy Procedia* **2014**, *54*, 376–388. [CrossRef]
5. Nabnean, S.; Nimnuan, P. Experimental performance of direct forced convection household solar dryer for drying banana. *Case Stud. Therm. Eng.* **2020**, *22*, 100787. [CrossRef]
6. Elieser, T. Mathematical modeling and simulation of a solar agricultural dryer with back-up biomass burner and thermal storage. *Case Stud. Therm. Eng.* **2018**, *12*, 149–165. [CrossRef]
7. Mehmet, D.; Erdem, A.; Ebru, K.A. Numerical and experimental analysis of heat and mass transfer in the drying process of the solar drying system. *Eng. Sci. Technol. Int. J.* **2021**, *24*, 236–246. [CrossRef]
8. Petros, D.; Mesele, H.; Amanuel, K.; Asfafaw, H.; Mekonnen, G.; Maarten, V. Design, development and CFD modeling of indirect solar food dryer. *Energy Procedia* **2019**, *158*, 1128–1134. [CrossRef]
9. Guofeng, Y.; Liang, H.; Xing, L.; Li, X.; Wenxue, T.; Zhifeng, W. Experimental investigation of a solar dryer system for drying carpet. *Energy Procedia* **2018**, *70*, 626–633. [CrossRef]
10. Nimnuan, P.; Nabnean, S. Experimental and simulated investigations of the performance of the solar greenhouse dryer for drying cassumunar ginger (*Zingiber cassumunar* Roxb.). *Case Stud. Therm. Eng.* **2020**, *22*, 100745. [CrossRef]
11. Mohammed, A.H.; Klaus, G.; Mohammad, S.H. Mathematical model for a heat pump dryer for aromatic plant. *Procedia Eng.* **2013**, *56*, 510–520. [CrossRef]
12. Hosain, D.; Abbas, R.A.; Ali, A.; Mohsen, A.; Gholamhassan, N.; Jalal, K. Study of the drying kinetics of pepper. *J. Saudi Soc. Agric. Sci.* **2014**, *13*, 130–138. [CrossRef]
13. Mourad, A.; Salah, B.M.; Imed, Z.; Ahmed, B. Kinetic study of the convective drying of spearmint. *J. Saudi Soc. Agric. Sci.* **2014**, *13*, 1–7. [CrossRef]
14. Oluwole, A.O.; Oluyemisi, A.A.; Pelumi, E.O.; Hilary, I.O.; Adebowale, O.A.; Esivue, A.S. Drying kinetic of industrial cassava flour: Experimental data in view. *Data Brief* **2017**, *15*, 501–510. [CrossRef]
15. Ramallo, L.A.; Mascheroni, R.H. Quality evaluation of pineapple fruit during drying process. *Food Bioprod. Process.* **2012**, *90*, 275–283. [CrossRef]
16. Shiva, K.M.; DurgaPrasad, B.; Basavaraj, M. An Experimental Study on Drying Kinetics of Guava Fruit (*Psidium guajava* L.) By Thin Layer Drying. *IOSR J. Environ. Sci. Toxicol. Food Technol.* **2015**, *1*, 74–80.
17. Tlatelpa, B.A.; Rico, M.R.; Castro, G.L.; Urquiza, G.; Calderón-Ramírez, M. Artificial Neural Networks (ANN) and Kalman Filter Algorithms to Predict Output Temperatures on a Heat Exchanger. *Int. J. Appl. Eng. Res.* **2018**, *17*, 13130–13135.
18. Hikmet, E.; Mustafa, I. Modelling of a vertical ground coupled heat pump system by using artificial neural networks. *Expert Syst. Appl.* **2009**, *36*, 10229–10238. [CrossRef]

19. González, G.R.; Rico, M.R.; Wolf, W.; Lübke, M.; Eiswirth, M.; Anderson, J.S.; Kevrekidis, I.G. Characterization of a two-parameter mixed-mode electrochemical behavior regime using neural networks. *Phys. D Nonlinear Phenom.* **2001**, *151*, 27–43. [CrossRef]
20. Kapil, V.; Panigrah, P.K. Artificial neural network control of a heat exchanger in a closed flow air circuit. *Appl. Soft Comput.* **2005**, *5*, 441–465. [CrossRef]
21. Ramírez, A.E.; Calderon, R.M.; Rico, M.R.; Gonzalez, F.C.; Parmananda, P. Detecting Bifurcations in an Electrochemical Cell Employing an Assisted Reference Model Strategy. *J. Phys. Chem.* **2012**, *117*, 535–540. [CrossRef]
22. Hikmet, E.; Mustafa, I.; Abdulkadir, S.; Mehmet, E. Performance prediction of a ground-coupled heat pump system using artificial neural networks. *Expert Syst. Appl.* **2008**, *35*, 1940–1948. [CrossRef]
23. Vinoth, K.K.; Paradeshi, L.; Srinivas, M.; Jayaraj, S. Parametric Studies of a Simple Direct Expansion Solar Assisted Heat Pump Using ANN and GA. *Energy Procedia* **2016**, *90*, 625–634. [CrossRef]
24. Luca, S.; Herbert, A.Z.; Jan, D.; Camilo, C.M.; Van Bael, J.; Adriano, S. Modeling the performance of a sorption thermal energy storage reactor using artificial neural networks. *Appl. Energy* **2019**, *253*, 113525. [CrossRef]
25. Ashly, M.T.; Nizar, A.; Subathra, M.S.; Lazarus, G.A. Analysing the Performance of a Flat Plate Solar Collector with Silver/Water Nanofluid Using Artificial Neural Network. *Procedia Comput. Sci.* **2016**, *93*, 33–40. [CrossRef]
26. Chaibi, Y.; Malvoni, M.; El, R.T.; Kousksou, T.; Zeraouli, Y. Artificial neural-network based model to forecast the electrical and thermal efficiencies of PVT air collector systems. *Clean. Eng. Technol.* **2021**, *4*, 100132. [CrossRef]
27. Gilles, N.; Christophe, P.; Said, D. Estimation of tilted solar irradiation using Artificial Neural Networks. *Energy Procedia* **2013**, *42*, 33–42. [CrossRef]
28. İlhan, C.; Okan, E.; Engin, G.; Ali, E.G. The prediction of photovoltaic module temperature with artificial neural networks. *Case Stud. Thermal Eng.* **2014**, *3*, 11–20. [CrossRef]
29. Anurag, K.; Nikhil, R.; Pijush, S. Prediction of heat transfer rate of a Wire-on-Tube type heat exchanger: An Artificial Intelligence approach. *Procedia Eng.* **2013**, *64*, 74–83. [CrossRef]
30. Haritonova, L. On Neural Modeling of Heat Exchange in Heat Exchangers (Recuperators) with the Systems of Plane-Parallel Impingement Jets for Machine Building and Metallurgical Productions. *Procedia Eng.* **2017**, *206*, 1002–1008. [CrossRef]
31. Promvonge, P.; Eiamsa, A.S.; Wongcharee, K.; Chuwattanakul, V.; Samruaisin, P.; Chokphoemphun, S.; Nanan, K.; Eiamsa, A.P. Characterization of heat transfer and artificial neural networks prediction on overall performance index of a channel installed with arc-shaped baffle turbulators. *Case Stud. Therm. Eng.* **2021**, *26*, 101067. [CrossRef]
32. Xie, G.N.; Wang, Q.W.; Zeng, M.; Luo, L.Q. Heat transfer analysis for shell-and-tube heat exchangers with experimental data by artificial neural networks approach. *Appl. Therm. Eng.* **2007**, *27*, 1096–1104. [CrossRef]
33. Yasar, I. A new approach for the prediction of the heat transfer rate of the wire-on-tube type heat exchanger—use of an artificial neural network model. *Appl. Therm. Eng.* **2003**, *23*, 243–249. [CrossRef]
34. Gabriele, B.; Vincenzo, T.; Nadia, U. Neural Networks Implementation for Analysis and Control of Heat Exchange Process in a Metal Foam Prototypal Device. *Procedia CIRP* **2017**, *62*, 518–522. [CrossRef]
35. Guptaa, A.K.; Kumara, P.; Sahooa, R.K.; Sahub, A.K.; Sarangia, S.K. Performance measurement of plate fin heat exchanger by exploration: ANN, ANFIS, GA, and SA. *J. Comput. Des. Eng.* **2017**, *4*, 60–68. [CrossRef]
36. Yasar, I.; Akif, K.; Cem, P. Performance prediction for non-adiabatic capillary tube suction line heat exchanger: An artificial neural network approach. *Energy Convers. Manag.* **2005**, *46*, 223–232. [CrossRef]
37. Mohammad, K.; Vali, R.S.; Reza, A.C.; Ebrahim, T.; Abbaspour, G.Y.; Iman, G. ANFIS and ANNs model for prediction of moisture diffusivity and specific energy consumption potato, garlic and cantaloupe drying under convective hot air dryer. *Inf. Process. Agric.* **2018**, *5*, 372–387. [CrossRef]
38. Momenzadeh, L.; Zomorodian, A.; Mowla, D. Applying Artificial Neural Network for Drying Time Prediction of Green Pea in a Microwave Assisted Fluidized Bed Dryer. *J. Agr. Sci. Technol.* **2012**, *14*, 513–522. [CrossRef]
39. Shamsollah, A.; Kosari, M.A.; Mohammad, B. Prediction of wheat moisture content at harvest time through ANN and SVR modeling techniques. *Inf. Process. Agric.* **2020**, *7*, 500–510. [CrossRef]
40. Tohsing, K.; Janjai, S.; Lamlert, N.; Mundpookhier, T.; Chanalert, W.; Bala, B.K. Experimental performance and artificial neural network modeling of solar drying of litchi in the parabolic greenhouse dryer. *J. Renew. Energy Smart Grid Technol.* **2018**, *13*, 83–95.
41. Krischer, K.; Rico, M.R.; Kevrekidis, I.G.; Rotermund, H.H.; Ertl, G.; Hudson, J.L. Model identification of a spatiotemporally varying catalytic reaction. *AIChE J.* **1993**, *1*, 89–98. [CrossRef]
42. Khataeea, A.R.; Kasiri, M.B. Artificial neural networks modeling of contaminated water treatment processes by homogeneous and heterogeneous nanocatalysis. *J. Mol. Catal. A Chem.* **2010**, *331*, 86–100. [CrossRef]
43. Arzu, S.S. Performance analysis of single-stage refrigeration system with internal heat exchanger using neural network and neuro-fuzzy. *Renew. Energy* **2011**, *36*, 2747–2752. [CrossRef]
44. Jian, Z.; Fariborz, H. Development of Artificial Neural Network based heat convection algorithm for thermal simulation of large rectangular cross-sectional area Earth-to-Air Heat Exchangers. *Energy Build.* **2010**, *42*, 435–440. [CrossRef]
45. Reza, B.; Masoud, R. Prediction of heat transfer and flow characteristics in helically coiled tubes using artificial neural networks. *Int. Commun. Heat Mass Transf.* **2012**, *39*, 1279–1285. [CrossRef]
46. Gongnan, X.; Bengt, S.; Qiuwang, W.; Linghong, T. Performance predictions of laminar and turbulent heat transfer and fluid flow of heat exchangers having large tube-diameter and large tube-row by artificial neural networks. *Int. J. Heat Mass Transf.* **2009**, *52*, 2484–2497. [CrossRef]

47. Soheil, A.; Carey, J.S.; Robert, W.B. Application of neural networks to predict the transient performance of a Run-Around Membrane Energy Exchanger for yearly non-stop operation. *Int. J. Heat Mass Transf.* **2012**, *55*, 5403–5416. [CrossRef]
48. Unal, A.; Aydin, K.M.; Feridun, O.A. Estimation of heat transfer in oscillating annular flow using artificial neural networks. *Adv. Eng. Softw.* **2009**, *40*, 864–870. [CrossRef]
49. Romero, V.M.; Cerezo, E.; Garcia, M.; Sanchez, M.H. Simulation and validation of vanilla drying process in an indirect solar dryer prototype using CFD Fluent program. *Energy Procedia* **2014**, *57*, 1651–1658. [CrossRef]
50. Suherman, S.; Hadiyanto, H.; Evan, E.S.; Shesar, A.R.; Aditya, R.P. Towards an optimal hybrid solar method for lime-drying behavior. *Heliyon* **2020**, *6*, e05356. [CrossRef] [PubMed]

AgriEngineering

MDPI

Article

Artificial Neural Networks for Modeling and Optimizing Egg Cost in Second-Cycle Laying Hens Based on Dietary Intakes of Essential Amino Acids

Walter Morales-Suárez, Luis Daniel Daza and Henry A. Váquiro *

Faculty of Agronomic Engineering, Universidad del Tolima, Ibagué 730006, Colombia; wmoraless@ut.edu.co (W.M.-S.); lddazar@ut.edu.co (L.D.D.)
* Correspondence: havaquiro@ut.edu.co

Abstract: Egg production is a significant source of animal protein for human consumption. Feed costs significantly impact the profitability of egg production, representing more than 70% of the variable costs. This study evaluated the effect of dietary intakes of three essential amino acids (EAAs) on the egg cost for H&N Brown second-cycle laying hens. The hens were fed for 20 weeks with 23 diets that varied in their lysine, methionine + cystine, and threonine contents. These amino acids were derived from both dietary and synthetic sources. Zootechnical results were used to calculate the feed cost per kilogram of egg (FCK), considering the cost of raw materials and the diet composition. Multivariate polynomial models and artificial neural networks (ANNs) were validated to predict FCK as a function of the EAAs and time. The EAA intakes that minimize FCK over time were optimized using the best model, a cascade-forward ANN with a softmax transfer function. The optimal scenario for FCK (0.873 USD/kg egg) at 20 weeks was achieved at 943.7 mg lysine/hen-day, 858.3 mg methionine + cystine/hen-day, and 876.8 mg threonine/hen-day. ANNs could be a valuable tool for predicting the egg cost of laying hens based on the nutritional requirements. This could help improve economic efficiency and reduce the feed costs in poultry companies.

Keywords: animal nutrition; egg laying; feed costs; mathematical model; multivariate analysis; poultry

Citation: Morales-Suárez, W.; Daza, L.D.; Váquiro, H.A. Artificial Neural Networks for Modeling and Optimizing Egg Cost in Second-Cycle Laying Hens Based on Dietary Intakes of Essential Amino Acids. *AgriEngineering* **2023**, *5*, 1832–1845. https://doi.org/10.3390/agriengineering5040112

Academic Editors: Ray E. Sheriff and Chiew Foong Kwong

Received: 13 September 2023
Revised: 3 October 2023
Accepted: 5 October 2023
Published: 12 October 2023

1. Introduction

According to FAO estimations, the world population may increase to 9.8 billion by 2050 [1]. This accelerated growth has aroused worldwide concern, mainly due to the difficulty of producing more food without affecting food security and of making it accessible to the entire population. Among the possible strategies that can be implemented to increase food production worldwide, egg production is emerging as a viable alternative to advance progress toward the second Sustainable Development Goal (end hunger, achieve food security and improved nutrition, and promote sustainable agriculture) [2].

From a nutritional point of view, eggs are an essential source of proteins, phospholipids, vitamins (e.g., thiamin, riboflavin, niacin, vitamin B12, vitamin A, vitamin E, vitamin D, and vitamin K), minerals (e.g., iron, selenium, calcium, magnesium, phosphorus, potassium, sodium, and zinc) and antioxidants, and their consumption could cover part of the daily needs of these types of nutrients [3,4]. Furthermore, this product presents advantages over other protein sources, primarily stemming from its low commercial cost, which has increased its consumption and production worldwide. It is estimated that between 2018 and 2020, egg production increased by more than 8% worldwide [5]. Despite the growth of this industry and the egg's nutritional value, some factors limit its production, such as costs associated with animal feed, which can exceed 70% of variable costs [6]. An effective strategy to reduce egg production costs is the implementation of second production cycles in birds. By adopting this method, different benefits are obtained, such as increased laying cycles and egg size, which provide responsiveness to fluctuations in costs associated with production supplies [7]. For second-cycle production to be efficient, the nutritional requirements of laying birds must be defined. However, these requirements vary according to genetic lines, production volume, environmental conditions, and the age of the birds [8].

Amino acids are critical components in the nutrition and productivity of laying hens. Therefore, a deficit or excess of these compounds can be configured as a disadvantage from an economic point of view [9]. Among the amino acids that are part of the diet of laying hens, lysine (Lys), sulfur amino acids (methionine and cysteine; Met + Cys), and threonine (Thr) are the most relevant. Lys is an essential amino acid (EAA) for the maintenance, growth, and production of birds, with the primary function of participating in body protein synthesis [9,10]. As plant-based protein sources have low levels of Met + Cys, these are the first limiting amino acids for poultry diets [9]. The third limiting amino acid in poultry diets is Thr [11], which is related to maintaining the integrity of the intestinal barrier and the production of antibodies that play an essential role in bird immunity [9]. Several studies have evaluated the optimal ratio of amino acids in the diet of laying hens to improve productivity or egg size [12,13]. However, only some studies have focused on minimizing the feed cost per kilogram of egg (FCK) by evaluating the nutritional requirements associated with amino acids in the diet.

The challenges and trends in poultry require investment in technology and innovation, especially in data analysis, to improve the industry's efficiency, profitability, sustainability, and competitiveness [14]. Data analysis can support planning and cost reduction in egg production by estimating the productive behavior of laying hens under various nutritional, sanitary, environmental, and economic conditions [10,12,15,16].

Predictive models are valuable tools to address the challenges of poultry production in terms of business management and cost reduction [17–19]. They can provide benefits such as enhancing the control and monitoring of farm indicators to identify health or nutrition issues; conducting comparative analysis between different production conditions to assess the performance and profitability of companies and identify areas for improvement or correction; and supporting companies to adapt to market or environmental changes, estimate scenarios, and plan actions aligned with their objectives.

In this study, two models based on ANNs were applied to assess the dependence of the feed cost per kilogram of egg (FCK) on the dietary intakes of the three main EAAs for laying hens. The model with the best fit was used to minimize FCK and to determine the relative importance of Lys, Met + Cys, and Thr in H&N Brown second-cycle laying hens (SCLHs) under field conditions.

2. Materials and Methods

2.1. Fieldwork, Experimental Design, and Diet Formulation

This study was conducted on a commercial farm in San Pedro, Antioquia (Colombia). The university bioethics committee approved the protocols and experimental procedures on the laying hens (Act 03/2017). For the fieldwork, H&N Brown SCLHs aged 91 weeks were used (1380 hens). The animals were housed in California-type cages under environmental conditions and distributed in a layer shed after being subjected to proper molting and resting protocols [12]. In each cage, 12 hens were placed (576 cm^2/bird), and each cage was fitted with a PVC side feeder and two nipple drinkers. Five cages were used as replicates for each treatment (60 birds per treatment).

The diet formulation per treatment and the feed consumption were used to express the design matrix as dietary intakes of Lys (iLys), Met + Cys (iMetCys), and Thr (iThr) (Table 1) according to the methodology and results of Morales-Suárez et al. [12]. This methodology combined levels of the three EAAs from a comprehensive literature review using a central composite design. The resulting 23 diets (treatments) were prepared weekly by adding L-lysine (98.5% feed grade), DL-methionine (99% feed grade), and L-threonine (98.5% feed grade) to a diet based on corn and soybean meal (Supplementary Table S1). Thus, the Lys, Met + Cys, and Thr levels in the experimental diets were adjusted by considering the contributions of the synthetic amino acids and food ingredients. The layers were fed at 115 g/hen-day for 20 weeks.

Table 1. Essential amino acid intakes of H&N Brown SCLHs over 20 weeks and diet cost.

Diet	iLys * (mg/hen-day)	iMetCys * (mg/hen-day)	iThr * (mg/hen-day)	Diet Cost (USD/kg)
1	727.0 ± 10.0	661.9 ± 9.1	552.1 ± 7.6	0.3974
2	732.4 ± 9.6	666.8 ± 8.8	875.2 ± 11.5	0.4015
3	728.4 ± 18.0	852.1 ± 21.0	712.4 ± 17.6	0.4036
4	728.5 ± 17.2	1042.3 ± 24.6	553.2 ± 13.0	0.4058
5	736.3 ± 18.6	1053.5 ± 26.7	879.8 ± 22.3	0.4099
6	809.6 ± 11.3	736.6 ± 10.3	614.6 ± 8.6	0.4015
7	812.7 ± 18.0	739.5 ± 16.4	805.8 ± 17.9	0.4040
8	828.4 ± 22.7	982.6 ± 26.9	628.9 ± 17.2	0.4065
9	810.2 ± 11.3	961.9 ± 13.4	803.3 ± 11.2	0.4089
10	951.7 ± 17.8	673.7 ± 12.6	723.7 ± 13.6	0.4033
11	947.4 ± 19.0	861.7 ± 17.3	559.4 ± 11.2	0.4055
12	938.4 ± 13.6	853.6 ± 12.4	713.6 ± 10.3	0.4075
13	943.7 ± 12.2	858.3 ± 11.1	876.8 ± 11.3	0.4095
14	935.1 ± 12.9	1040.2 ± 14.3	711.0 ± 9.8	0.4117
15	1076.9 ± 23.4	751.2 ± 16.3	626.8 ± 13.6	0.4061
16	1066.3 ± 15.7	743.9 ± 10.9	810.7 ± 11.9	0.4085
17	1058.8 ± 14.6	963.9 ± 19.5	616.3 ± 12.5	0.4111
18	1058.5 ± 14.6	963.6 ± 13.3	804.7 ± 11.1	0.4135
19	1164.7 ± 30.3	674.3 ± 17.7	562.5 ± 14.7	0.4052
20	1143.1 ± 15.8	661.9 ± 9.1	868.8 ± 12.0	0.4092
21	1154.2 ± 18.4	858.7 ± 13.7	771.9 ± 11.4	0.4114
22	1153.3 ± 20.7	1049.5 ± 18.8	557.0 ± 10.0	0.4135
23	1159.5 ± 16.9	1055.1 ± 15.4	881.2 ± 12.3	0.4176

* Values expressed as means ± standard deviations (n = 100). Source: Morales-Suárez et al. [12].

In this way, the experiment considered the dependence of FCK on four factors: iLys, iMetCys, iThr, and time.

2.2. Zootechnical and Economic Results

The feed conversion ratio (CR) was computed weekly as the ratio between kilograms of feed intake and kilograms of eggs.

The feed cost per kilogram of egg (FCK) was calculated considering the cost of the diets (Table 1) calculated from the local cost of the raw materials in USD in March 2021 (Table 2). For each treatment, the FCK in USD/kg egg was estimated weekly by multiplying the cost per kilogram of feed and CR.

2.3. Mathematical Modeling

The FCK was modeled as a function of dietary intake of EEAs (iLys, iMetCys, iThr) and time (t) using multivariable polynomial models and ANNs.

The second- (Equation (1)) and third-order (Equation (2)) polynomial models were evaluated by means of stepwise regression to include only statistically significant terms at a 95% confidence level [20].

$$\mathrm{FCK} = \beta_0 + \sum_{j=1}^{4} \beta_j x_j + \sum_{j=1}^{3} \sum_{k>j}^{4} \beta_{jk} x_j x_k + \sum_{j=1}^{4} \beta_{jj} x_j^2 \quad (1)$$

$$\mathrm{FCK} = \beta_0 + \sum_{j=1}^{4} \beta_j x_j + \sum_{j=1}^{3} \sum_{k>j}^{4} \beta_{jk} x_j x_k + \sum_{j=1}^{4} \beta_{jj} x_j^2 + \sum_{j=1}^{2} \sum_{k>j}^{3} \sum_{l>k}^{4} \beta_{jkl} x_j x_k x_l \sum_{j=1}^{4} \sum_{k\neq j}^{4} \beta_{jkk} x_j x_k^2 + \sum_{j=1}^{4} \beta_{jjj} x_j^3 \quad (2)$$

Table 2. Cost of diet ingredients.

Ingredient	Cost (USD/kg)
Corn	0.29
Soybean meal	0.53
Palm oil	0.93
Corn gluten	0.92
Calcium carbonate	0.05
Monocalcium phosphate	0.62
Bentonite	0.06
Choline chloride	1.15
Salt	0.08
Vitamin premix	11.78
Mineral premix	6.42
L-lysine *	1.73
DL-Methionine *	2.56
L-threonine *	1.50
Tryptophan	9.96
Arginine	15.37
Sodium bicarbonate	0.44
Valine	4.20

* Synthetic amino acids used as dietary supplements.

Here, FCK is the feed cost per kilogram of egg (USD/kg egg); x terms represent the EEA dietary intakes (iLys, iMetCys, iThr) and the time; and β terms are the parameters associated with linear, quadratic, cubic, and cross-product terms.

Additionally, feed-forward (Equation (3)) (Figure 1a) and cascade-forward (Equation (4)) (Figure 1b) ANN architectures with one hidden layer were used to evaluate the dependence of the FCK on the independent factors. These network architectures have been used due to their prediction capability for complex multivariate nonlinear problems and their modeling capability for diverse natural phenomena that are very difficult to handle via classical techniques [21]. Five to eleven neurons and four transfer functions (hyperbolic tangent, log sigmoid, radial basis, and softmax) (Equations (5)–(8), respectively) were assessed for each ANN architecture [22].

$$\mathrm{FCK} = \mathrm{w_{ho}} \times \mathrm{f} + \mathrm{b_o} \tag{3}$$

$$\mathrm{FCK} = \mathrm{w_{ho}} \times \mathrm{f} + \mathrm{w_{io}} \times \mathrm{x} + \mathrm{b_o} \tag{4}$$

$$\mathrm{f} = \frac{2}{1 + \mathrm{e}^{[-2(\mathrm{w_{ih}} \times \mathrm{x} + \mathrm{b_h})]}} - 1 \tag{5}$$

$$\mathrm{f} = \frac{1}{1 + \mathrm{e}^{[-(\mathrm{w_{ih}} \times \mathrm{x} + \mathrm{b_h})]}} \tag{6}$$

$$\mathrm{f} = \mathrm{e}^{[-(\mathrm{w_{ih}} \times \mathrm{x} + \mathrm{b_h})^2]} \tag{7}$$

$$\mathrm{f} = \frac{\mathrm{e}^{(\mathrm{w_{ih}} \times \mathrm{x} + \mathrm{b_h})}}{\sum \mathrm{e}^{(\mathrm{w_{ih}} \times \mathrm{x} + \mathrm{b_h})}} \tag{8}$$

Here, FCK is the output estimation, w_{ho} is the weights between the hidden and the output layers, f is the transfer function, w_{io} is the weights between the input and the output layers in the cascade-forward architecture, x is the input predictor (iLys, iMetCys, iThr, t), b_o is the bias of the output layer, w_{ih} is the weights between the input and the hidden layers, and b_h is the biases of the hidden layer.

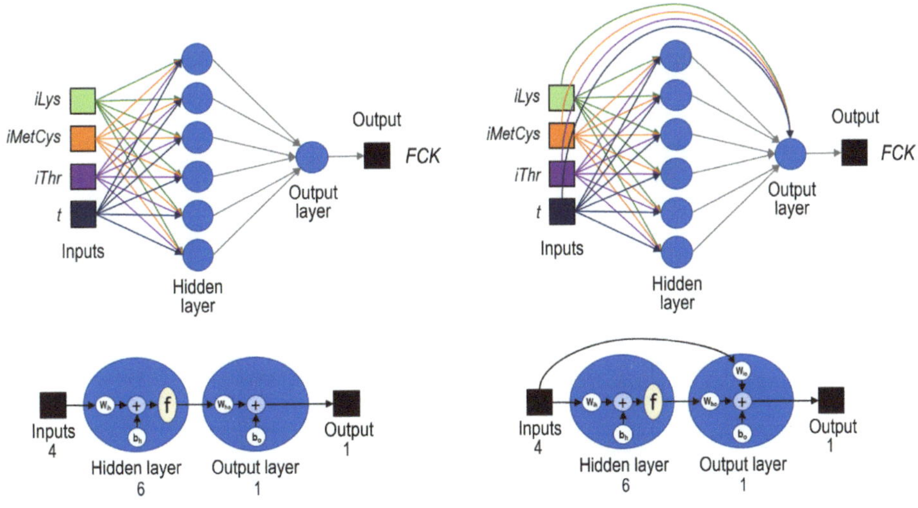

Figure 1. Feed-forward (**a**) and cascade-forward (**b**) architectures used in the artificial neural network modeling. The examples illustrate four-input, one-output neural networks with six neurons in a hidden layer.

2.4. Training, Validation, and Statistical Analysis

MATLAB® R2019a (The MathWorks Inc., Natick, MA, USA) was used for the analysis of two sets of data: the training group with 1380 data records (three treatment replicates) and the validation group with 920 data records (two treatment replicates).

The statistically significant parameters of the multivariate polynomial models (Equations (1) and (2)) were obtained through the MATLAB® function "stepwisefit". The parameters of the ANN models (biases and weights) were identified using the "trainbr" function of MATLAB®, which avoids overfitting in the training process through a Bayesian regularization algorithm [23].

The accuracy, goodness of fit, and deviation of the models were evaluated by examining the root-mean-square error (RMSE) (Equation (9)), the adjusted coefficient of determination (R^2_{adj}) (Equation (10)), and the bias (Equation (11)), respectively. Meanwhile, the quality of the models was determined by employing the Akaike Information Criterion (AIC) (Equation (12)) [24]. The models with the best fit were evaluated for the adequacy of the estimates using a residual analysis.

$$\mathrm{RMSE} = \sqrt{\frac{\sum_{i=1}^{n}\left(\mathrm{FCK}_i^* - \mathrm{FCK}_i\right)^2}{n}} \tag{9}$$

$$R^2_{adj} = 1 - S^2_{yx}/S^2_y \tag{10}$$

$$\mathrm{bias} = \frac{1}{n}\sum_{i=1}^{n}\left(\mathrm{FCK}_i^* - \mathrm{FCK}_i\right) \tag{11}$$

$$\mathrm{AIC} = n\log\left[\det\left(\frac{1}{n}\sum_{i=1}^{n}\varepsilon\varepsilon^T\right)\right] + 2n_p + n[\log(2\pi) + 1] \tag{12}$$

Here, FCK_i^* and FCK_i represent the observed and predicted values, respectively; S_y and S_{yx} are the standard deviations of the sample and model estimations, respectively; ε is the prediction error; and n and n_p are the numbers of data points and estimated parameters, respectively. A high value of R^2_{adj} and low values of RMSE and AIC were used to select the best model.

2.5. Optimization

An optimization problem was formulated to find the iLys, iMetCys, and iThr values that minimize FCK (Equation (13)) at 4, 8, 12, 16, and 20 weeks. The optimization was solved for each of the weeks mentioned above using the MATLAB® function "fmincon" and constraining the limits to the maximum and minimum values of iLys, iMetCys, and iThr used in the experiment (Table 1).

$$\text{minimize FCK(iLys, iMetCys, iThr)such that} \begin{cases} \text{iLys}_{\text{min}} \le \text{iLys} \le \text{iLys}_{\text{max}} \\ \text{iMetCys}_{\text{min}} \le \text{iMetCys} \le \text{iMetCys}_{\text{max}} \\ \text{iThr}_{\text{min}} \le \text{Thr} \le \text{iThr}_{\text{max}} \end{cases} \quad (13)$$

3. Result

3.1. Egg Cost and Zootechnical Results

The observed FCK values were between 0.998 and 1.559 USD/kg egg in week 8, between 0.949 and 1.214 USD/kg egg in week 12, between 0.928 and 1.090 USD/kg egg in week 16, and between 0.919 and 1.041 USD/kg egg in week 20 (Table 3). FCK generally decreases over time since feed costs are amortized as the hens' egg production increases over the production cycle.

The lowest FCK corresponds to diet 13 at weeks 8, 12, 16, and 20, followed by diet 12 at weeks 12, 16, and 20 and diet 3 at week 20 (Table 3). In these diets, the iLys values ranged from 728.4 mg/hen-day (diet 3) to 943.7 mg/hen-day (diet 13); the iMetCys values ranged between 852.1 mg/hen-day (diet 3) and 858.3 mg/hen-day (diet 13); and the iThr values ranged between 712.4 mg/hen-day (diet 3) and 876.8 mg/hen-day (diet 13) (Table 1).

The observed CR was between 2.02 and 2.36 kg feed/kg egg, comparable to the values for commercial hens at 20 weeks of production [25]. Diet 13 presented CR values between 3.40 ± 0.19 kg feed/kg egg (week 4) and 2.18 ± 0.16 kg feed/kg egg (week 20) (Table 4).

Egg production ranged from 83.1% to 85.3%, which agrees with the values reported for commercial hens at 20 weeks of production in the Hy-Line Brown management guide [25]. Diet 13, the treatment with the lowest FCK, presented the following egg production rates: $72.1 \pm 4.7\%$ (week 4); $80.9 \pm 5.1\%$ (week 8); $81.2 \pm 9.1\%$ (week 12); $77.1 \pm 6.5\%$ (week 16); and $80 \pm 4.9\%$ (week 20) [12]. At week 20, diet 13 showed 105.1 ± 4.2 hen-housed eggs, an average of 5.25 eggs per week.

3.2. Curve Fitting and Statistical Criteria

FCK was modeled using two different approaches (Table 5). In the first approach, the multivariate polynomial models presented RMSE values of 0.0818 and 0.0841 and R^2_{adj} values of 0.8642 and 0.8519 for the training and validation, respectively.

The best polynomial model for FCK was the third-order equation in Equation (14), where the independent variables iLys, iMetCys, iThr, and t are represented as x_1, x_2, x_3, and x_4, respectively.

$$\begin{aligned} \text{FCK} = &-3.0975 + 0.1523x_1 - 0.1945x_2 + 7.009 \times 10^{-2}x_3 - 0.6893x_4 + 5.901 \times 10^{-5}x_1x_2 + 1.653 \times 10^{-4}x_2x_3 \\ &-2.323 \times 10^{-4}x_1x_3 + 6.095 \times 10^{-4}x_4x_1 + 4.990 \times 10^{-4}x_4x_2 - 3.589 \times 10^{-4}x_4x_3 - 1.073 \times 10^{-4}x_1^2 \\ &+1.192 \times 10^{-4}x_2^2 - 4.103 \times 10^{-5}x_3^2 + 2.060 \times 10^{-2}x_4^2 + 9.065 \times 10^{-8}x_1^2x_3 - 2.760 \times 10^{-7}x_1^2x_4 \\ &-3.276 \times 10^{-8}x_2^2x_1 - 9.762 \times 10^{-8}x_2^2x_3 - 2.298 \times 10^{-7}x_2^2x_4 + 4.207 \times 10^{-8}x_3^2x_1 \\ &+1.174 \times 10^{-7}x_3^2x_4 - 1.952 \times 10^{-7}x_1x_2x_4 + 1.131 \times 10^{-7}x_2x_3x_4 + 9.898 \times 10^{-8}x_1x_3x_4 \\ &+1.710 \times 10^{-8}x_1^3 - 5.055 \times 10^{-9}x_2^3 - 4.380 \times 10^{-4}x_4^3 \end{aligned} \quad (14)$$

The best results for FCK using ANN models were obtained using a cascade-forward architecture, nine neurons in the hidden layer, and a softmax transfer function (Table 5).

The fitting results presented RMSE values of 0.042 and 0.043 and R^2_{adj} values of 0.9534 and 0.9564 for the training and validation, respectively.

Table 3. Feed cost per kilogram of egg (USD/kg egg) of H&N Brown SCLHs for the diets over 20 weeks.

Diet	Week 8	Week 12	Week 16	Week 20
1	1.33 ± 0.04	1.15 ± 0.06	1.06 ± 0.06	1.02 ± 0.05
2	1.56 ± 0.03	1.21 ± 0.03	1.09 ± 0.03	1.04 ± 0.04
3	1.16 ± 0.02	1.02 ± 0.03	0.96 ± 0.03	0.94 ± 0.03
4	1.15 ± 0.01	1.03 ± 0.02	0.99 ± 0.02	0.96 ± 0.03
5	1.18 ± 0.02	1.05 ± 0.03	0.98 ± 0.02	0.95 ± 0.02
6	1.43 ± 0.02	1.16 ± 0.03	1.06 ± 0.03	1.02 ± 0.03
7	1.23 ± 0.04	1.06 ± 0.04	0.98 ± 0.04	0.94 ± 0.03
8	1.41 ± 0.03	1.15 ± 0.04	1.04 ± 0.04	0.99 ± 0.04
9	1.21 ± 0.01	1.07 ± 0.02	1.01 ± 0.03	0.98 ± 0.03
10	1.22 ± 0.03	1.08 ± 0.04	1.00 ± 0.03	1.01 ± 0.09
11	1.16 ± 0.02	1.06 ± 0.03	0.99 ± 0.02	0.97 ± 0.02
12	1.12 ± 0.05	1.01 ± 0.02	0.96 ± 0.02	0.94 ± 0.02
13	1.00 ± 0.04	0.95 ± 0.02	0.93 ± 0.04	0.92 ± 0.04
14	1.17 ± 0.02	1.05 ± 0.01	1.00 ± 0.02	0.97 ± 0.03
15	1.09 ± 0.03	1.02 ± 0.03	0.97 ± 0.03	0.95 ± 0.03
16	1.25 ± 0.01	1.11 ± 0.01	1.03 ± 0.02	1.00 ± 0.02
17	1.21 ± 0.02	1.04 ± 0.02	0.98 ± 0.02	0.95 ± 0.03
18	1.28 ± 0.05	1.09 ± 0.03	1.03 ± 0.03	1.00 ± 0.03
19	1.36 ± 0.04	1.14 ± 0.01	1.05 ± 0.02	1.01 ± 0.01
20	1.34 ± 0.02	1.16 ± 0.03	1.06 ± 0.04	1.02 ± 0.04
21	1.35 ± 0.03	1.14 ± 0.04	1.05 ± 0.04	1.02 ± 0.04
22	1.48 ± 0.03	1.20 ± 0.03	1.07 ± 0.03	1.03 ± 0.04
23	1.41 ± 0.04	1.21 ± 0.02	1.09 ± 0.02	1.04 ± 0.02

Values expressed as means ± standard deviations (n = 5).

Table 4. Feed conversion ratio (kg feed/kg egg) of H&N Brown SCLHs for the diets over 20 weeks.

Diet	Week 8	Week 12	Week 16	Week 20
1	3.43 ± 0.23	2.58 ± 0.12	2.16 ± 0.33	2.22 ± 0.06
2	3.62 ± 0.20	2.62 ± 0.14	2.03 ± 0.15	2.03 ± 0.11
3	2.89 ± 0.11	2.37 ± 0.11	2.08 ± 0.25	2.25 ± 0.17
4	2.88 ± 0.18	2.20 ± 0.14	2.09 ± 0.26	2.18 ± 0.24
5	3.45 ± 0.44	1.88 ± 0.08	2.15 ± 0.28	1.95 ± 0.14
6	5.05 ± 0.53	2.17 ± 0.07	2.00 ± 0.23	2.03 ± 0.10
7	3.50 ± 0.36	2.00 ± 0.11	1.99 ± 0.30	2.09 ± 0.22
8	4.20 ± 0.29	2.16 ± 0.09	1.99 ± 0.18	2.08 ± 0.10
9	2.87 ± 0.26	2.30 ± 0.15	1.89 ± 0.07	2.17 ± 0.23
10	3.25 ± 0.30	2.34 ± 0.17	2.27 ± 0.23	2.09 ± 0.17
11	3.59 ± 0.26	2.10 ± 0.04	2.11 ± 0.19	2.23 ± 0.19
12	3.05 ± 0.30	2.01 ± 0.05	2.03 ± 0.18	2.14 ± 0.23
13	2.48 ± 0.20	2.08 ± 0.16	2.11 ± 0.21	2.18 ± 0.27
14	3.40 ± 0.19	2.12 ± 0.05	2.05 ± 0.14	2.17 ± 0.11
15	2.59 ± 0.30	2.26 ± 0.16	2.14 ± 0.22	2.59 ± 1.33
16	3.10 ± 0.10	2.43 ± 0.07	2.19 ± 0.21	2.24 ± 0.23
17	3.30 ± 0.22	2.25 ± 0.03	2.00 ± 0.05	2.18 ± 0.14
18	3.50 ± 0.13	2.15 ± 0.09	2.03 ± 0.14	2.29 ± 0.18
19	3.84 ± 0.30	2.35 ± 0.11	2.18 ± 0.08	2.05 ± 0.15
20	4.07 ± 0.53	2.36 ± 0.14	2.16 ± 0.10	2.18 ± 0.19
21	3.48 ± 0.21	2.34 ± 0.17	2.21 ± 0.19	2.08 ± 0.08
22	4.39 ± 0.16	2.35 ± 0.14	2.14 ± 0.26	2.05 ± 0.08
23	3.60 ± 0.18	2.39 ± 0.15	2.11 ± 0.04	2.02 ± 0.10

Values expressed as means ± standard deviations (n = 5).

Table 5. Modeling results for the feed cost per kilogram of egg (FCK) of H&N Brown SCLHs.

Model	Specifications	Training RMSE	Training R^2_{adj}	Validation RMSE	Validation R^2_{adj}	*p*-Value	Bias	AIC
Multivariate polynomial								
Second order	12 parameters	0.108	0.7663	0.111	0.7485	0.001	2.38×10^{-5}	19028
Third order	28 parameters	0.082	0.8642	0.084	0.8519	0.001	7.11×10^{-5}	18621
ANN								
Feed-forward network	Softmax transfer function Nine hidden neurons 55 parameters	0.047	0.9418	0.045	0.9537	0.001	2.38×10^{-7}	17714
Cascade-forward network	Softmax transfer function Nine hidden neurons 59 parameters	0.042	0.9534	0.043	0.9564	0.001	1.23×10^{-7}	17608

RMSE is the root-mean-square error; R^2_{adj} is the adjusted coefficient of determination; AIC is Akaike's information criterion; *p*-value is the probability to test the validity of the null hypothesis that the residuals came from a normal distribution, according to the Lilliefors test.

The ANN models showed better-fitting results than other modeling approaches, presenting higher R^2_{adj}, lower RMSE, and lower AIC values. In addition, the ANN models predicted the behavior of FCK using a simple network architecture (Equations (4) and (7)), which would facilitate the subsequent analysis and optimization. The parameters (weights and biases) of the best ANN model are presented in Table 6.

Table 6. Weights and biases of the ANN model for feed cost per kilogram of egg (FCK) of H&N Brown SCLHs.

$$w_{ih} = \begin{bmatrix} 1.5971 & 0.8748 & -2.4899 & 1.1157 \\ 19.078 & 4.7011 & 1.5910 & -3.5728 \\ 18.684 & 4.4364 & 1.7995 & 0.4581 \\ 4.4308 & -4.2488 & -0.8286 & 0.1034 \\ -17.374 & 1.1351 & 3.4314 & -2.8917 \\ 0.5897 & 0.7214 & -2.6382 & 1.6316 \\ -10.577 & -11.288 & 1.3771 & 1.9248 \\ -15.701 & 2.4790 & 0.9823 & 2.3833 \\ -0.7277 & 1.1889 & -3.2246 & -1.1524 \end{bmatrix}$$

$$w_{ho} = \begin{bmatrix} -0.00138 \\ 0.00223 \\ -0.00131 \\ 0.00121 \\ -0.00079 \\ 0.00090 \\ -0.00114 \\ -0.00127 \\ 0.00265 \end{bmatrix}^T \quad w_{io} = \begin{bmatrix} 5.35 \times 10^{-5} \\ 1.61 \times 10^{-5} \\ -1.56 \times 10^{-6} \\ -3.69 \times 10^{-5} \end{bmatrix}^T \quad b_h = \begin{bmatrix} 8.9212 \\ -3.6435 \\ 3.2475 \\ 0.1961 \\ -8.2080 \\ 5.1132 \\ -8.5495 \\ 0.1083 \\ 2.8147 \end{bmatrix} \quad b_o = 0.00111$$

w_{ih} is the weights between the input and the hidden layers; w_{ho} is the weights between the hidden and the output layers; b_h is the biases of the hidden layer; b_o is the bias of the output layer.

3.3. Optimization

The minimum FCK during 20 weeks of production was obtained with mean amino acid levels of 909.5 mg/hen-day of iLys, 830.71 mg/hen-day of iMetCys, and 881 mg/hen-day of iThr (Table 7).

Table 7. Optimal intakes of lysine, methionine + cysteine, and threonine between 8 and 20 weeks of production for H&N Brown SCLHs considering the feed cost per kilogram of egg (FCK) as the optimization criterion.

Week	Minimum FCK (USD/kg egg)	Optimal Intakes (mg/hen-day)		
		Lysine	Methionine + Cysteine	Threonine
8	1.040	965.91	864.45	881
12	0.963	965.91	860.48	881
16	0.930	979.18	856.52	881
20	0.873	727.00	741.39	881
Mean		909.5	830.71	881

4. Discussion

4.1. Egg Cost and Zootechnical Results

The iLys requirement of 943.7 mg/hen-day found in diet 13, corresponding to the treatment with lower FCK (Table 4), is similar to that reported by Schneider [26], who indicated 942 mg/hen-day for maximum production, and is above the values observed by Rostagno et al. [27], Schmidt et al. [28], and Kakhki et al. [29] of 807 mg/hen-day, 824 mg/hen-day, and 848 mg/hen-day, respectively.

The iMetCys requirement of 858.3 mg/hen-day found in diet 13 is above the values of 747.46 mg/hen-day, 786.51 mg/hen-day, 791 mg/hen-day, and 811 mg/hen-day reported for maximum production by Macelline et al. [9], Polese et al. [30], Schmidt et al. [28], and Rostagno et al. [27], respectively.

The iThr requirement of 876.8 mg/hen-day observed in diet 13 is above the values reported for maximum production by Agustini et al. [31], Schmidt et al. [32], and Rostagno et al. [27] of 516.26 mg/hen-day, 525.04 mg/hen-day, and 621 mg/hen-day, respectively.

For the best FCK treatment (diet 13), the CR was 2.16 $\pm$ 0.14 kg feed/kg egg (Table 4) for H&N Brown SCLHs at 20 weeks of production (111 weeks of age). This result is lower than that obtained by Sariozkan et al. [33], who found 2.2 to 2.4 kg feed/kg egg at week 20 of production in Hy-Line Brown SCLHs between 81 and 92 weeks of age. The feed conversion ratio result for diet 13 found at 16 weeks (2.18 $\pm$ 0.17 kg feed/kg egg) is higher than the 1.97 kg feed/kg egg reported for Isa Brown hens at 16 weeks (103–106 weeks of age) [34] and 2.04 kg feed/kg of egg reported for H&N Brown SCLHs at 14 weeks of production (81–95 weeks of age) [35].

A good level of amino acids in the diet, especially sulfur amino acids (Met + Cys), is essential in the second production cycle since they directly influence egg size [36]. The difference in egg sizes could also contribute to making diet 13 the most cost-effective treatment. The percentage of eggs weighing 60 g or more (94%) was higher than that of diet 2 (92%) [12], which presented the highest FCK.

Finally, the results of diet 13 regarding egg production and CR at week 20 [12] show the importance of each EAA in meeting the nutritional needs of H&N Brown SCLHs. From 8 weeks of egg production, the productive levels increased with the consumption of Lys in the diet because this EAA is considered physiologically crucial for the maintenance, growth, and production of laying hens, and its primary function is the synthesis of muscle protein. In addition, egg production responds to high levels of Lys, possibly due to the increase in the concentration of plasma albumin, which is the main protein required by the body for the synthesis of egg protein in the oviduct [36].

4.2. ANN Model to Analyze Egg Cost as a Function of EAAs and Time

The ANN models accurately described the nonlinear behavior for H&N Brown SCLHs compared to the other models. These models also enabled effective prediction of FCK based on iLys, iMetCys, iThr, and time (Table 5). The fitting results were better for both the training and validation data sets. The AIC values also showed that the ANN models presented a better generalization than the polynomial models.

The ANN models were appropriate for estimating FCK, as confirmed by the random and balanced distribution of the residuals around zero, with 98.2% of them ranging from -0.12 to 0.12 USD/kg egg (Figure 2), despite the non-normality of the residuals (p-value < 0.05) (Table 5).

According to previous results [12], the ANN models were more accurate in describing FKC than egg production based on EAAs and time in H&N Brown SCLHs. Although this is the first study of its kind, applying ANNs to simultaneously analyze the influence of nutritional factors and time on feed costs, other studies have reported that ANN models are more precise than other nonlinear regression models in estimating zootechnical parameters [37,38].

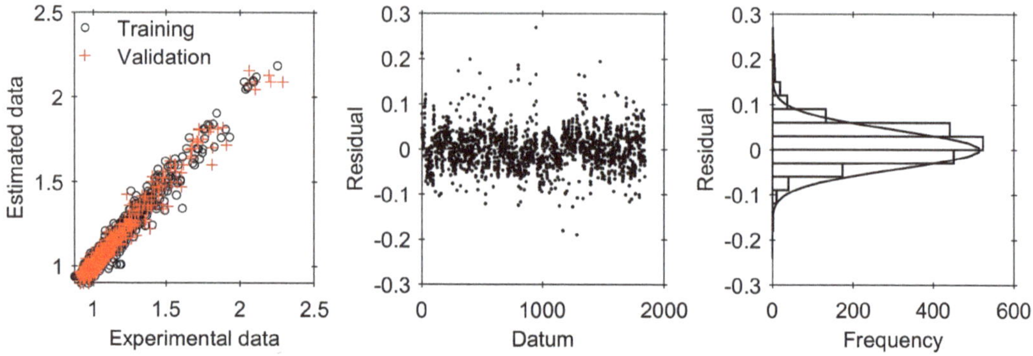

Figure 2. Residual analysis of the ANN model for the feed cost per kilogram of egg (FCK) of H&N Brown second-cycle laying hens (92–111 weeks old).

The versatility of ANNs allows for estimating the weekly, overall, or final nutritional requirements in terms of costs for the second production cycle. The best FCK estimations covered intermediate levels of EAAs among those evaluated in the experimental design (Table 1). Thus, the ANN model estimated levels of iLys for low FCK values at 20 weeks (Figure 3) comparable to the 942 mg/hen-day recommended by Schneider [26] for Shaver Brown SCLHs. On the contrary, other authors recommend lower levels: 713 mg/hen-day for hens in the post-molting phase [25], or 751 mg/hen-day [9], 848 mg/hen-day [29], and 807 mg/hen-day for laying hens in the first production cycle [27].

Combinations of iLys, iMetCys, and iThr to minimize FCK at different weeks can be found in Figure 3. For week 8, FCK values less than 1.076 USD/kg egg can be achieved with 943–1000 mg/hen-day of iLys, 800–1000 mg/hen-day of iMetCys, and 717–881 mg/hen-day of iThr. For week 12, the recommended levels are 943–1000 mg/hen-day of iLys, 853–961 mg/hen-day of iMetCys, and 881 mg/hen-day of iThr, which can result in FCK values less than 0.973 USD/kg egg. For week 16, the adequate levels are 943–1000 mg/hen-day of iLys, 853–961 mg/hen-day of iMetCys, and 881 mg/hen-day of iThr, which can lead to FCK values less than 0.948 USD/kg egg. For week 20, several possible combinations of EAA intakes can produce FCK values below 0.922 USD/kg egg, as shown in Figure 3. For example, some options could be 728 mg/hen-day of iLys, 661–991 mg/hen-day of iMetCys, and 552 mg/hen-day of iThr, or 727 mg/hen-day of iLys, 700–900 mg/hen-day of iMetCys, and 881 mg/hen-day of iThr.

The limiting amino acid concept is based on the interruption of protein synthesis due to amino acid deficiency in the provided diet. For birds, methionine is the first limiting amino acid, not only due to the high demand for cysteine to form feathers and methionine to form ovalbumin but also due to the composition of the diet generally used for this species [39]. As glycogenic agents, methionine and cysteine are metabolized to pyruvic acid via succinyl-CoA and participate in gluconeogenesis—the process of making glycogen in the liver and muscle—improving the body recovery of the birds [36].

The estimated iThr at 20 weeks is also different from the 562 mg/hen-day for Shaver Brown hens (75–90 weeks of age) [31], 462 mg/hen-day for Lohmann Brown SCLHs (79–95 weeks of age) [32], and 499 mg/hen-day for Hy-Line commercial layers [25]. These reported levels are lower than the 876.8 mg/hen-day of iThr recommended in this study for H&N Brown SCLHs.

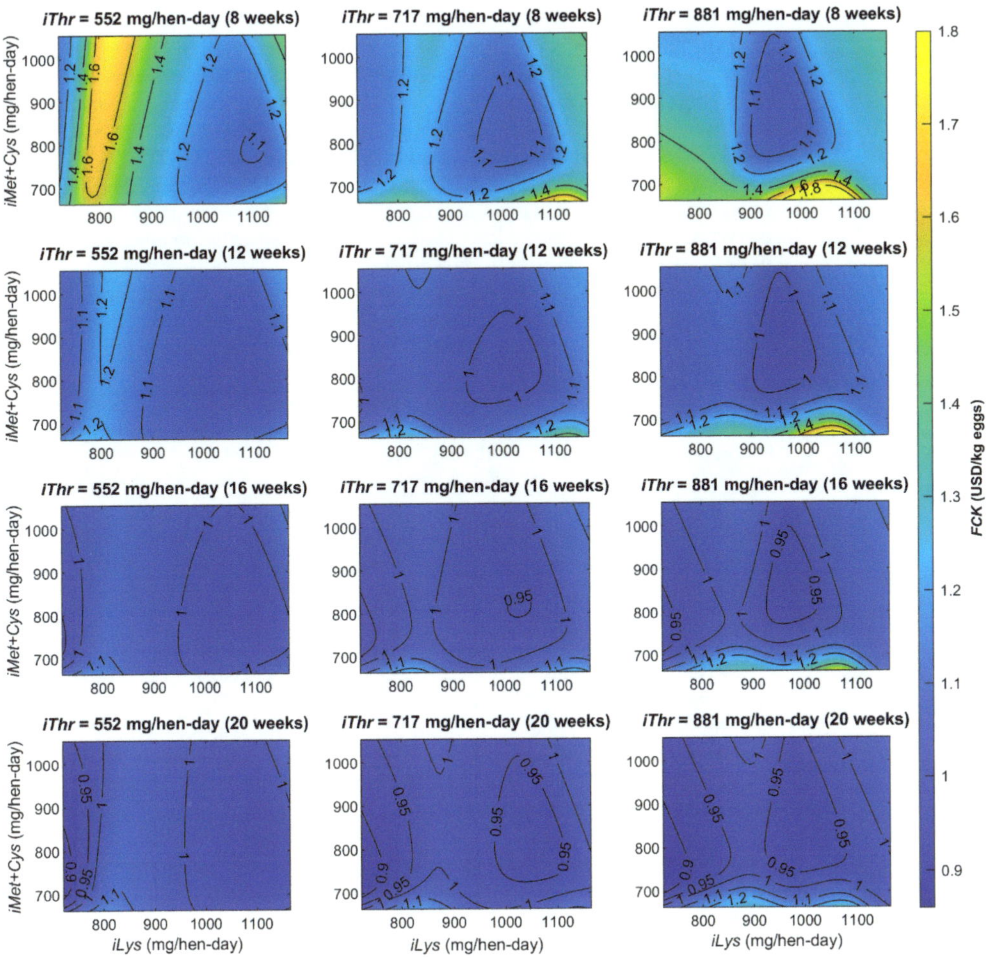

Figure 3. Dependence of the feed cost per kilogram of egg (FCK) on lysine (iLys), methionine + cysteine (iMetCys), and threonine (iThr) intakes at 8, 12, 16, and 20 weeks (99, 103, 107, and 111 weeks of age) for H&N Brown SCLHs.

Threonine becomes more relevant the older the hens are because its maintenance ratio increases [40]. Therefore, adequate Thr supplementation allows the bird to express its maximum production potential. Hence, its nutritional demand is important to consider in egg production, albeit indirectly [41]. The large membrane carriers are found in the ileum, suggesting that this is a crucial site of Thr uptake. Likewise, the intestinal mucosa develops by growing taller and denser villi, which means more epithelial cells and improved digestion and absorption of nutrients [42].

The ANN model shows how each EAA affects different FCK outcomes (Figure 3), indicating that it is necessary to validate scenarios of technical and economic interest to the producers and studies where other nutritional requirements can be used for different production stages.

The optimal levels of Lys, Met + Cys, and Thr for SCLHs have been inconsistent across different studies despite being the most critical EAAs for laying hens. This study demonstrated the importance of the interactions among the EAAs for predicting the egg cost in laying hens, using ANN models as a novel approach.

4.3. Optimization

The minimum FCK decreased between weeks 8 and 20 from 1.040 to 0.873 USD/kg egg. Digestible iLys and iMetCys levels diminished between 8 and 20 weeks from 965.91 mg/hen-day to 727 mg/hen-day and from 864.45 to 741.39 mg/hen-day, respectively. Digestible iThr levels were 881 mg/hen-day for all weeks analyzed (Table 7). At 20 weeks of production, the iMetCys/iLys ratio was 102%, and the iThr/iLys ratio was 121%.

H&N Brown SCLHs showed Thr requirements of 881 mg/hen-day, higher than those of laying hens in the first production cycle, to achieve excellent egg production and feed conversion results with the lowest FCK. Thr supplementation enables animals to make use of proteins and minor amino acids in feed [31]. Protein synthesis and body protein turnover depend on Thr. In addition, Thr and serine are essential for feather formation, accounting for 20% of EAA requirements [40]. Mucin production, necessary for intestinal health, nutrient absorption, and antibody production, also requires Thr [43].

The iThr/iLys ratio at 20 weeks of 121.18% for H&N Brown SCLHs is higher than the EAA estimation of 77% for laying hens during the first cycle of production [27]. The iMetCys/iLys ratio at 20 weeks of 101.98% for H&N Brown SCLHs is also higher than the 98% reported by Rostagno et al. [27].

The mean optimal requirements (Table 7) are close to the EAA intakes in diet 13. On diet 13, the observed FCK between weeks 8 and 20 decreased by 7.92% (Table 3), corresponding to an increment of 2.45% in egg production [12] and a decrease of 3.36% in CR (Table 4). When comparing the observed and estimated FCK, we found percentage errors of less than 2% between weeks 12 and 20, reflecting the accuracy of the ANN model.

5. Conclusions

According to this study, the feed cost per kilogram of egg for H&N Brown second-cycle laying hens is better described by artificial neural networks than by multivariate polynomial models. Artificial neural networks can also estimate essential amino acid intakes to minimize feed costs, offering excellent egg production and conversion ratio results.

The requirements of lysine, methionine + cysteine, and threonine in H&N Brown second-cycle laying hens to minimize the feed cost per kilogram of egg differ from those defined to maximize egg production when models based on artificial neural networks are used. This modeling methodology will allow producers to design profitable and productive diets according to the changing scenarios of the poultry sector.

Supplementary Materials: The following supporting information can be downloaded at: https://www.mdpi.com/article/10.3390/agriengineering5040112/s1, Table S1: Ingredient and nutrient composition of experimental diets (g/kg as-fed basis).

Author Contributions: Conceptualization, W.M.-S. and H.A.V.; methodology, W.M.-S. and H.A.V.; software H.A.V.; validation, W.M.-S., L.D.D. and H.A.V.; formal analysis, W.M.-S., L.D.D. and H.A.V.; investigation, W.M.-S., L.D.D. and H.A.V.; resources, W.M.-S. and H.A.V.; data curation, W.M.-S., L.D.D. and H.A.V.; writing—original draft preparation, W.M.-S., L.D.D. and H.A.V.; writing—review and editing, W.M.-S., L.D.D. and H.A.V.; visualization, W.M.-S., L.D.D. and H.A.V.; supervision, W.M.-S. and H.A.V.; project administration, W.M.-S. and H.A.V.; funding acquisition, W.M.-S. and H.A.V. All authors have read and agreed to the published version of the manuscript.

Funding: This research was funded by the Departamento del Tolima and the Ministry of Science, Technology, and Innovation of Colombia through the "Convocatoria 755/2016-Formación de Capital Humano de Alto Nivel para el Departamento de Tolima", and the University of Tolima project 430120.

Institutional Review Board Statement: The study was approved by the Bioethics Committee of the University of Tolima (Act 03/2017).

Data Availability Statement: Data are included within the article.

Acknowledgments: The authors acknowledge the technical assistance of the "Postharvest and Quality Control" laboratory of the University of Tolima and the poultry farm "San Carlos".

Conflicts of Interest: The authors declare no conflict of interest. The funders had no role in the design of the study; in the collection, analyses, or interpretation of data; in the writing of the manuscript; or in the decision to publish the results.

References

1. FAO. *The Future of Food and Agriculture–Alternative Pathways to 2050*; Food and Agriculture Organization of the United Nations: Rome, Italy, 2018.
2. Morris, S.S.; Beesabathuni, K.; Headey, D. An Egg for Everyone: Pathways to Universal Access to One of Nature's Most Nutritious Foods. *Matern. Child Nutr.* **2018**, *14*, e12679. [CrossRef] [PubMed]
3. Kuang, H.; Yang, F.; Zhang, Y.; Wang, T.; Chen, G. The Impact of Egg Nutrient Composition and Its Consumption on Cholesterol Homeostasis. *Cholesterol* **2018**, *2018*, 6303810. [CrossRef] [PubMed]
4. Marventano, S.; Godos, J.; Tieri, M.; Ghelfi, F.; Titta, L.; Lafranconi, A.; Gambera, A.; Alonzo, E.; Sciacca, S.; Buscemi, S. Egg Consumption and Human Health: An Umbrella Review of Observational Studies. *Int. J. Food Sci. Nutr.* **2020**, *71*, 325–331. [CrossRef] [PubMed]
5. FAO. FAOSTAT Statistical Database—Crops and Livestock Products. Available online: http://www.fao.org/faostat/en/# (accessed on 22 August 2023).
6. Wongnaa, C.A.; Mbroh, J.; Mabe, F.N.; Abokyi, E.; Debrah, R.; Dzaka, E.; Cobbinah, S.; Poku, F.A. Profitability and Choice of Commercially Prepared Feed and Farmers' Own Prepared Feed among Poultry Producers in Ghana. *J. Agric. Food Res.* **2023**, *12*, 100611. [CrossRef]
7. Filho, R.L.A.; Milbradt, E.L.; Okamoto, A.S.; Silva, T.M.; Vellano, I.H.B.; Gross, L.S.; Oro, C.S.; Hataka, A. *Salmonella* Enteritidis Infection, Corticosterone Levels, Performance and Egg Quality in Laying Hens Submitted to Different Methods of Molting. *Poult. Sci.* **2019**, *98*, 4416–4425. [CrossRef]
8. Junqueira, O.M.; de Laurentiz, A.C.; da Silva Filardi, R.; Rodrigues, E.A.; Casartelli, E.M.C. Effects of Energy and Protein Levels on Egg Quality and Performance of Laying Hens at Early Second Production Cycle. *J. Appl. Poult. Res.* **2006**, *15*, 110–115. [CrossRef]
9. Macelline, S.P.; Toghyani, M.; Chrystal, P.V.; Selle, P.H.; Liu, S.Y. Amino Acid Requirements for Laying Hens: A Comprehensive Review. *Poult. Sci.* **2021**, *100*, 101036. [CrossRef] [PubMed]
10. Mehri, M. Optimization of Response Surface and Neural Network Models in Conjugation with Desirability Function for Estimation of Nutritional Needs of Methionine, Lysine, and Threonine in Broiler Chickens. *Poult. Sci.* **2014**, *93*, 1862–1867. [CrossRef]
11. Min, Y.N.; Liu, S.G.; Qu, Z.X.; Meng, G.H.; Gao, Y.P. Effects of Dietary Threonine Levels on Growth Performance, Serum Biochemical Indexes, Antioxidant Capacities, and Gut Morphology in Broiler Chickens. *Poult. Sci.* **2017**, *96*, 1290–1297. [CrossRef]
12. Morales-Suárez, W.; Ospina-Rojas, I.C.; Méndez-Arteaga, J.J.; do Nascimento Ferreira, A.E.; Váquiro-Herrera, H.A. Multivariate Modeling Strategies to Predict Nutritional Requirements of Essential Amino Acids in Semiheavy Second-Cycle Hens. *Rev. Bras. Zootec.* **2021**, *50*, e20200262. [CrossRef]
13. Soares, L.; Sakomura, N.K.; Dorigam, J.C.d.P.; Liebert, F.; Sunder, A.; do Nascimento, M.Q.; Leme, B.B. Optimal In-Feed Amino Acid Ratio for Laying Hens Based on Deletion Method. *J. Anim. Physiol. Anim. Nutr.* **2019**, *103*, 170–181. [CrossRef] [PubMed]
14. Astill, J.; Dara, R.A.; Fraser, E.D.G.; Roberts, B.; Sharif, S. Smart Poultry Management: Smart Sensors, Big Data, and the Internet of Things. *Comput. Electron. Agric.* **2020**, *170*, 105291. [CrossRef]
15. Faridi, A.; Gitoee, A.; Donato, D.C.Z.; France, J.; Sakomura, N.K. Broiler Responses to Digestible Threonine at Different Ages: A Neural Networks Approach. *J. Anim. Physiol. Anim. Nutr.* **2016**, *100*, 738–747. [CrossRef] [PubMed]
16. Galeano-Vasco, L.; Galvan, I.M.; Aler, R.; Cerón-Muñoz, M.F. Forecasting Egg Production Curve with Neural Networks. *Arch. Zootec.* **2018**, *67*, 81–86. [CrossRef]
17. Ahmad, H.A. Egg Production Forecasting: Determining Efficient Modeling Approaches. *J. Appl. Poult. Res.* **2011**, *20*, 463–473. [CrossRef] [PubMed]
18. Ojo, R.O.; Ajayi, A.O.; Owolabi, H.A.; Oyedele, L.O.; Akanbi, L.A. Internet of Things and Machine Learning Techniques in Poultry Health and Welfare Management: A Systematic Literature Review. *Comput. Electron. Agric.* **2022**, *200*, 107266. [CrossRef]
19. De Almeida, L.G.B.; Oliveira, E.B.; Furian, T.Q.; Borges, K.A.; da Rocha, D.T.; Salle, C.T.P.; Moraes, H.L.d.S. Artificial Neural Networks on Eggs Production Data Management. *Acta Sci. Vet.* **2020**, *48*, 1732. [CrossRef]
20. Ameer, K.; Bae, S.-W.; Jo, Y.; Lee, H.-G.; Ameer, A.; Kwon, J.-H. Optimization of Microwave-Assisted Extraction of Total Extract, Stevioside and Rebaudioside-A from *Stevia rebaudiana* (Bertoni) Leaves, Using Response Surface Methodology (RSM) and Artificial Neural Network (ANN) Modelling. *Food Chem.* **2017**, *229*, 198–207. [CrossRef]
21. Wang, S.; Chung, F.-L.; Wang, J.; Wu, J. A Fast Learning Method for Feedforward Neural Networks. *Neurocomputing* **2015**, *149*, 295–307. [CrossRef]
22. Morales-Suárez, W.; Elliott, S.; Váquiro-Herrera, H.A. Multivariate Analysis of Sources of Polyunsaturated Fatty Acids, Selenium, and Chromium on the Productive Performance of Second-Cycle Laying Hens. *Rev. Bras. Zootec.* **2022**, *51*, e20210204. [CrossRef]
23. Daza, L.D.; Homez-Jara, A.; Solanilla, J.F.; Váquiro, H.A. Effects of Temperature, Starch Concentration, and Plasticizer Concentration on the Physical Properties of Ulluco (*Ullucus tuberosus* Caldas)-Based Edible Films. *Int. J. Biol. Macromol.* **2018**, *120*, 1834–1845. [CrossRef] [PubMed]
24. Mathworks. *MATLAB User's Guide*; The Mathworks Inc.: Natick, MA, USA, 2019.

25. Hy-Line. *Management Guide for Hy-Line Brown Commercial Layers*; Hy-Line International: Des Moines, IO, USA, 2014.
26. Schneider, S.E. Exigência de Lisina Digestível Para Poedeiras Semipesadas. Master's Thesis, Universidade Estadual do Oeste do Paraná, Marechal Cândido Rondon, Paraná, Argentina, 2011.
27. Rostagno, H.S.; Texeira Albino, L.F.; Hannas, M.I.; Lopes Donzele, J.; Sakomura, N.; Perazzo, F.G.; de Oliveira Brito, C. *Tablas Brasileñas Para Aves y Cerdos: Composición de Alimentos y Requerimientos Nutricionales*, 4th ed.; Universidad Federal de Viçosa: Viçosa, MG, Brazil, 2017; ISBN 9788581791227.
28. Schmidt, M.; Gomes, P.C.; Teixeira, L.F.; Nunes, R.V.; Brumano, G. Níveis Nutricionais de Metionina+ Cistina Digestível Para Poedeiras Semipesadas No Segundo Ciclo de Produção. *Rev. Bras. Zootec.* **2009**, *38*, 1962–1968. [CrossRef]
29. Kakhki, R.A.M.; Golian, A.; Zarghi, H. Effect of Dietary Digestible Lysine Concentration on Performance, Egg Quality, and Blood Metabolites in Laying Hens. *J. Appl. Poult. Res.* **2016**, *25*, 506–517. [CrossRef]
30. Polese, C.; Nunes, R.V.; Vilela, C.G.; Murakami, A.E.; Agustini, M.A.B.; Takahashi, S.E.; Vilela, V.O.; Souza, C.; Schneider, S.E. Quantidade de Metionina+ Cistina Digestível Para Poedeiras Semipesadas de 75 a 91 Semanas de Idade. *Arq. Bras. Med. Vet. Zootec.* **2012**, *64*, 1682–1690. [CrossRef]
31. Agustini, M.A.B.; Nunes, R.V.; Vilela, C.G.; Takahashi, S.E.; Vilela, V.O.; Bueno, R.S.; Polese, C. Níveis Nutricionais de Treonina Digestível Para Poedeiras Semipesadas de 75 a 90 Semanas de Idade. *Semin. Cienc. Agrar.* **2014**, *35*, 3449–3456. [CrossRef]
32. Schmidt, M.; Gomes, P.C.; Rostagno, H.S.; Albino, L.F.T.; Nunes, R.V.; Cupertino, E.S.; Gomes, P.C.; Albino, L.F.T.; Cupertino, E.S.; Nunes, R.V.; et al. Exigências Nutricionais de Treonina Digestível Para Poedeiras Semipesadas No Segundo Ciclo de Produção. *Rev. Bras. Zootec.* **2010**, *39*, 1099–1104. [CrossRef]
33. Sariozkan, S.; Kara, K.; Güçlü, B.K. Applicability of Non-Feed Removal Programs to Induce Molting Instead of the Conventional Feed Withdrawal Method in Brown Laying Hens. *Braz. J. Poult. Sci.* **2016**, *18*, 535–542. [CrossRef]
34. Morales, W.; Rodríguez, V.; Verjan, N. Parámetros Productivos y Económicos de Gallinas Ponedoras ISA Brown En Segundo Ciclo de Producción Suplementadas Con Aminoácidos No Esenciales. *Rev. Investig. Vet. Perú* **2018**, *29*, 533–543. [CrossRef]
35. Gongruttananun, N.; Guntapa, P.; Saengkudrua, K.; Boonkaewwan, C. Effects of a Short-Term Molt Using Cassava Meal, Broken Rice, or Corn Meal on Plasma Thyroxin Concentrations, Organ Weights and Intestinal Histopathology in Older (95 Wk) Laying Hens. *Agric. Nat. Resour.* **2017**, *51*, 526–531. [CrossRef]
36. Domingues, C.H.d.F.; Sgavioli, S.; Praes, M.F.F.M.; Santos, E.T.; Castiblanco, D.M.C.; Petrolli, T.G.; Duarte, K.F.; Junqueira, O.M. Lisina e Metionina+ Cistina Digestíveis Sobre o Desempenho e Qualidade de Ovos de Poedeiras Comerciais: Revisão. *Pubvet* **2016**, *10*, 448–512. [CrossRef]
37. Safari-Aliqiarloo, A.; Faghih-Mohammadi, F.; Zare, M.; Seidavi, A.; Laudadio, V.; Selvaggi, M.; Tufarelli, V. Artificial Neural Network and Non-Linear Logistic Regression Models to Fit the Egg Production Curve in Commercial-Type Broiler Breeders. *Eur. Poult. Sci.* **2017**, *81*. [CrossRef]
38. Akilli, A.; Gorgulu, O. Comparative Assessments of Multivariate Nonlinear Fuzzy Regression Techniques for Egg Production Curve. *Trop. Anim. Health Prod.* **2020**, *52*, 2119–2127. [CrossRef]
39. Carvalho, T.S.M. Metionina+ Cistina Digestível Na Dieta de Poedeiras Comerciais e Sua Influência Sobre o Desempenho, Qualidade e Perfil Aminoácidico Dos Ovos e Avaliação Econômica. Master's Thesis, Universidade Federal de Minas Gerais, Belo Horizonte, Brazil, 2017.
40. Lelis, G.; Tavernari, F.d.C.; Calderano, A.; Costa, F.A.D.; Albino, L.F.T.; Rostagno, H.; Brito, C.; de Araújo, W.A.G.; Júnior, V.R. Digestible Threonine-to-Lysine Ratio in Diets for Brown Egg-Laying Hens. *J. Appl. Poult. Res.* **2019**, *28*, 145–152. [CrossRef]
41. Barbosa, L.M.; da Costa, P.R.d.S.; Silva, L.G.L.; de Freitas, P.V.D.X.; de Almeida Brainer, M.M.; Modesto, K.P. Digestible Threonine: Lysine Ratios for Light Laying Hens from 29 to 45 Weeks. *Rev. Bras. Ciências Agrárias* **2017**, *12*, 256–260. [CrossRef]
42. Castro, M.R.; Pinheiro, S.R.F.; Oliveira, R.G.; Abreu, L.R.A.; Mota, L.F.M.; Miranda, J.A. Digestible Threonine to Lysine Ratios for Meat-Type Quails. *Braz. J. Poult. Sci.* **2020**, *22*, eRBCA-2017-0636. [CrossRef]
43. Fouad, A.M.; Zhang, H.X.; Chen, W.; Xia, W.G.; Ruan, D.; Wang, S.; Zheng, C.T. Estimation of L-Threonine Requirements for Longyan Laying Ducks. *Asian Australas. J. Anim. Sci.* **2017**, *30*, 206. [CrossRef] [PubMed]

MDPI AG
Grosspeteranlage 5
4052 Basel
Switzerland
Tel.: +41 61 683 77 34

AgriEngineering Editorial Office
E-mail: agriengineering@mdpi.com
www.mdpi.com/journal/agriengineering

www.ingramcontent.com/pod-product-compliance
Lightning Source LLC
Chambersburg PA
CBHW041410210926
51850CB00013B/3227